AF902537

NEW APPROACH OF INDOOR AND OUTDOOR LOCALIZATION SYSTEMS

Edited by **Fouzia Boukour Elbahhar** and **Atika Rivenq**

New Approach of Indoor and Outdoor Localization Systems
http://dx.doi.org/10.5772/2751
Edited by Fouzia Boukour Elbahhar and Atika Rivenq

Contributors

Manuel Perez-Ruiz, Shrini K. Upadhyaya, María Laura Mateo, María Virginia Mackern, Stephan J.G. Gift, F. Elbahhar, B. Fall, A. Rivenq, M. Heddebaut, R. Elassali, Alberto Giretti, Alessandro Carbonari, Massimo Vaccarini, Miika Valtonen, Timo Vuorela, Stefano Maddio, Alessandro Cidronali, Gianfranco Manes, Abdullah Al-Ahmadi, Tharek Abd. Rahman, Xiaoying Kong, Li Liu, Heung-Gyoon Ryu, Aleksander Nawrat, Karol Jędrasiak, Krzysztof Daniec, Roman Koteras, Xin Xu, Kai Zhang, Haidong Fu, Shunxin Li, Yimin Qiu, Xiaofeng Wang

Published by InTech

Janeza Trdine 9, 51000 Rijeka, Croatia

Copyright © 2012 InTech

All chapters are distributed under the Creative Commons Attribution 3.0 license, which allows users to download, copy and build upon published articles even for commercial purposes, as long as the author and publisher are properly credited, which ensures maximum dissemination and a wider impact of our publications. However, users who aim to disseminate and distribute copies of this book as a whole must not seek monetary compensation for such service (excluded InTech representatives and agreed collaborations). After this work has been published by InTech, authors have the right to republish it, in whole or part, in any publication of which they are the author, and to make other personal use of the work. Any republication, referencing or personal use of the work must explicitly identify the original source.

Notice

Statements and opinions expressed in the chapters are these of the individual contributors and not necessarily those of the editors or publisher. No responsibility is accepted for the accuracy of information contained in the published chapters. The publisher assumes no responsibility for any damage or injury to persons or property arising out of the use of any materials, instructions, methods or ideas contained in the book.

Publishing Process Manager Oliver Kurelic
Technical Editor InTech Prepress, Novi Sad
Cover InTech Design team

Additional hard copies can be obtained from orders@intechopen.com

New Approach of Indoor and Outdoor Localization Systems,
Edited by Fouzia Boukour Elbahhar and Atika Rivenq
p. cm.
ISBN 978-953-51-0775-0

We are IntechOpen, the world's leading publisher of Open Access books

Built by scientists, for scientists

4,000+
Open access books available

116,000+
International authors and editors

120M+
Downloads

151
Countries delivered to

Our authors are among the
Top 1%
most cited scientists

12.2%
Contributors from top 500 universities

WEB OF SCIENCE™

Selection of our books indexed in the Book Citation Index in Web of Science™ Core Collection (BKCI)

Interested in publishing with us?
Contact book.department@intechopen.com

Numbers displayed above are based on latest data collected.
For more information visit www.intechopen.com

For EU product safety concerns:
IN TECH d.o.o., Prolaz Marije Krucifikse Kozulić 3, 51000 Rijeka, Croatia,
info@intechopen.com or visit our website at intechopen.com.

Meet the editors

Dr. Fouzia Elbahhar Boukour was born in 1975. She received the M.S and Ph.D degrees form the University of Valenciennes (France) in 2000 and 2004, respectively. She is actually employed as researcher at IFSTTAR/ LEOST, Villeneuve d'Ascq, France. Fouzia Boukour is author or coauthor of more than 50 research papers, including journal articles, book chapters and conference proceedings. She participates at many national and European projects dedicated to transport applications. She serves also as a reviewer for several journals and international Conference. She is involved in signal processing especially Ultra wide band technology. Her major research interests are land transportation like Communication Vehicle to Vehicle and vehicle to Infrastructure, localisation.

Professor Atika Rivenq was graduated Engineer from The ENSIMEV School in 1993, received the M.S. degree in 1993 and then her Ph.D. degree in 1996 from the University of Valenciennes (France). She is Professor in electronics at this university with the IEMNDOAE Lab. Her main activity is in signal processing applied to intelligent transport systems (localisation, collision avoidance radar, cruise control associated to GPS). She participates to many national and European projects dedicated to guided-train communication and inter-vehicles communication especially using UWB technology.

Contents

Preface

The ability to localise people and assets will be driving not only emerging Localization services, but also mobile advertising, safety and security applications. In fact, location based services are becoming an indispensable part of life.

For outdoor application, Global positioning system GPS requires communication with at least four GPS satellites, and offers location accuracy of several meters. It is used mainly for outdoor location-based application, because its accuracy could be degraded significantly in indoor scenarios.

In additional, the wide adoption of Global Positioning System in mobile devices, combined with wireless sensor networks has solved many problems of outdoor Localization.

However, it concerns only outdoors application. There are numerous areas of ubiquitous computing and application, which require the knowledge of indoors user position.

Over the past years, significant effort has been dedicated to the development of indoor Localization systems. The results depend essentially on the environment of the concerning application, the implementation cost and the used technologies. In fact, indoor positioning faces additional difficulties as compared to outdoor positioning. Attenuation and multipath reflections of the line-of-sight (LOS) signal (or direct path) by walls, floors, and ceilings of tunnels are the main factors prohibiting typical GPS receivers to operate in indoor environments. Most of the time, the sum of multipath signals is stronger than the direct path signal, thereby preventing the receiver from accurately calculating the time of arrival.

Despite the effort to improve these results, the existing indoor positioning systems present real challenges for the researchers.

This book addresses the state of the art technology and the results from cutting edge research for indoor and outdoor Localization applications. In fact, the writing of this book was prompted by the fact that actually many applications require high Localization accuracy for indoor and outdoor environment.

In this book, the chapters are organised in three main topic areas:

Section 1: "Satellite Systems for Positioning" consists of chapters which cover whole satellite system (GPS, GNSS) and some applications using this system. It consists of three chapters. The first chapter is related to the satellite Localization system especially The GNSS system. In fact, the authors present some results in terms of precision related the application the GNSS receivers in agricultural applications. The second chapter presents new concepts to calculate the coordinates of GNSS stations, existing in some countries, in the international reference systems using information and applying a methodology of rigorous processing and adjusting. This technique has proved to be advantageous when the objective is to introduce the Reference System simultaneously with the precision and parameters estimation. The last chapter of this section deals with the use of GPS technology in the determination of the one-way speed of light.

Section 2: "Wireless Sensor Network Localization Techniques", is dedicated to the Localization applications using the wireless sensor networks. The section consists of five chapters. The first chapters are related the applications of UWB Localization system especially applied to building construction environments and the transport applications. The third chapter is related to the capacitive transmit-mode measurement methods that can be used to passively track an individual indoors in three dimensions and non-intrusive. The fourth chapter is dedicated to the concepts for an indoor Localization system suitable for a Wireless Sensor Network in a GPS-denied scenario way. The last chapter is dedicated to the off-the-shelf positioning system using IEEE802.11 wireless local area network technology without any additional devices.

Section 3: "Coupled GPS and other sensors", consists of three chapters. The first two chapters present some results for Inertial Navigation System/GPS association and some applications of this coupling. The last chapter discusses some interesting results concerning the development and implementation of a "System On Programmable Chip" SOPC based on portable positioning systems PPS, which is capable of handling the data-fused positioning and can provide user a feasible way to obtain their position anywhere.

The editor would like to thank all authors for their contribution, and all those people who directly or indirectly helped make this work. We hope the information included will be useful for Communication & Localization Engineers, Signal Processing Engineers and the academic researchers occupied in the field of indoor and outdoor Localization systems.

Fouzia Boukour Elbahhar
Chargée de Recherche, IFSTTAR, Institut Français des Sciences et Technologies des Transports, de l'Aménagement et des Réseaux,
France

Atika Rivenq
University of Valenciennes,
France

Section 1

Satellite Systems for Positioning

Chapter 1

GNSS in Precision Agricultural Operations

Manuel Perez-Ruiz and Shrini K. Upadhyaya

Additional information is available at the end of the chapter

http://dx.doi.org/10.5772/50448

1. Introduction

Today, there are two Global Navigation Satellite Systems (GNSS) that are fully operational and commercially available to provide all-weather guidance virtually 24 h a day anywhere on the surface of the earth. GNSS are the collection of localization systems that use satellites to know the location of a user receiver in a global (Earth-centered) coordinate system and this has become the positioning system of choice for precision agriculture technologies. At present North American Positioning System known as Navigation by Satellite Timing and Ranging Global Position System (NAVSTAR GPS or simply GPS) and Russian Positioning System known as *Globalnaya Navigatsionnaya Sputnikovaya Sistema* or Global Navigation Satellite System (GLONASS) both qualify as GNSS. Two other satellite localization systems, Galileo (European Union) and Compass (Chinese), are expected to achieve full global coverage capability by 2020. Detailed information on GNSS technology is plentiful, and there are many books that provide a complete description of these navigation systems [9-11]. But the focus of this chapter is on the applications of GPS in agricultural operations. These applications include positioning of operating machines, soil sampling, variable rate application and vehicle guidance.

The basic principle of operation on which GNSS systems is based is often referred to as resection (also called triangulation), and it involves estimating the distances from at least three satellites orbiting the Earth along different and sufficiently separated trajectories to determine the position of an object in 2-D along with the uncertainty in measurement. Typically, each GPS satellite continuously transmits at least two carrier waves consisting of two or more codes, and a navigation message. GNSS receivers measure the time it takes for the signal to travel from the transmitter on the satellite to the receptor in the receiver antenna and use that time to calculate the distance (or range) between them. To perform a positioning or navigation task, a GNSS receiver must lock onto the signals from at least three satellites to calculate a two-dimensional (2D) position (latitude and longitude). If four or more satellites are in view, the receiver can determine three-dimensional (3D) position (latitude, longitude, and altitude) of the user.

INTECH
open science | open minds

© 2012 The Author(s). Licensee InTech. This chapter is distributed under the terms of the Creative Commons Attribution License http://creativecommons.org/licenses/by/3.0), which permits unrestricted use, distribution, and reproduction in any medium, provided the original work is properly cited.

Bibliography on GNSS systems is rich, and there are many monographs that provide a full description of these navigation systems [9-11]. However, the authors have decided to focus this chapter on the precision agricultural applications of GNSS receivers due to its popularity in this field in recent years.

2. GPS system

The North American GPS consists of 24 operational satellites (+some spares) in six orbits (A-F). Normally 4 to 10 satellites can be seen anywhere in the world with an elevation mask of 10 degrees. These orbits are nearly circular with an elevation of 20,200 km and an eccentricity of less than 1%. The orbital period is 11 hours and 58 minutes. This means that these satellites go around the Earth two times a day. The orbits are inclined at 55 degrees to the equatorial plane. The satellites have orbital speeds of about 3.9 km/s in an Earth-centered non rotating coordinate frame of reference. This system was completed in 1993 and became fully operational in 1995.

The current GPS consists of three major segments- space, control and user. The space segment consists of 24 operational satellites plus additional spares (- 8 at present). Control segment consists of worldwide network of tracking stations and a Master Control Station (MCS) to track the satellites in order to predict their exact locations, almanac and ephemeris, obtain data related to satellite integrity, satellite clocks, atmospheric data, etc., and upload the information to GPS satellites. The user segment consists of GPS receivers.

Figure 1 shows a typical GPS satellite with L- and S- band antennas. These satellites transmit positioning signals using L- band and S- band is used for uploading almanac and ephemeris data to the satellites from uplink stations. Table 1 lists the current GPS constellation. Note that the system consists of IIR, IIRM, and IIF satellites (-These are different generation satellites with different specifications and capabilities). Each satellite is recognized by a pseudo random number (PRN) or a space vehicle number (SVN). Note that PRN is not at all random and is generated by a complex mathematical algorithm to identify a given satellite. Table 1 includes the launch date of each satellite and the orbit in which it is located. Since timing is the key for receiver position determination as will be described later, each satellite is equipped with three to four atomic clocks.

The control segment consists of 12 tracking stations with the master control station located in Colorado Springs, CO, USA. All stations are unmanned and they transmit data to the master control station by satellite communication. Four of the 12 stations have the uplink capabilities and can upload almanac, ephemeris and other relevant information to satellites.

As mentioned earlier, GPS receivers constitute the user segment. These can consist of simple and inexpensive receivers costing only about $100 to 150 or very expensive receivers costing thousands of dollars that provide high positioning accuracy. The positioning accuracy of inexpensive receivers may be about 10 m without any correction and can be improved to about 3m with satellite based wide area correction. The more expensive receivers can provide centimeter level positioning accuracy. While most receivers use pseudo-ranging technique to determine their location, more expensive receivers employ carrier phase measurement to provide centimeter level accuracy.

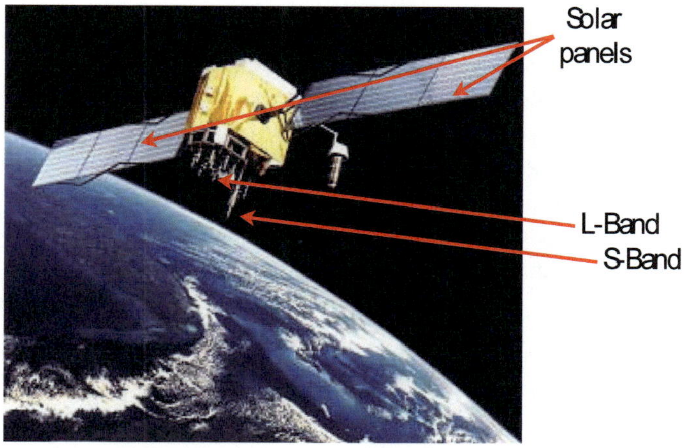

Figure 1. A typical IIF GPS satellite with L-and S- band antennas. (Source: http://www.kowona.de /en/gps/)

Satellites	PRN	SVN	Launch Date	Plane
IIR-2	13	43	23 Jul 1997	F3
IIR-3	11	46	07 Oct 1999	D5
IIR-4	20	51	11 May 2000	E1
IIR-5	28	44	16 Jul 2000	B3
IIR-6	14	41	10 Nov 2000	F1
IIR-7	18	54	30 Jan 2001	E4
IIR-8	16	56	29 Jan 2003	B1
IIR-9	21	45	31 Mar 2004	D3
IIR-10	22	47	21 Dec 2003	E2
IIR-11	19	59	20 Mar 2004	C3
IIR-12	23	60	23 Jun 2004	F4
IIR-13	2	61	06 Nov 2004	D1
IIR-14M	17	53	26 Sep 2005	C4
IIR-15M	31	52	25 Sep 2006	A2
IIR-16M	12	58	17 Nov 2006	B4
IIR-17M	15	55	17 Oct 2007	F2
IIR-18M	29	57	20 Dec 2007	C1
IIR-19M	7	48	15 Mar 2008	A4
IIR-20M*		49	24 Mar 2009	
IIR-21M	05	50	17 Aug 2009	E3
IIF-1	25	62	28 May 2010	B2
IIF-2	01	63	16 Jul 2011	D2

Satellite is no longer in service

Table 1. GPS constellation as of January 2012 (Source: ftp://tycho.usno.navy.mil/pub/gps/gpsb2.txt)

3. GLONASS system

GLONASS (Global Navigation Satellite System) was developed by former Soviet Union in 1980s almost in parallel with the United States and is now operated for the Russian government by the Russian Space Force. The original GLONASS constellation was completed in 1995, but then the unstable economic situation following the collapse of the former Soviet Union led to the deterioration of this satellite constellation. In December 2011 the GLONASS achieved full global coverage for the second time (27 satellites, 24 operational and 3 in reserve). These satellites are located in medium Earth orbits (MEO) at 19,100 km altitude with a 64.8 degrees inclination and a period of 11 hours and 15 minutes. This constellation operates in three orbital planes, with 8 evenly spaced satellites in each.

4. Galileo system

Galileo is a programme for a global navigation satellite system and it is currently being built by the all European Union countries and the European Space Agency (ESA). Recognizing the importance of satellite navigation, positioning, and timing in different fields, a civilian European system was conceived and developed in the early 1990s. It started with the European contribution to the first generation of GNSS (GNSS-1), the EGNOS program, and continues with the generation of GNSS-2, the Galileo program. The goal is for it to be completely functional by 2020 and will provide coverage to the Polar Regions. When developed, the Galileo system will consist of 30 satellites (27 operational + 3 active spares), positioned in three circular medium Earth orbit (MEO) planes inclined at 56 degrees to the equatorial planes at an elevation of 23,222 km altitude above the Earth and an orbital period of 14 hours and 5 minutes.

5. BeiDou-COMPASS system

The BeiDou Satellite Navigation and Positioning System is being developed by China. This system was designed to provide positioning, fleet-management, and precision-time dissemination to Chinese military and civil users. At present, it has 10 satellites and covers the Asia-Pacific region. Unlike other GNSS, which use MEO (altitudes between 19,000-23,000 km), BeiDou located its satellites in geostationary orbit, approximately 36,000 km above sea level in the plane of equator. However, the Beidou system is being currently upgraded under the name COMPASS to achieve full GNSS capability by 2020. When completed this system is expected to have 35 satellites in 21, 150 km orbits inclined at 55.5 degrees to the equatorial plane and an orbital period of 12 hr and 36 min.

In addition to the above systems that either have or expected to have GNSS capability, two other regional systems also provide position measurement over a limited region. Indian Regional Navigational Satellite System (IRNSS) is planned to have seven geostationary (GEO) satellites and is expected to provide 20 m accuracy within India and 2000 km of its neighbourhood. The Japanese Quasi-Zenith Satellite System (QZSS) is primarily a communication system with navigational capability. It consists of three highly inclined, geosynchronous satellites. At least one satellite is over Japan at all times.

In the future the combined use of the GNSS systems will increase the overall performance, robustness of satellite navigation for the benefit of all potential users.

6. GPS signal and structure

As mentioned earlier, GPS signal consist of two carrier waves ((L_1 = 1575.42 MHz or 19 cm and L_2 =1227.60 MHz or 24.4 cm), two or more digital codes (Coarse Acquisition code or C/A on L_1 and P-code on both L_1 and L_2), and a navigation message. Civilians have access to C/A code on L_1 only. P-code is encrypted with an unknown W-code resulting in a Y-code [i.e., P (Y)] and is not available to civilians (for military purpose only). This is called antispoofing. Use of P-code can provide very accurate estimation of position (precise positioning service, PPS) as ionospheric distortion can be completely eliminated using L_1 and L_2 signals. However, use of C/A code only cannot provide very accurate estimation of position (standard positioning service, SPS). The newer satellites transmit two additional codes (L_2 CM – civilian moderate and L_2 CL –civilian long). These additional codes will be helpful in minimizing errors due to atmospheric effects. The navigational message contains information about almanac, ephemeris, clock correction, satellite health, atmospheric correction etc. This is added to both C/A and P-Code. These codes contain information about the satellite identity (PRN) and timing information. These codes are then added on to L_1 (both C/A and P codes) and L_2 (P-code only). Figure 2 is a schematic diagram of the GPS signal.

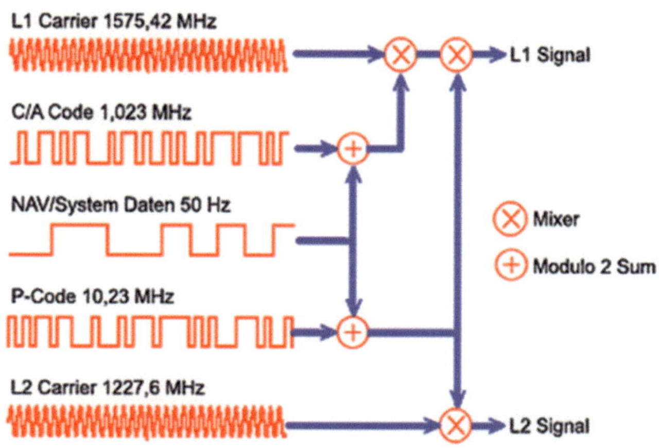

Figure 2. Composition of the signals from GPS satellites (Source: http://www.kowoma.de/en/gps/signals.htm)

7. GPS positioning principle

Resection is the principle used for locating the position of an object on the surface of the earth. Figure 3a shows that location A of an object on the surface of the earth can be easily determined if travel times of signals from two satellites to a receiver located at position A are measured. Travel times can be multiplied by the speed of electromagnetic wave (speed of light = 299,729,458 m/s) to determine the distances between the satellites and the receiver. Resection principle can then be used to locate the receiver position A (i.e., point of intersection of two circles drawn with satellites as the center of the circle and respective distances to point A as radii – basically triangulation). Note that the other point of intersection of these circles is an unacceptable solution (why?).

Since the distances involved are never really measured, but are estimated from the times required for the signal to travel from satellites to the object, the process is called pseudoranging. Note that any small error in time measurement may lead to relatively large error in position estimation as speed of light is very high. As indicated in figure 3b, any error in time measurements can locate the receiver at position B rather than the real location A and we would have no estimation of the magnitude of this error. However, if the timing signal is measured from a third satellite, a curved triangular region (B-B-B) can be determined within which the receiver should be located as shown in figure 3c. Thus it is critical to get signals from at least three satellites to get an estimate of position location and relative accuracy of that measurement. Since all GPS position measurements are performed in 3-D, signals from at least four satellites are needed to obtain the position fix (latitude, longitude, and altitude) and a measure of relative accuracy of that measurement.

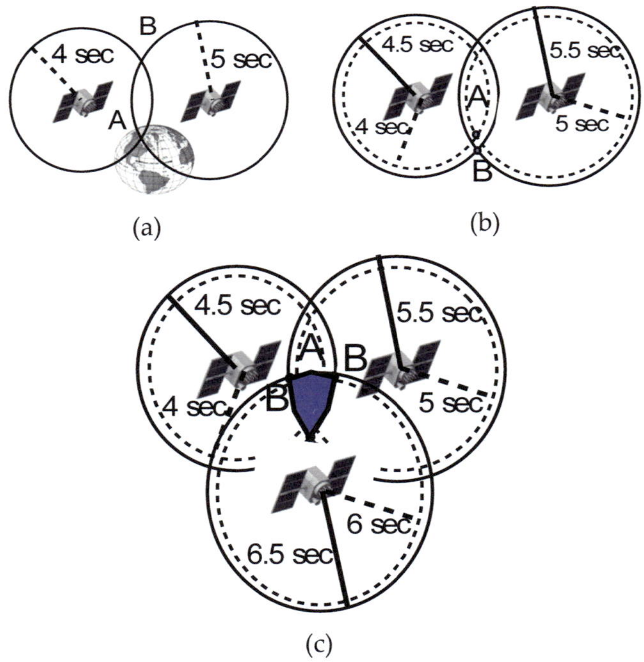

Figure 3. Determination of 2-D position of object A using pseudoranging principle – (a) no error in time measurement, (b) effect of error in measuring time, and (3) region of uncertainty (or estimate of error) when timing signals from three satellites are utilized for determining position in 2-D.

Thus, if (x_0, y_0, z_0) are the unknown Cartesian coordinates of the receiver and (x_i, y_i, z_i) are the respective coordinates of the ith satellite, the distance of the receiver from the satellite, d_i, can be obtained by:

$$d_i^2 = [(c(t_i - e_t)]^2 = (x_i - x_o)^2 + (y_i - y_o)^2 + (z_i - z_o)^2 \tag{1}$$

where c is the speed of light and e_t is error in measuring time. If time measurements are available from four different satellites (i=1,2,3,4), then we can write four nonlinear algebraic equations in four unknowns (x_0, y_0, z_0, and e_t) that can be solved. Note that positions of satellites are known because each satellite transmits ephemeris as a part of the navigation message. If time measurements from more than four satellites are available, least square minimization is used to obtain the best estimation of the receiver location and associated measurement error.

8. Carrier phase measurement

An alternate way to measure the location of a receiver is to measure phase angle of the signal received by the GPS receiver. As shown in figure 4, a signal from the satellite will complete an integer number of cycles (say, N) and a portion of the waveform as it arrives at the receiver. The receiver measures only the partial waveform or phase of the signal. It does not have any idea of the integer number of cycles between itself and the satellite. This is known as integer ambiguity. If the receiver tracks the satellite over time, it is possible to keep track of the phase change from the start. This information along with an optimization procedure can be used to solve for the unknown integer N. This is called ambiguity resolution. Since a fraction of the waveform or phase angle can be measured and wavelength of L_1 carrier wave is 19 cm, this technique can provide millimeter precision compared with a few meters for C/A code measurements. To achieve this level of accuracy a local base station or a virtual base station is necessary to provide an accurate reference. This system is often referred to as real-time kinematic GPS or RTK GPS.

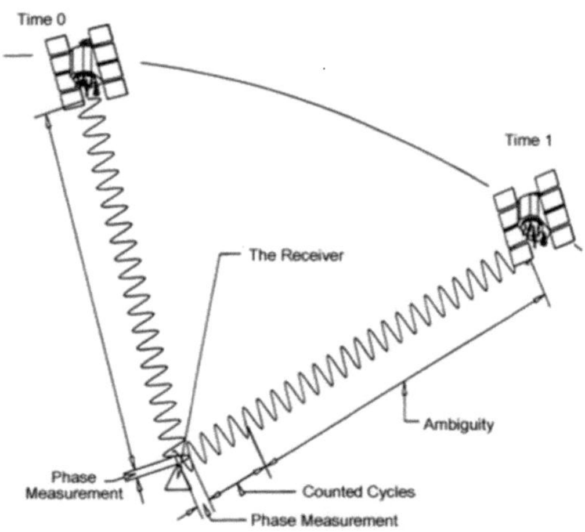

Figure 4. Principle of carrier phase measurement (Source: http://nptel.iitm.ac.in/courses)

9. Real-time differential GPS correction

Very high GPS accuracy can be achieved using post processing. However, for real-time applications that require on-the-go corrections, a differential GPS (DGPS) is preferred. A straightforward manner of accomplishing this is to use two GPS receivers (a rover and a base) that track same satellites, so that many of the errors can be minimized and higher accuracy can be obtained in real-time. Figure 5 provides a schematic diagram of the principle involved in differential correction. Since the position of the base station is known accurately, the error in estimating the location of the base station using satellite signals can be determined. This correction information can be communicated to the field GPS receiver (i.e., the rover) by a radio link and this information can be used increase its accuracy. However, the deployment of two GPS receivers for agricultural applications could be expensive in many instances. An alternative, to reduce the cost without degrading the positional accuracy, is to use one of the available differential correction services. If the GPS users obtain one of this available services, only one receiver can be used as a rover and no base receiver would be required.

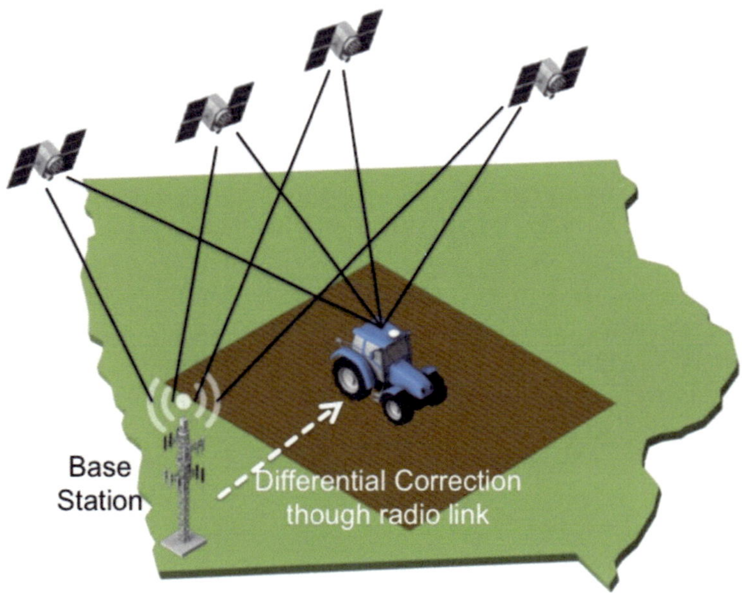

Figure 5. Schematic diagram of a differential correction technique (Courtesy of TRIMBLE)

Agricultural use of GPS has significantly expanded due to the increased availability of differential correction. Today there are various types of differential correction services that are readily available to the user. These are:

- DGPS radio beacons (e.g., US Coastguard DGPS beacons along major waterways): These services can provide sub-meter DGPS accuracy. Reliable coverage is available on land, sea and air. This service is free and available in more than 40 countries; however, the correction signals degrade as you move away form the beacon location.

- Space-Based Augmentation System (SBAS): It is satellite based system that provides regional correction signals (e.g., Wide Area Augmentation System (WAAS) within North America, European Geostationary Navigation Overlay Service (EGNOS) within Europe, Multi-Functional Satellite Augmentation System (MSAS) within Japan and Southeast Asia, and GPS and geo-augmented navigation (GAGAN) within India) over a wide-area (L-band DGPS) through the use of additional satellite-broadcast messages. All of these systems work similarly and are compatible with each other, however, the accuracies of these free satellite based systems vary. They consist of reference stations distributed over a wide area, master stations to process and upload data, and geostationary satellites to transmit the correction signals to users. The WAAS service within USA is fully operational for safety–critical operations such as aircraft navigation and is specified at 7 m accuracy. Agricultural users have found WAAS to be a reliable source of correction, with an accuracy of better than 3 m and a much better pass-to-pass accuracy [12]. The two major commercial L-band satellite based correction providers are Fugro (OmniSTAR service) and Deere (Starfire service). OmniSTAR provides almost complete worldwide coverage. The Starfire service is based on the NASA Jet Propulsion Laboratory correction system. Both of these commercial service providers have a high-accuracy service that uses dual-frequency receivers and antennas for performance in the decimeter range (100–300 mm). OmniStar and Starfire are subscription services.
- Dedicate-use RTK base station and RTK networks: Real-time kinematic (RTK) systems establish the most accurate solution for GNSS applications, producing typical errors of less than 2 cm. This level of precision is not needed for general site-specific farming, but it does permit treatment of specific small location such as a plant-specific operation and is essential for precision guidance [6,13], controlled traffic farming [7,8], mechanical intra-row weed control or thinning of crop plants [14]. In this method, a base station is located at a known point close to where the vehicle operates and communicates with the rover through a radio transmitter. Two disadvantages of RTK-GPS solutions are: (i) the requirement that a base station be located within 10 km at all times which limits its use when farms are large or spread out and (ii) high capital cost. An alternative to the local base station that is becoming increasingly popular is the Virtual Reference Station (VRS) that essentially creates a virtual reference point near the rover using a network of RTK base stations. The VRS service is available for a fee from vendors such as OmniSTAR.

10. Applications of GNSS in agriculture

Satellite-based localization solutions have become quite mature and the GNSS receivers have found numerous applications in agriculture. These receivers are a key part of the precision agriculture technologies as position information is a prerequisite for site-specific crop management. However, not all tasks that need to be performed in precision agriculture need the same level of positioning accuracies [6,12]. Some precision agriculture operations such as yield monitoring, soil sampling or variable rate applications, can be performed using submeter accuracy differential GPS (DGPS) as errors below 1 m are acceptable for these applications. Other tasks like mechanical intra-row weed control, thinning of crop plants, precise planting or autonomous navigation within tight rows demand decimeter- or even centimeter- level accuracy. A solution to this demanding requirement can be found in real-time kinematic GPS (RTK-GPS) which was discussed earlier.

Despite the fact that there exist different global positioning satellite systems, GPS and GLONASS are the only two fully operational GNSS system today. As discussed before, these two systems are similar. However, North American GPS has been in continuous use since the middle of 1990s and many of the agricultural applications have been developed using this system. Applications of GPS for agricultural purposes have exploded in the recent years and the literature is rich with may very interesting examples. In the following discussion we limit our attention to six specific applications with which the authors are very closely involved:

i. Yield monitoring
ii. Compaction profile sensing
iii. Tree planting site-specific fumigant application
iv. RTK GPS based plant mapping
v. Precise weed management system
vi. Robotic applications

10.1. Yield monitors

The ability to continuously monitor and map yield at harvest and observe its spatial variability is a key step in implementing site-specific crop management. Spatial variation in site-specific yield data within a field frequently reflects the variation in soil, plant, and environmental characteristics. Farmers, consultants and researchers have utilized yield monitors to map yield of many crops. However, majority of precision agriculture practice adoption has occurred in grains, oilseeds and cotton. Generally speaking, cereal grain combines use physical sensors to measure grain flow (i.e. impact sensor), whereas cotton yield monitors use microwave or near-infrared sensors to measure amount of cotton. GPS device is a key part of the yield monitor as position data is critical to determine spatial variability in crop yield. Other sensors such as forward speed sensor (i.e. radar, ultrasonic sensor or magnetic pickup on the transmission drive shaft), crop moisture sensor and header height sensor are also mounted on the combine. With all of these sensors and instrumentations, it is possible to map spatial variability in yield data and create yield maps and track field performance from year to year. These maps are very useful to create different management zones for various inputs within a field.

The yield monitor shown in figure 5 was used in a study conducted at University of Cordoba, Spain. The yield monitor (model PF3000, AgLeader Technology Inc.) was mounted on a four-row cotton harvester (model 9965, Deere & Company Ltd.) and calibrated. It consists of four optical cotton flow sensors located in the ducts and a GPS receiver (Trimble model AgGPS 132) with Onmistar differential corrections [15]. The goal of this study was to investigate the relationship between yield variability and spatial variability of some soil properties for 6 ha irrigated cotton field located in southern Spain. The soil samples were taken from the 0-20 cm depth in the Spring on a 20x20 m grid (133 samples). The soil was analyzed for a range of properties including texture (sand, silt and clay), organic matter (OM), phosphorus (P), potassium (K) [16]. Kriged maps of each soil property and crop yield were generated using Surfer software. The main cause of the spatial variability in cotton yield in this field was found to be due to the spatial variation in soil texture (in particular, the sand and clay percentages) as seen in figure 6. Soil texture variation influenced water content distribution and consequently the uniformity of plant stand [17].

Figure 6. Cotton yield monitor mounted and used at University of Cordoba, Spain. The combine is equipped with 4 optical flow sensors, a DGPS receiver and a display.

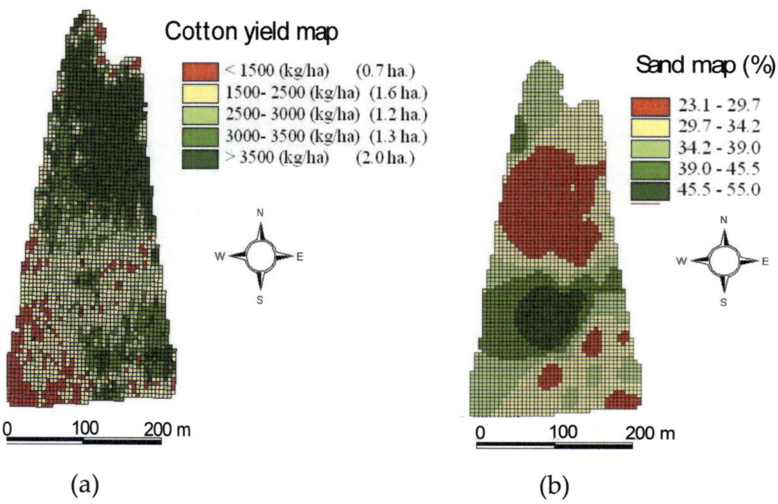

Figure 7. Kriged maps of: (a) cotton yield and (b) sand content in the test field located in southern Spain.

10.2. Compaction profile sensor

Soil properties and environmental conditions are generally regarded as the main causes contributing to variability in crop yield within a field. The research conducted at the University of California Davis in a processing tomato field has indicated that variability in water infiltration rate caused by variability in soil compaction is a major factor affecting processing tomato yield (Figure 8). Soil compaction is often measured using an ASABE (American Society of Agricultural and Biological Engineers) standard cone penetrometer (force per unit area of a penetrating standard cone known as cone index). However, cone index (CI) is a point measurement that exhibits high variability, and is labour intensive and time consuming to measure if a huge amount of data needs to be obtained to map a large field. To overcome these limitations a compaction profile sensor shown in figure 9 was developed. The device consists of five 5.1 cm long, active cutting elements that are directly connected to five octagonal load cells and can measure cutting resistance of soil directly ahead of the cutting elements. These active cutting elements are isolated from each other by 2.5 cm long dummy elements. Moreover, a dummy element of length 8 cm was attached above the topmost active element. This long dummy element was included since an earlier study had indicated that the soil cutting data from the top layer was unreliable due to depth fluctuations and potentially a different mechanism of soil failure (i.e., crescent versus lateral soil failure). This device was capable of getting soil cutting resistance data over the depth profile of 7.5 to 45.7 cm below the surface. A sub-meter accuracy DGPS receiver that used coastguard beacon differential correction was included with this system to provide position information. In addition, a radar (model RADAR II, DICKEY-john Corporation, Illinois, USA), was employed to measure ground speed.

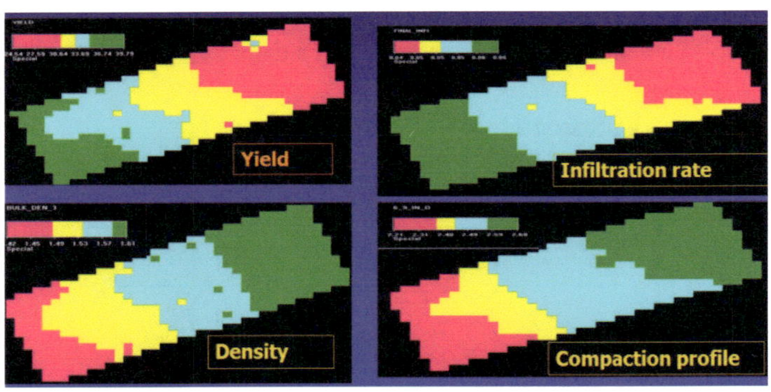

Figure 8. Spatial variability in crop yield, infiltration rate, bulk density of soil, and soil compaction level for a section of a processing tomato field located in Winters, CA, USA.

The compaction profile sensor was calibrated and then tested in agricultural fields in California and in the Midwest. ASABE standard cone penetrometer was also tested in the same fields. The force acting on the unit area of the cutting element, termed CIE (cone index equivalent), was related to CI values at the same depth and depth of operation of the cutting element. Figure 9 shows the plot of the measured CIE values versus the predicted CIE values based on a multiple linear relationship given by the following equation:

$$CIE_i = 0.15CI_i + 2.244d_i + 0.69CI_i x\ d_i \quad (2)$$

where d_i is the depth of operation of the ith cutting element (i=1, 2,....5) and CIE and CI are the corresponding cone index equivalent and cone index values respectively.

(a) (b)

Figure 9. The compaction profile sensor developed at UC Davis. The figure on the left (a) provides an overview of the system when it is mounted on a toolbar. The figure on the right (b) provides internal construction details of the sensing elements.

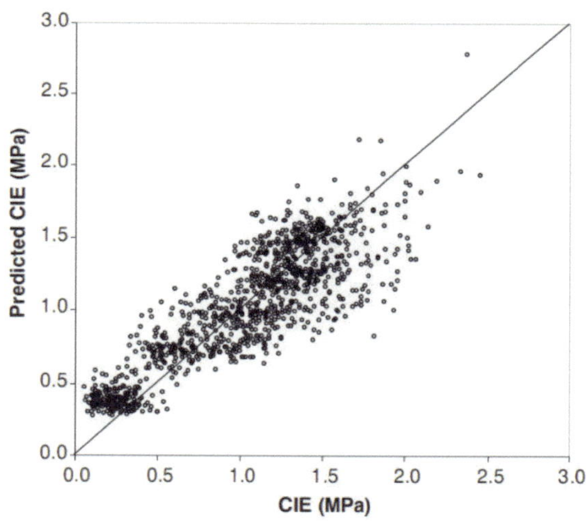

Figure 10. Comparison between predicted CIE (cone index equivalent) values and measured CIE values obtained during the field tests in the Midwetern United States [Source: Andrade-Sanchez et al. (2008)].

The map of the soil compaction level estimated from the force on the cutting element located between 15 to 22.5 cm deep layer of soil is shown in the lower right hand side of figure 8. The soil compaction map for this layer correlated very well with the yield map (upper left hand side). Complete description of the compaction profile sensor and its application for mapping soil compaction profile can be found in Andrade-Sanchez et al. (2008).

10.3. Tree planting site-specific fumigant application

A major concern when one replaces an old orchard with a new one is the incidence of replant disease. For example, when young almond trees are planted at sites from where the old almond or stone fruit trees have been removed, the new plants get stunted or even get killed due to a poorly defined soil borne disease complex called replant disease. Although the exact cause of this disease is not well-understood, pre-plant, site-specific application of small amount of fumigant such as methyl bromide (MB), chloropicrin (CP), 1,3-Dichloropropane (1,3-D), or two-way mixtures of CP with MB or (1,3-D) can control the incidence of replant disease. Therefore, it is a common practice to apply fumigants to the soil over 2-4 m wide continuous strips centered over the future tree rows.

However, the researchers at USDA/UC Davis have found that application of a small amount (0.2 kg/site) of fumigant in the vicinity of future tree planting sites can control replant disease effectively. While continuous application requires about 168 kg/ha of fumigant, tree planting site-specific application requires only about 40 to 70 kg/ha of the fumigant based on the tree spacing along the row. This is a 58 to 76% reduction of chemical load on the environment and cost. Therefore tree planting site-specific application of fumigants is not only economical, but also beneficial to the environment.

However, manual tree planting-site-specific fumigant application is very labor intensive and handling of fumigants poses some risk. Accurately locating tree-planting site is a time consuming process. However, with the advent of high performance GPS (HPGPS), computer technology can be used to apply the right amount of fumigants at the right location. This type of GPS system has an accuracy of about 20 cm rather than the coastguard beacon based DGPS that has sub-meter accuracy. This higher accuracy is necessary to turn on and off the fumigant applicators over a two meter long strip at the center of which a tree would be planted (Note that a sub-meter accuracy DGPS system may introduce very large error in the treatment zone).

A shank type fumigant applicator on loan from TriCal Inc. (figure 11a) was retrofitted with a precision fumigant applicator (PFC). This unit communicated with a rate controller to obtain the actual application rate. PFC was uploaded with a tree-planting grid (tree map) developed using a specially writing software. The inputs to the software consisted of HPGPS coordinates of the four corners of the orchard, row spacing, tree spacing along the row, fumigant application zone length, and pattern of planting (i.e., rectangular versus diamond shape). The PFC receives the location information from the HPGPS receiver and information from an inclination sensor located near the solenoid valve (to determine if the shanks are in raised or lowered position). The PFC would turn on a solenoid valve and apply fumigants, if the HPGPS unit located the shank within the treatment zone and the shanks were lowered into the ground. If the applicator was in the raised position or if the shanks were outside the treatment zone, PFC would close the solenoid valve. The system was calibrated using road and field tests and was found to work well (less than 15 cm error in applying the fumigant in the treatment zone if appropriate look ahead value was used).

The system has been used to treat tree-planting sites in several orchards over the past five years.

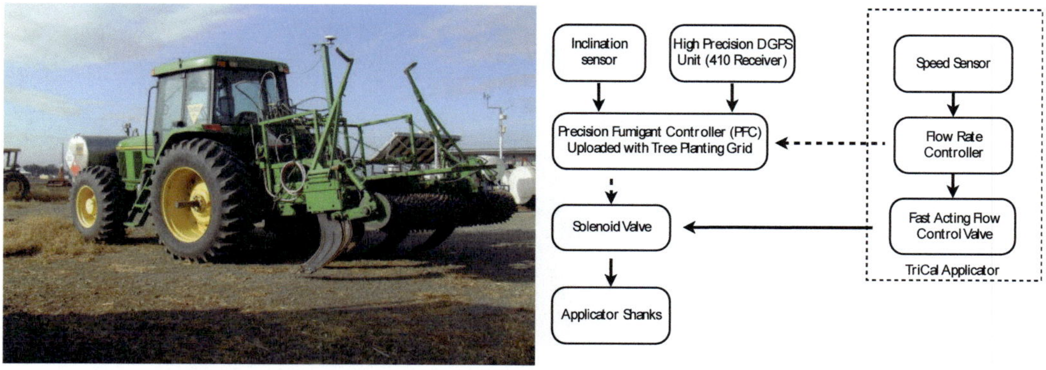

(a) (b)

Figure 11. A shank type fumigant applicator (a) retrofitted with a precision fumigant controller and a HPGPS positioning unit, (b) Schematic of the fine-tuned site-specific fumigant application system.

10.4. RTK-GPS based plant mapping

In recent years, application of centimeter accuracy RTK-GPS has received a lot of attention because of its ability to provide extremely precise location information. These highly precise RTK-GPS based systems, such as automated tractor steering systems, have become very popular in recent years. Figure 12a shows an auto guidance tractor equipped with RTK-GPS forming the bed to plant processing tomatoes. Figure 12b shows the same tractor being used to install drip tape about 12.5 cm away from the plant line and 12.5 cm below the soil surface, while figure 12c shows planting processing transplants along the centerline of the bed. Figure 12d shows mechanical cultivation using cultivator tines spaced about 5 cm away from the plant line, while figure 12e shows the resultant cultivated field. Figure 12f shows a deep cultivation operation in the same field along the centerline of the bed after the tomato crop was removed. Although these auto guidance systems use cm accuracy RTK-GPS system, the overall pass-to-pass accuracy of the tractor is expected to be about 2.5 cm. What is very interesting is that when the cultivator blades were placed 5 cm away from the plant line and the auto guidance tractor was operated at 11 km/h, vary little plant damage occurred. Moreover, when deep cultivation was done along the centerline of the bed following tomato harvest, no damage occurred to buried drip tape placed 12.5 cm away from the plant line [18].

While RTK GPS based autoguidance tractors have become a commercial reality, extremely interesting additional applications exist for this type of system in production agriculture. One such possibility is the ability to create a plant map using RTK GPS by monitoring the seeds or transplants while they are being planted. The availability of precision mapping technologies for crop plants enables a new opportunity for plant specific treatment systems where the resources for plant care are tailored to the needs of individual plants rather than providing the same level or resources to all plants in the field irrespective of the need or potential for utilization [19,20].

Figure 12. RTK GPS based autoguidance system used for various cultural operations – (a) bedding, (b) drip tape installation setup, (c) transplanting processing tomatoes, (d) cultivation, (e) plants following cultivation, and (f) deep tillage after plant removal.

One such possibility is the ability to create a plant map using RTK-GPS by monitoring the seeds or transplants while they are being planted. Such a plant map can then be utilized for subsequent intra-row, weed-specific cultivation or chemical application. Figure 13 shows a 4-row, Salvo 650 vacuum planter retrofitted with a RTK-GPS unit developed by [13]. The system consists of two microcomputers one of which monitors the seeds using optical sensors (four sensors - one per row) as they are being planted and records the event along with the time (i.e., row number in which the seed was seen and the time at which the event occurred). Moreover, it obtains the RTK-GPS coordinates and records them along with the time tag. The second microcomputer monitors the first microcomputer and displays the planter performance information on a monitor mounted in the cab. The time tag allows to determine the exact location where a given seed was dropped into the ground facilitating the creation of a seed planting map. The actual plant map is expected to be slightly different due to system dynamics. The study conducted at UC Davis has shown that the difference between RTK-GPS based expected seed location versus actual plant position in the field was in the range of 3.0 to 3.8 cm.

Figure 13. A RTK GPS based seed monitoring system retrofitted onto a 4-row, Salvo 650 vacuum planter

However, the Ehsani's system utilized an additional RTK-GPS system dedicated to the planter, in addition to any RTK-GPS auto guidance systems present on the tractor, greatly increasing the capital cost of the crop mapping operation. Perez-Ruiz et al. (2012) [14] developed a centimeter-level accuracy plant mapping system for transplanted row crop which utilized a single RTK-GPS auto guidance system mounted on the tractor, and not the planter, thereby reducing the capital cost of the system (A similar approach was investigated by Mr. Mark Mattson, a Graduate Student Researcher working under Dr. Upadhyaya's guidance at UC Davis in 2002). They developed an instrumented hitch orientation sensor that allowed for accurate real-time monitoring of the position of the transplanting sled in relationship to the tractor. When combined with tractor mounted RTK GPS coordinate data, a transplant map could be created by sensing transplant placement during planting. Field tests using this system showed that the mean RMS accuracy of the system was 2.67 cm in the along-track direction where 95% of the crop plants were located within a circular radius of 5.58 cm from the mapped location. These results showed that it was possible to use a single RTK-GPS system mounted on the tractor for GPS location mapping of planting events occurring on the tractor-drawn transplanter without the need for an independent RTK-GPS system located on the transplanter or planter. Figure 12 shows the crop plant locations determined by the automatic GPS mapping transplanter during planting (orange triangle). The inset photo shows the manual RTK-GPS survey measurement of the plant location obtained during ground truthing. The ground truth points (black circles) were overlaid on the automatically generated map for comparison.

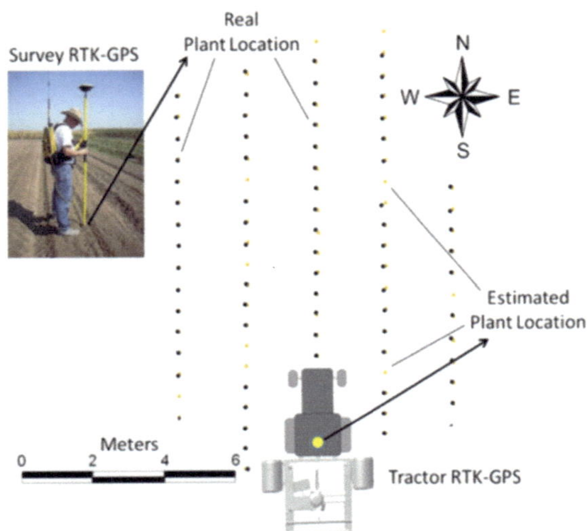

Figure 14. Automatically generated crop geoposition map

10.5. Precision weed management system

Improved mechanical methods of weed control have been motivated by an increased consumer demand for organic produce, consumer and regulatory demands for a reduction in environmentally harmful herbicide use, and a decrease in the availability of farm workers willing to perform manual agricultural tasks such as hand weeding. Extensive research has been conducted to address this issue and alternate techniques have been developed to control weeds in the plant line [21-23].

Since the RTK GPS based seed or transplant map was close to the actual plant map, it was hypothesized that a simple greenness sensor could be used to look for plants and when a plant is detected its coordinates could be compared with the coordinates of plants in the plant map. If there is no corresponding plant on the plant map, then it can be assumed to be a weed and an appropriate herbicide could be applied kill the weed. Such weed-specific chemical application can reduce the amount of chemical by 24-51% thus reducing cost and protecting the environment from the harmful effect of the chemical. Employing this principle, we designed and built, at University of Sevilla, a fully automatic electro-hydraulic side-shift frame for row center positioning controlled by RTK GPS location information to perform a precise mechanical (between row) and narrow herbicide band spray (over the crop row) weed control. Figure 15 shows the frame, placed between the tractor and implement, that allowed centering the narrow band treatments (10 cm) of herbicide above the rows and parallel to the crop rows with a minimum lateral drift (cross track error).

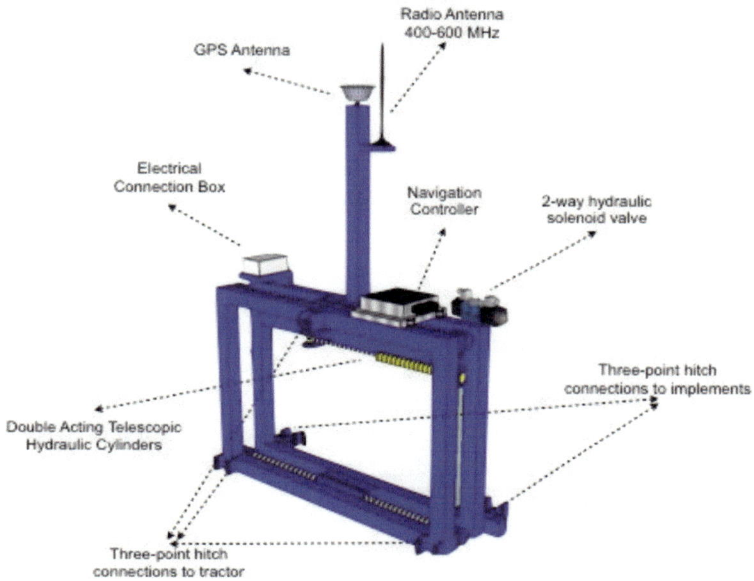

Figure 15. Schematic diagram showing the side-shift frame system developed for row position centering controlled with a RTK-GPS geopositioning system.

This new system, equipped with RTK GPS technology, was used for targeted herbicide application to weeds along crop rows, without reducing the efficacy of the intra-row chemical control treatment, while providing savings of approximately 50% of herbicide. The savings in applied chemical not only reduced production costs but also reduced the environmental impact caused by the chemical. Moreover, use of this system led to reduction of labor required to hand weed on the average from 15.3 hours per hectare for the conventional treatment and 13.2 hours for the improved plant/weed specific treatment. Complete elimination of herbicide application while achieving a high percentage of elimination weeds is a very attractive proposition and is critical for organic growers to reduce production cost. However, it is a very challenging task. An interesting approach is to use plant-specific mechanical cultivation based on a RTK-GPS based plant map. Dr. Slauhgter and his research group at the department of Biological and Agricultural Engineering (UC Davis) have developed an automatic intra-row, automatic weeding system using cultivator knives that remove the weeds along the plant lines of transplanted processing tomato crop using RTK-GPS based plant map obtained during transplanting operation. Field test results indicated that this RTK-GPS based automatic weeding system did not damage any plants while performing intra-row cultivation at travel speeds of 0.8 and 1.6 km/h. Additional information of automatic intra-row weeding system can be found in [14].

(a) (b) (c)

Figure 16. A RTK-GPS based robotic cultivation system-(a) A cultivator retrofitted with a RTK-GPS system in operation, and (b) The cultivating tines open to protect the plant based on the plant map and closed to get rid of weeds in between plants in the crop row.

10.6. Intelligent system applications

The introduction of semiautomatic systems in combine harvesters a few decades back was one of the first steps towards automation. Today, full automatic, robotic systems have been incorporated into many different agricultural operations - from harvesting to intelligent application of herbicides. These new systems in the agricultural sector present new challenges such as safety, user education and training, and machine actuation. Of these, safety is the most important as actuation often requires sensory information before mechanical execution. Robotic systems require sophisticated hardware and software in order to allow the adaptation to changing environments and accomplish exigent missions in a safe and efficient way. Two different approaches have become essential characteristic of intelligent vehicles system: combining local information with global localization to enhance autonomous navigation, and integrating inertial systems with GNSS for vehicle automation [24].

Many researchers have spent a significant amount of effort in recent years to solve pressing and challenging problems facing agriculture today. For instance, the DEMETER Project (USA, Carnegie Mellon University, 2000) has led to the development of a new generation self-propelled hay harvester for agricultural operation. The goal was to provide a "Program-Execute" so that an expert harvester operator merely has to harvest a given field once ("programming the field") allowing a less-skilled operator play back the programmed field ("executing the field") at a later date. A major disadvantage of this approach from the point of view of safety was the potential loss of large and heavy machines in the field leading to very dangerous situations. This could be avoided by using a fleet of small machines that are deployed simultaneously and less dangerous to the people and the surrounding while being equally effective in completing the operation.

There have been some attempts to configure colonies of robots that could be used for agricultural activities, such as the project entitled “Cognitive Colonies” (USA, Carnegie Mellon University, 2001). This project consists of ten small robots constituting a “model” facility. These robots will form a colony whose sole purpose is the generation of a map of this area. After an initial period during which basic distributed mapping operation is created, the sponsors will be asked to "disable" robots of their choice and observe the reaction of the colony to this loss. The results of this project can be applied to configure groups of robots that work collaboratively in accomplishing a task changing the configuration depending on the number of operative robots.

In the past couple of decades, precision agriculture has emerged as a promising field to increase crop productivity. For instance, the Robotic Weeding Project under development at the Department of Agricultural Engineering, University of Aarhus, Denmark, is devoted to building an autonomous vehicle with a vision system capable of controlling a grid-dosing sprayer system. Likewise, the project "Autonomous Agricultural Spraying" (USA, Carnegie Mellon University, 2007) devoted to make agricultural spraying significantly cheaper, safer and more environmentally friendly through automation, such that a single operator, from a remote location, can oversee the nighttime operation of at least four spraying vehicles.

In the last two years, projects such as RHEA (Robot Fleets for Highly Effective Agriculture and Forestry Management) have emerged to develop a fleet of heterogeneous –land and aerial-robots to carry surveillance and actuation system over the mission field. The land units will be based on medium-sized autonomous vehicle with onboard equipment for navigation and application of treatments. Thanks to the integration of GNSS and sensors each robot controller can receive its desired trajectories from the mission planner and its current position from its own geographical positioning system. Every robot controller can compute its own control signals for traction and steering in order to track the desired trajectory avoiding obstacles – trees, bushes, rocks, holes, protrusions, animals, humans, etc.- in the path.

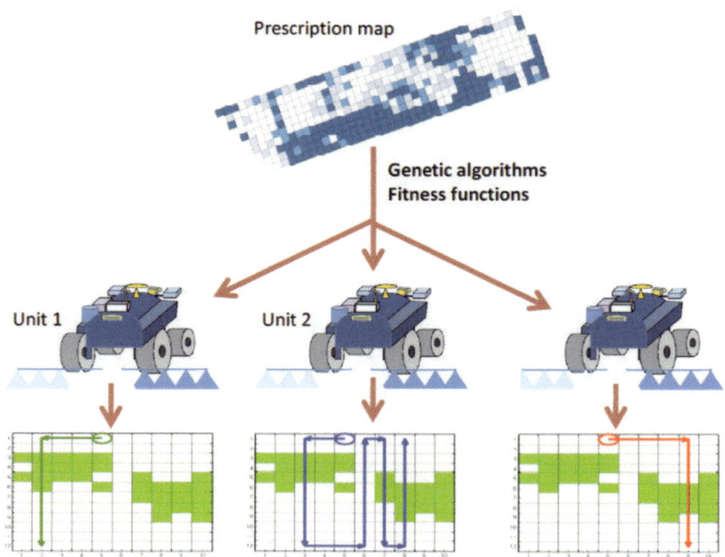

Figure 17. Distribution of three ground units to conduct a spraying operation following a predefined prescription map. Path plan for each individual unit marked with a different color. Green cells indicate the presence of weeds. Circles indicate the starting position

11. Conclusion and outlook

The future applications of GNSS in precision agriculture operations know no bounds. This type of agriculture, where the positioning along with additional data on the vehicle status, soil properties, crop health, and fertilizer requirements provide the knowledge base for decision making and management to improve productivity, safety, and quality while reducing cost and environmental impact. The central concept of precision agriculture is to apply only the inputs - what you need where you need and when you need - and this can only be done if large amount of geo-refenced data are available to make informed management decisions.

Agricultural applications such as yield monitoring, variable rate application, plant mapping, precise weed management, etc. require many sensors to acquire data from the field, but these data can only be linked together through a map by means of the location information provided by the GPS or any other GNSS receiver. With this type of precision agriculture data, the prescriptions maps can be created for planning future farming tasks.

One benefit of GNSS receivers over GPS-only is the increased number of satellite available for location calculations by the receiver. This is possible because GNSS-compatible equipment can use navigation satellites from other networks outside the GPS system. In addition, reliability is increased in areas where GPS receivers cannot operate or provide poor accuracy. The higher end GNSS receivers can currently observe up to around 72 satellites and are capable of accommodating additional satellites as more satellite-based systems become operational. More benefits of GNSS receiver include; 1) A shorter warm-up time (known as "time to first fix"); 2) Reduced delay in recomputing a position if satellite signals are temporarily blocked by obstructions (reaquisition time); 3) The ability to compute a position where is difficult for a GPS receiver operation, specially near tree rows, building, big obstacles, etc. In order to clarify and avoid confusion among agricultural users it would be worthwhile to remember that by design the GNSS receivers are compatible with GPS; however, GPS receivers are not necessarily compatible with GNSS.

Currently, the scientific community is devoting great efforts to avoid the GNSS signal interruption caused by shading of the GNSS antenna by terrain or obstacles (e.g. trees, buildings, implements, etc.) or by interference from an external source to improve the accuracy of agricultural applications. The need to provide continuous location data or navigation during periods when the GNSS signal is interrupted is the impetus for integrating GNSS with various additional sensors (e.g. inertial sensor, dopplerometers, altimeters, odometers, etc.). The integration of GNSS products and services with sensors will expand the possibilities of agricultural use of this technology in the future even further.

Author details

Manuel Perez-Ruiz
Aerospace Engineering and Fluids Mechanics Department, University of Sevilla, Spain

Shrini K. Upadhyaya
Biological and Agricultural Engineering Department, University of California, Davis, USA

Acknowledgement

This work was partially financed by the European Union's Seventh Framework Programme [FP7/2007-2013] under Grant Agreement number 245986. The authors thank professor David Slaughter, Biological and Agricultural Engineering Department, UC Davis for his valuable suggestions during the writing process of this chapter.

12. References

[1] Griepentrog, H.W., Blackmore, B.S., Vougioukas, S. (2006). Positioning and navigation (Chapter 4.2). In A. Munack (Ed), CIGR handbook of agricultural engineering: Volume VI-information technology (pp. 195-204). St. Joseph, MI 49085, USA: ASABE.

[2] Bauer, W.D., Schefcik, M. (1994). Using differential GPS to improve crop yields. GPS World, 5(2), 38-41.

[3] Petersen, C. (1991). Precision GPS navigation for improving agricultural productivity. GPS World, 2(1), 38-44.

[4] Wilson, J.N. (2000). Guidance of agricultural vehicles-a historical perspective. Computers and Electronics in Agriculture, 25, 3-9.

[5] Pérez-Ruiz, M., Carballido, J., Agüera, J., Gil, J.A. (2011). Assessing GNSS correction signals for assisted guidance systems in agricultural vehicles. Precision Agriculture, 12, 639-652.

[6] Larsen, W.E., Nielsen, G.A., Tyler, D.A. (1994). Precision navigation with GPS. Computers and Electronics in Agriculture, 11, 85-95.

[7] Chamen, W.C.T., Watts, C.W., Leede, P.R., Longstaff, D.J. (1992). Assessment of a wide span vehicle (gantry), and soil and crop responses to its use in a zero traffic regime. Soil and Tillage Research, 24, 359-380.

[8] Chamen, W.C.T., Dowler, D., Leede, P.R., Longstaff, D.J. (1994). Design, operation and performance of a gantry system: experience in arable cropping. Journal of Agricultural Engineering Research, 59,45-60.

[9] Grewal, M.S., Weill, L.R., Andrews, A.P. (2011). Global Positioning Systems, Inertial Navigation, and Integration. Wiley, ISBN 0471-35032-X, New York.

[10] El-Rabbany, A. (2006). Introduction to GPS- The Global Positioning System. Second Edition. Artech House, Boston, 210 p.

[11] Misra, P., Enge, P. (2006). Global Positioning System: Signals, Measurements, and Performance, 2nd ed., Gamba-Jamuna Press, ISBN 0-9709544-1-7, Lincoln, MA.

[12] Heraud, J.A., Lange, A.F. (2009). Agricultural automatic vehicle guidance from horses to GPS: How we got here, and where we are going. ASABE Distinguished Lecture Series No. 33. ASABE, St. Joseph, MI 49085, USA.

[13] Ehsani, M.R., Upadhyaya, S.K. Mattson, M.L. (2004). Seed location mapping using RTK-GPS. Transactions of the ASABE, 47, 909-914.

[14] Pérez-Ruiz, M., Slaughter, D.C., Gliever, C.J., Upadhayaya, S.K. (2012). Automatic GPS-based intra-row weed knife control system for transplanted row crops. Computers and Electronics in Agriculture, 80, 41-49.

[15] Agüera, J., Pérez-Ruiz, M., Gil, J.A., Madueño, A., Zarco-Tejada, P., Blanco, G. (2003). Determining spatial variability of yield and reflectance of a cotton crop in the Guadalquivir Valley. In memory: 4th Conference on Precision Agriculture 347-348 pp.

[16] Spark, D.L., Page, A.L. Helmke, P.A., Loccpert, R.M., Sottanpour, P.N., Tabatai, M.A., Johnston, C.I, Sumner, M.E. (1996). Methods of soils analysis, part 3rd ed. chemical methods, Agron. eds No. 5, 3 ed., American Society of Agronomy, Madison.

[17] Bravo, C., Giráldez, J.V., Agüera, J., Pérez-Ruiz, M., Gónzalez, P., Ordóñez, R., Gil, J.A. (2004). Assessing and modelling spatial variability of soil properties and association with the cotton yield map. Book of Astracts. ISBN 90-76019-258. AgEng 2004, Leuven. pp. 916-917.

[18] Abidine, A.Z., Heidman, B.C., Upadhyaya, S.K., Hills, D.J. (2004). Autoguidance system operated at high speed cuases almost no tomato damage. California Agriculture, 58, 44-47.

[19] Chancellor, W.J. (1981). Substituting information for energy in agriculture. Transactions of the ASAE, 24, 802-807.

[20] Chancellor, W.J., Goronea, M.A. (1993). Effects of spatial variability of nitrogen, moisture, and weeds on the advantages of site-specific applications for wheat. Transactions of the ASAE, 37, 717-724.

[21] Jørgensen, R.N. Sørensen, C.G., Maagaard, J., Havn, I., Jensen, J., Søgaard, H.T., Sørensen, L.B., 2007. HortiBot: A system design of a robotic tool carrier for high-tech plant nursing. Agricultural Engineering International: CIGR Ejournal Manuscript ATOE 07 006. Vol. IX: 13pp.

[22] Nørremark, M., Griepentrog, H.W., Nielsen, J., Søgaard. H. T., 2008. The development and assessment of the accuracy of an autonomous GPS-based system for intra-row mechanical weed control in row crops. Biosyst. Eng. 101, 396-410.

[23] Van Evert, F. K., Samson, J., Polder, G., Vijn, M.,Van Dooren, H., Lamaker, A., Van Der Heijden, G. W. A. M., Van der Zalm, T., Lotz, L. A., 2011. A robot to detect and control broad-leaved dock (Rumex obtusifolius L.) in grassland. J. Field Robot. 28, 264-277.

[24] Rovira, F., Zhang, Q., Hansen, A.C. (2010). Mechantronics and Ingelligent Systems for Off-road Vehicles. London: Springer.

Chapter 2

Adjustment Methodology in a Regional Densification of a Terrestrial Reference Frame

María Laura Mateo and María Virginia Mackern

Additional information is available at the end of the chapter

http://dx.doi.org/10.5772/50548

1. Introduction

The measurement of geodetic networks by satellite positioning is becoming more and more common, given the advantages of this technology to solve the main problems of geodesy:

- Planimetric geo-reference to a geocentric system
- Independence of intervisibility between points
- Increase of relative precision.

The purpose of working under the Network scheme is to have superabundant observations that allow adjustment by estimating parameter (coordinates) values and precisions.

Given the versatility of the method of indirect or parametric observations, it has become the most widely used as a method of geodetic adjustment.

The Adjustment Method by Non-Conditioned Indirect Observations, also known as Method of Coordinate Variation, or Parametric Method has been widely accepted in Geodesy. by means of the use of observation equations, in which the parameters are the coordinates.

This adjustment made under the application of the method of minimum squares, estimates parameters as well as their precisions.

It allows to incorporate equations of coordinates, which are essential when introducing the reference frame and generating an external control.

In the adjustment, an appropriate management of the coordinates requires:

- The introduction of fiducial coordinates with a precision better than the precision of the network to be adjusted.
- The adoption of appropriate weights to these coordinates, in order not to create unnecessary deformations.

INTECH
open science | open minds

© 2012 The Author(s). Licensee InTech. This chapter is distributed under the terms of the Creative Commons Attribution License http://creativecommons.org/licenses/by/3.0), which permits unrestricted use, distribution, and reproduction in any medium, provided the original work is properly cited.

These objectives, as will be shown in this chapter, are strongly correlated.

The strict frame introduction by fixing the fiducial points could often result in an unfavorable option because it induces loss of precision to the network, caused by internal deformation. On the contrary it is possible to generate a free or quasi free adjustment omitting the introduction of the datum in order not to cause deformation to the network. Finally, it is possible to achieve a condition of balance by adopting the appropriate strategy, the optimum quantity and distribution of fiducial points. This solution must achieve the required precisions and the densification of the selected reference frame [2].

2. Adjustment strategy

2.1. Internal precision

Before the datum input it is necessary to estimate the internal precision of the network to be adjusted. Knowledge of network internal precision will allow the definition of tolerances when datum is introduced.

This analysis is possible because of the existence of superabundant observations that is to say in the presence of degrees of freedom. In the satellite positioning techniques the adjustment over observations made in different periods or days is very important since these observations are affected by factors that change through time such as the tropospheric and ionospheric delay, oceanic and terrestrial tides, etc. This is the reason why it is necessary to measure the points of the networks in different sessions (2 or more are recommended). The analysis of network precision is carried out on such points of over-occupation.

A "free" adjustment to the network is performed (assigning a very low a-priori weight to the coordinates of the points). This adjustment combines the total group of sessions in the so called multisession adjustment. After the application of similarity transformation (Fig.1), this result is compared with the network solution of each daily session. The internal precision of the free network is estimated from the quantification of the residuals obtained (Fig.2).

The methodology for performing the analysis on the internal accuracy is summarized in the following items and Fig. 2 outlines the procedure.

1. A free weekly adjustment is performed incorporating each of daily networks.
2. Transformation parameters are estimated between the weekly combined network and each daily networks.
3. The transformation parameters are applied to each of the daily solutions.
4. The residuals are calculated by comparing the coordinates arising from the previous processing and the coordinates of the weekly solution.
5. The obtained residuals were analyzed according to predetermined tolerances. If it is necessary stations are reduced.

The knowledge of the internal precision of the network makes possible to perform the tie of the network to a certain reference frame.

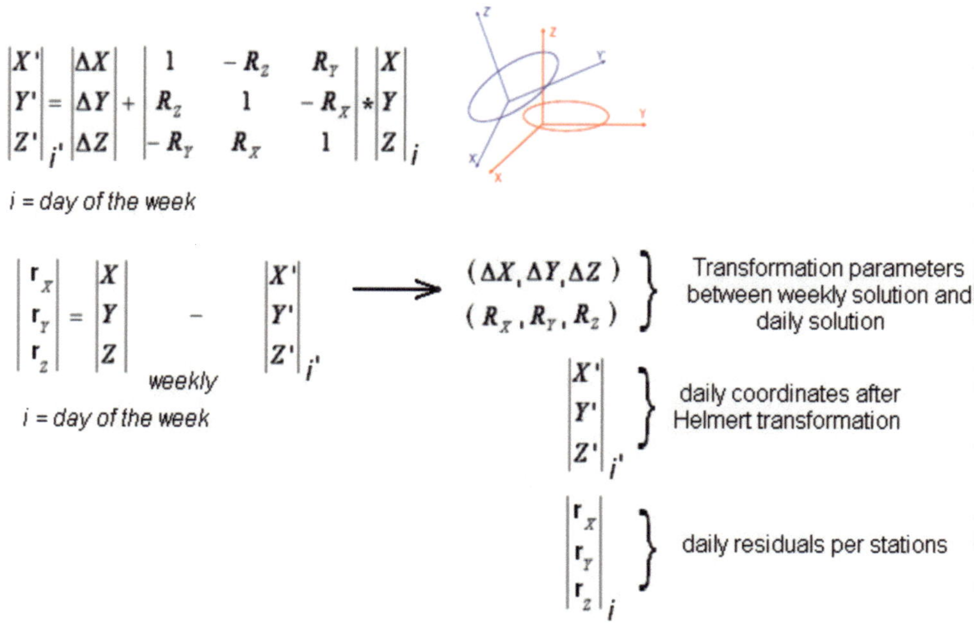

Figure 1. Similarity transformation.

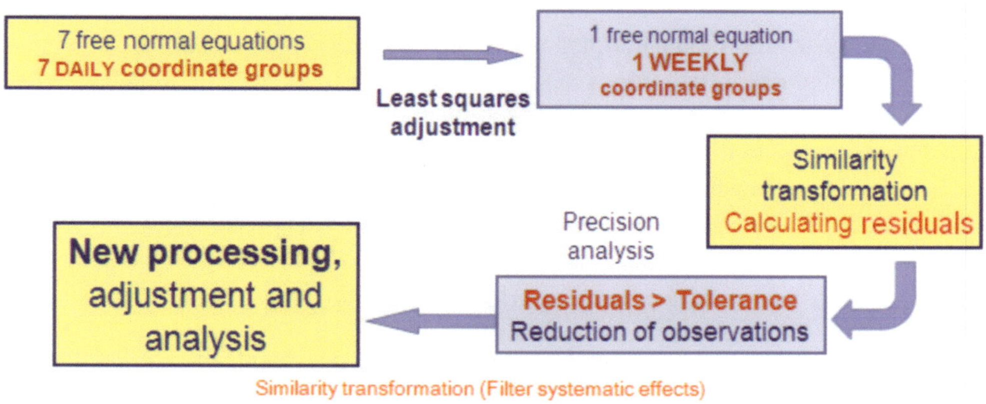

Figure 2. Analysis of the internal precision of the network.

2.2. Introduction to the reference frame

Once the analysis of the internal precision of the network to be adjusted is carried out, the second adjustment, "weighted adjustment" takes place, in which the control points are selected and a-priori errors are set.

The datum input strategies are mainly based on the adhesion or not to the equations of coordinates for fiducial points. An appropriate weight is added to them in order to link to a greater or lesser extent the network vectors to the positions determined by the values of the coordinates mentioned.

The most common strategies in the adjustment of normal equations, with the objective to link the network to a certain reference frame are the followings [5][6]:

- Minimum constraint solution: this kind of solution adjusts the network by performing a roto-translation in the measured and processed network towards the reference frame desired to be used as control or Datum, materialized in the fiducial coordinate group. Similarity parameters are applied, calculated from the coordinates of fiducial points and their corresponding of the free network.
- Coordinates constrained: this strategy allows you to force the coordinates of the points taken as fiducial network to their corresponding coordinates in the reference frame. The latter should be assigned a weight according to their precisions.
- Coordinates fixed: with this option the coordinates of the selected stations are completely fixed as fiducial, accepting that they do not receive any correction.

Entering the control coordinates with a certain weight produces a double effect: it ties the network to the reference system and consequently causes certain deformation. Such effects are strongly related. A greater weight at the fiducial coordinates introduces the frame with greater accuracy at the expense of greater distortion of the network. For these reasons there is no one adjustment strategy only. Different adjustments are made by varying both the adjustment strategy as the corresponding weights. Among the various solutions will choose the optimal strategy from a careful analysis of results [3].

The choice of the solution to be adopted is based on a double analysis, aiming at a balance between the reference frame input and the minimum network deformation. This is accomplished by controlling, on the one hand, the fact that the coordinates of the support points do not vary more than the error value with which they were determined, keeping the precision of the reference frame; and, on the other hand, that this adjustment does not create deformations to the network in the points that exceed its internal precision [3].

2.2.1. Strategy for the analysis of results in the introduction of the Datum

The proposal to analyze the solutions is based, on the one hand, on the estimation of similarity transformations between the coordinates obtained from free adjustment of the network and those obtained from weighted adjustment on the other hand. The purpose of the application of this transformation to the weighted network to be assimilated to the free network is to filter the systematic effects that may be affecting it, mainly the differences between reference systems. Only the residual effects that show the type of deformation generated by the weighted adjustment are left. This deformation can be quantified by means of the calculus of the residuals between the coordinates of the weighted network and the free network transformed [4]

In order to achieve balance the analysis of these deformations must be considered on one of the sides of the scale. On the other side of the scale it is necessary to quantify the precision of the frame over the adjusted network. This is observed in the magnitude of corrections of the estimated coordinates (new coordinates minus a-priori coordinates) in the fiducial points.

They should not exceed the error value with which they were determined in order to preserve the precision of the reference frame that they intend to densify.

The steps to be followed in this analysis methodology are mentioned below.

1. A quasi free adjustment is made, by assigning a very low a-priori weight to all point coordinates.
2. A weighted adjustment is made, by selecting a weighting strategy and assigning a-priori weights to the coordinates of the control points.
3. Seven similarity transformation parameters between the coordinates of the points of the free network and the weighted network are established.
4. The similarity transformation is applied to the free network in order to take it to the reference frame of the weighted network.
5. The residuals are calculated after the application of the transformation mentioned above.
6. The calculated residuals are analyzed.
7. The extent of the estimated corrections to the a-priori coordinates of the fiducial points is analyzed.
8. The process is repeated, modifying the adjustment strategy and the weights to the fiducial coordinates. It is also repeated changing the fiducial reference frame in case there is the possibility to choose more than one.
9. The optimum weighted adjustment is selected, based on the comparison of the results obtained for each case in points 6 and 7, or decisions are taken as regards the control network to be adopted.

Figure 3 outlines the proposed analysis methodology. Where it comes to finding the balance between the distortion introduced by the reference frame to be densified and accuracy on the same network densification.

3. Example - An actual case of application

In order to present this theory in practice, the Methodology used for the processing of a regional continuous GNSS network in South America (SIRGAS-CON-D-South), is described. In this point the models used in the processing, the adjustment methodology and the result analysis are described[7].

3.1. Densification network to be adjusted

As the main objective is to obtain a high precision network, it was necessary for the observations gathered by observation GNSS permanent stations to undergo a strict processing, modeling in a precise form the physical phenomena which affect the observations and estimating those parameters that cannot be modeled [5].

The observations of 98 continuous GNSS stations were gathered and precise ephemerides were used together with Earth Orientation Parameters (EOPs) provided by IGS (International GNSS Service.)

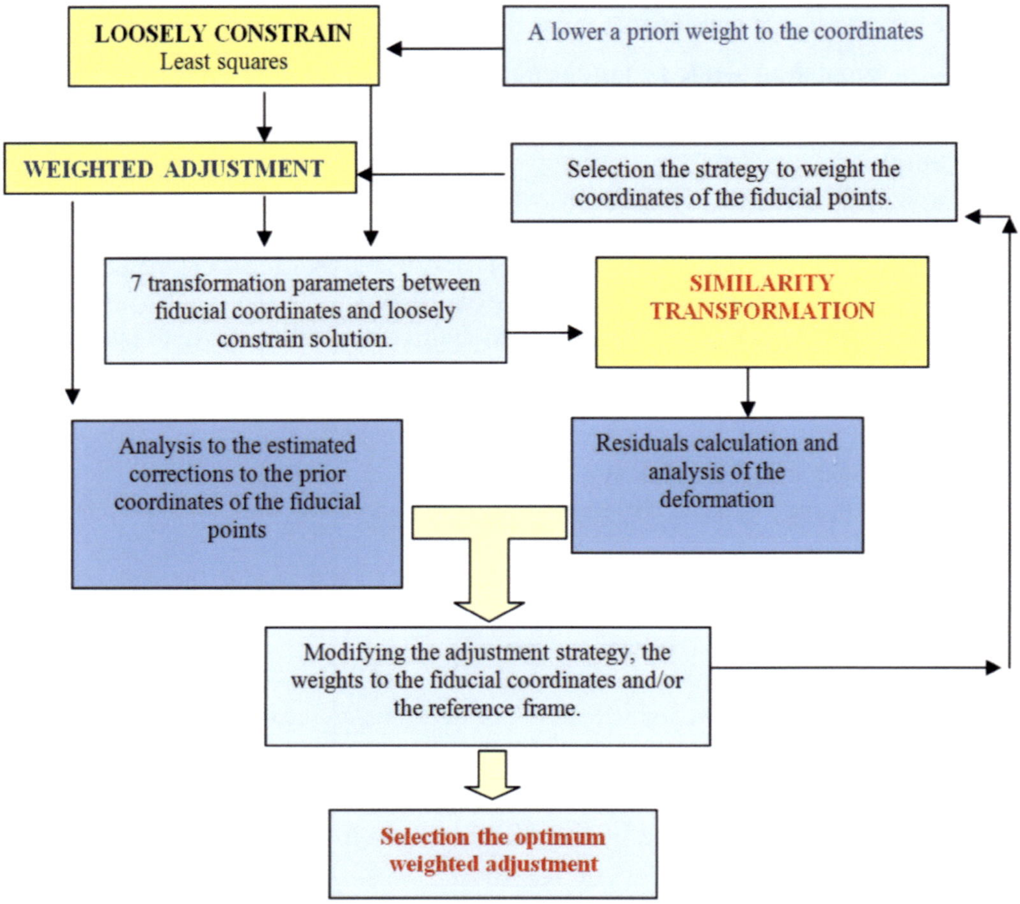

Figure 3. Methodology for analysis of results. Balance between accuracy and deformation.

The total 2.5 year-period of observation to be processed was divided into GPS weeks, witch provided data redundancy and over-occupation, in order to allow the corresponding accuracy analysis week by week and then the total that includes the whole period.

Other variables considered in the processing were: published satellite problems and maneuvers, as well as corrections to the antenna phase centers used. The latter were consistent with the offset of satellite antennas, on the one hand, and with the specific equipment (receiver, antenna, radome, off set, etc.) of each GNSS continuous station on the other. In addition, a-priori coordinates and velocities for all processed stations were incorporated to the processing.

Taking into account all these elements, a data pre-control and pre-processing were carried out, which detected outliers and inconsistencies in the data. In certain cases, it was necessary to reduce or cut data which did not comply with the required quality criteria.

Once the pre-processing was accomplished, the vectors network was constituted and also the parameters estimation was made. In this processing, models of ocean loading, gravitational attraction, nutation and precession were considered and improved by EOPs, calculated by IGS and only tropospheric parameters, ambiguities and corrections to a-priori coordinates were estimated.

The most important parameters introduced in the processing are detailed as follows [8]:

Observations	Double differenced
Sampling rate	30 sec
Elevation cutoff	03º
Baselines strategy	MAX-OBS
Observations weighting	cos Z
Orbits/EOP	IGS final - IGS08 and EOP week
A-priori Troposphere model	Niell dry component
Troposphere	Zenith delay estimated each 2 hours (12 daily corrections p/station) A-priori sigma applied with respect to Niell prediction model (wet component) -first parameter +/- 5 m absolute and +/- 10 cm relative
Ambiguities	QIF strategy, no ionosphere model applied
Ocean tide model	FES2004
Phase center variation	Absolute (IGS_08)
Coordinates and velocities	IGS05_R
Daily solution	NEQ files, free network solution (s=±1m)
Week solution	SINEX files, Free network solution (s=±1m)

Table 1. Parameters incorporated in the processing of GNSS observables

3.2. Free weekly adjustment, analysis of internal precision

The daily normal equations generated after parameter estimations, were subjected to a quasi-free weekly adjustment, with which the analysis of the internal precision of the network was performed. In this adjustment the normal equations obtained for each of the seven days of the week are combined in a single system of equations. All the point coordinates are constrained using a 1.00 meter a-priori sigma. In this way, the network is only tied to the frame established by the orbits at the moment of observation.

The solution of each of the 7 days was analyzed together with the combined weekly solution, starting from the comparison of coordinates after the application of similarity transformation. Residuals were calculated after the transformation and were verified to ensure that they were within the pre-established tolerances. This process was described in detail in point 2.1, Fig.1 and Fig.2.

It is worth noting that these tolerances of internal precision are determined in each case according to the repeatability analysis of coordinates. The latter was possible thanks to the observation redundancy and a particularly varied sampling according to the time (Fig. 4), which is relevant due to its influence on a vast quantity of factors taking part.

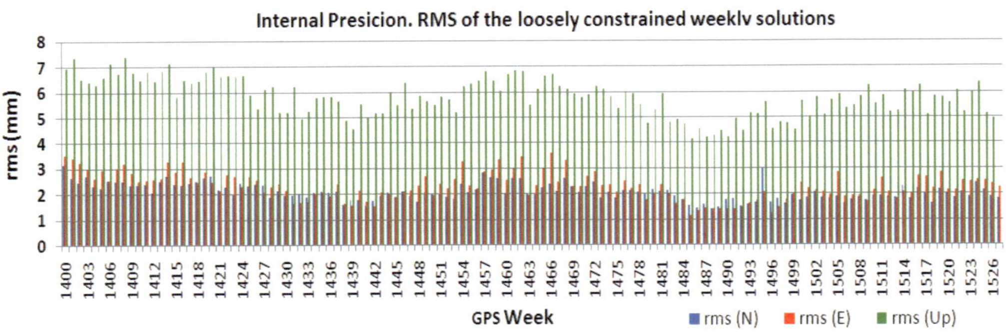

Figure 4. Internal precision. SIRGAS-CON-D-South network – Period: 127 weeks.

In the presented case these tolerances were considered for north and east components in 15 mm. and for a up component of 30 mm. For residuals larger than the tolerances, it was concluded that the main causes were: insufficient observations (shorter than 10 hours), too high offset of the receiver clock or too many outliers.

In some cases, it was necessary to partially or completely reduce the observations of such stations. Once the control was finished, the internal precision of the network in 2 mm. was estimated for the horizontal components and for up of 4 mm. (Fig. 4.)

Two of the three strategies mentioned were used: "Coordinates constrained" and "Minimum constraint solution".

The "Fixed Coordinates" method was discarded because it is a very precise network (2mm), as expressed in the previous paragraph, not counting to the date of this adjustment with fiducial coordinates that would present an order of precision better than the mentioned one.

3.3. Reference frame selection

Among available high precision global frames this selection was performed taking into account the precision quality of the network to be adjusted.

The availability and the distribution of processed points which had coordinates in one of the ITRF [9][10] defined by the IERS or in one of the frames defined by the IGS [11] were analized.

The epoch of reference frame was also taken into account, since it is necessary to update the coordinates to the date of the observations to be adjusted [12].

Of the most recent global frames to the date of the adjustment (2008.1), ITRF2005 and IGS05, IGS05 was selected, since it presents a good availability and distribution of fiducial points over the network to be adjusted (Fig.5). In this selection, the ITRF2005 was discarded because it had been calculated with relative antenna phase centre corrections whereas IGS05 was adopted because absolute antenna phase centre corrections had been used in its calculation.

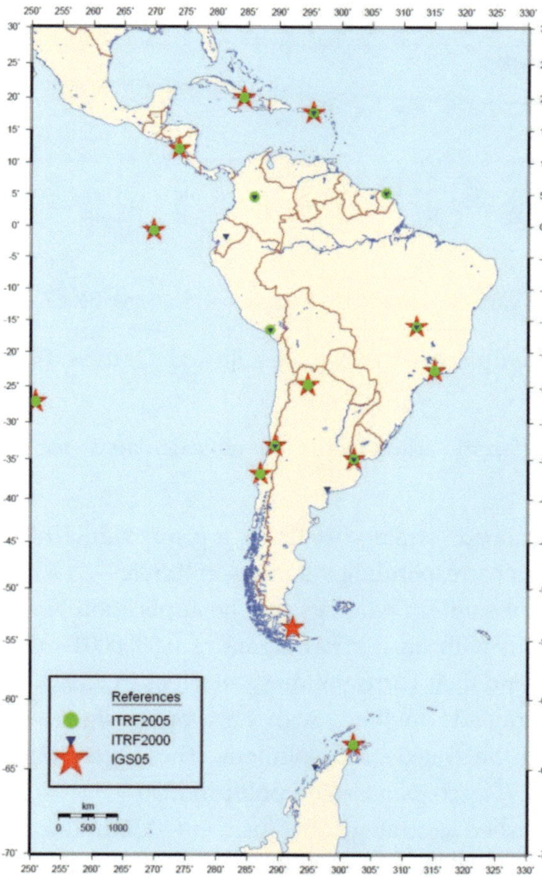

Figure 5. Stations with coordinates in ITRF00, ITRF05 and IGS05.

The possibility of adopting as frame IGS weekly solutions was also considered. IGS weekly coordinates refer to the IGS05. They coincide with the epoch of each week to adjust. In this case it was not necessary to update coordinates by velocities regardless of introducing error. The error introduced by the update coordinates for velocities is produced at the stations having in the time series of coordinates, nonlinear variations (mainly in the Up component). GNSS stations show significant seasonal position variations resulting from a combination of geophysical loading and systematic errors [2][5].

As an example (Fig. 6) shows the time series of the "up" component in the stations BRAZ and LPGS.

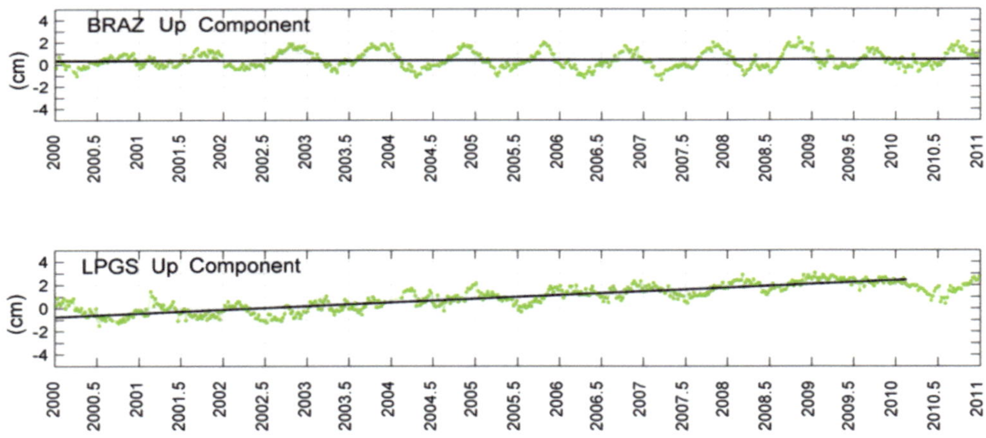

Figure 6. Seasonal position variations in the Up component. Stations: BRAZ and LPGS.

A total of 98 points made up the network to be adjusted. Of these 14 points were considered fiducial.

Summarizing, four different adjustments were calculated for each of the selected weeks:

1. "Coordinates constrained" strategy with an a-priori sigma of 0.00001 m. in fiducial coordinates and their corresponding velocities in IGS05.
2. "Minimum constraint solution" strategy, by the application Not-Net-Rotation and Not-Net-Translation, both with an a-priori sigma of 0.0000001 m without applying scale factor. Coordinates and their corresponding velocities in IGS05.
3. "Coordinates constrained" method, with an a-priori sigma of 0.00001 m, using as fiducial coordinates the weekly IGS solutions. These were obtained by a three-week delay in the site: ftp: // igscb.jpl.nasa.gov/pub/product/
4. Finally it was established as strategy of adjustment "Minimum constraint solution" with an a-priori sigma of 0.0000001 m without applying scale factor, by IGS weekly coordinates.

3.4. Analysis of deformation introduced by Datum

In order to identify which of two adjustment strategies used and which feasible group of coordinates used to introduce the datum was turning out to be most adapted to adjust the network object of this work, a comparison of results was made. 127 weeks of observations were included, from the 1400 GPS week, (November, 2006), up to the 1527 week, (March, 2009), two and a half years of observation).

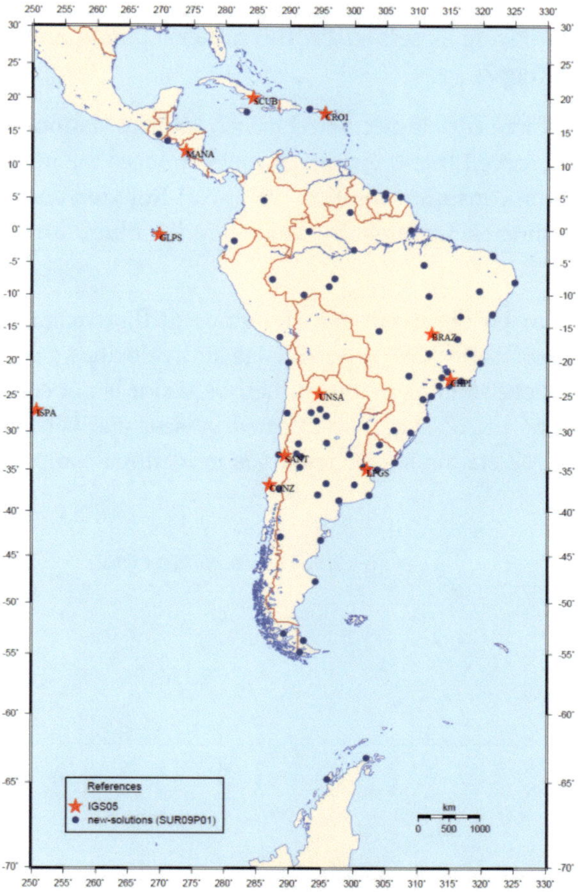

Figure 7. SIRGAS-CON-D-South and fiducial IGS stations.

The analysis was realized on the deformations caused to the loosely constrained network by the introduction of the datum. The deformations were calculated based on the quasi-free network corresponding to every week. It was considered that given the precision of this network, it was the best boss to analyze the deformations, since it is free of the systematic effects of rotation and translation introduced by the reference frame.

To estimate the deformations, each of the 4 solutions (obtained by applying the different two adjustment strategies and two frames), were compared, after the application of Helmert transformation to the quasi-free solution (2.2.1). It is to be noted that the parameters were calculated only with the groups of coordinates of the fiducial points, and then they were applied to the totality of the network for every week of calculation.

The residuals resulting from the totality of the comparisons were plotted for each of the stations, according to three components, north, east and up [5].

3.4.1. Analysis results

The major deformations were observed on the fiducial points and on the points of the contour of the network (Fig.7)

Fig. 8 and Fig. 9 show the residuals calculated in two fiducial stations, CONZ and BRAZ. In both graphs it can be observed that the strategy that produces the minimum deformation to the network is "Minimum constraint solution" (Not-Net-Rotation and Not-Net-Translation), both with IGS05 coordinates updated by velocities (in blue), and for the IGS Weekly coordinates (light blue).

In the same way, it can be observed that the strategy that major deformation offers is "Coordinates constrained" with IGS05 coordinates and velocities (red color). This response is mainly observed in those stations in which their behavior is not contemplated by a linear velocity. For example the BRAZ station is affected by seasonal variations principally in its Up coordinate. The CONZ station shows residuals in its three components, that increase as time passes (Fig. 8).

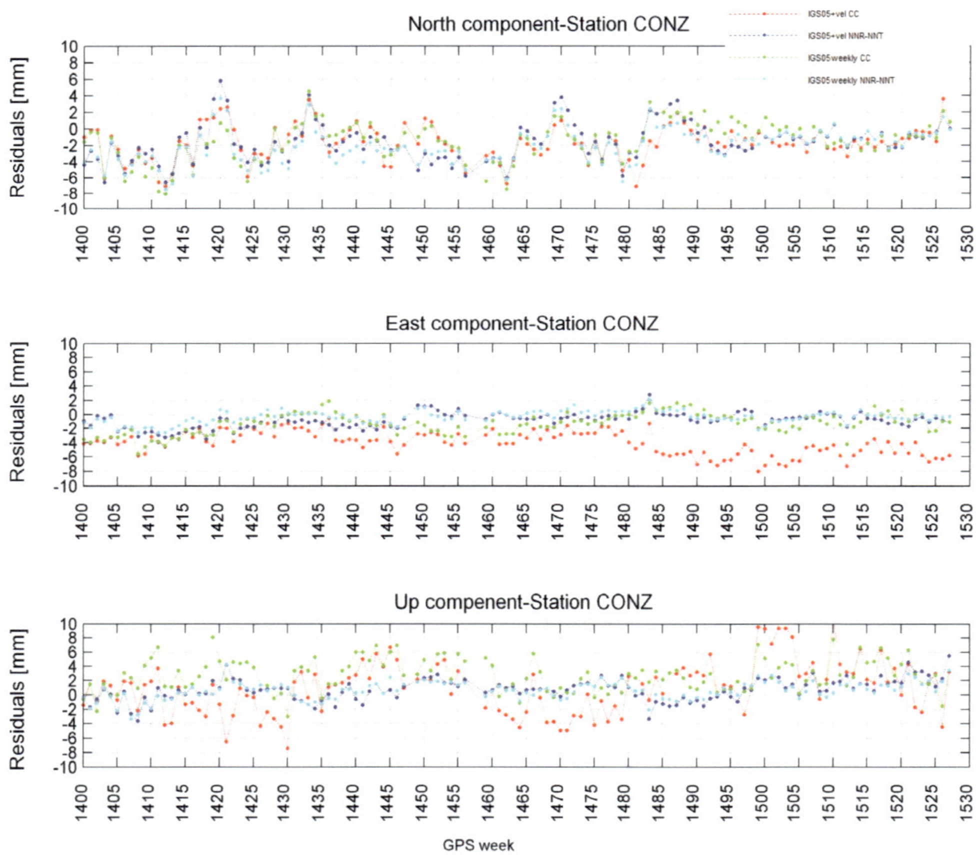

Figure 8. Residual of Helmert transformation in the fiducial station CONZ (Chile).

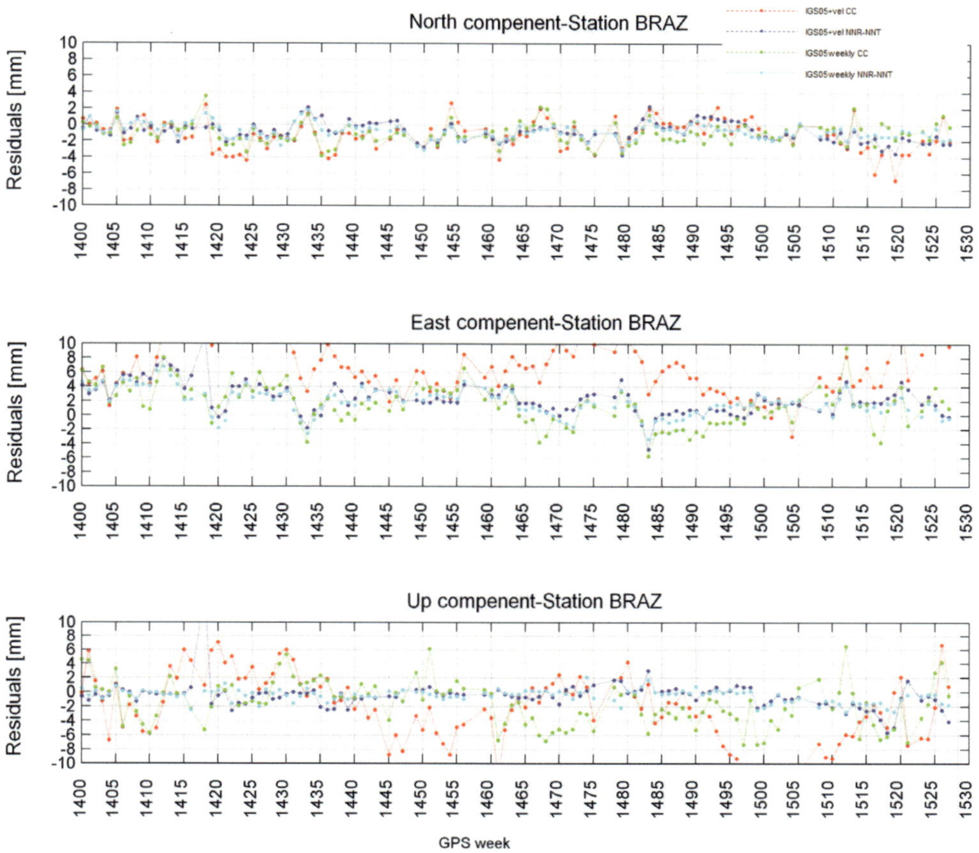

Figure 9. Residual of Helmert transformation in the fiducial station BRAZ (Brasil).

In the results, the behaviors observed in the variation of the coordinates it wasn´t linear. Which allowed to conclude in the need to calculate velocities determined in precise form and that they accompany on the real movement of the coordinates answering to a model defined with better precision.

In order to realize a quantitative analysis there were calculated the average values of these residuals by every station of reference for 127 analyzed weeks, in each of four strategies (table 2 and Fig.10).

Finally there was calculated the average value of these residuals for the whole network according to the strategy, with its corresponding standard deviation (table 1 and Fig. 11.). It is necessary to mention that the residuals in north and east components have combined in a residual called "Horizontal".

It was concluded that the strategy that minor deformation was causing to the network was “Minimum constraint solution” since this type of adjustment tries to support the quasi-free original network without deforming it and only torn it moves trying to accommodate to the frame of reference, producing of this form the minor residuals.

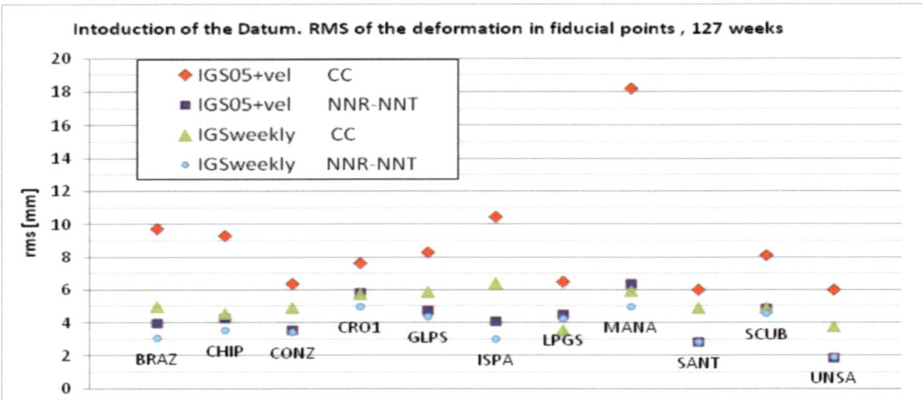

Figure 10. RMS in the reference points according to strategy of adjustment.

RMS (127 weeks)								
Station	IGS05+vel CC (mm)		IGS05+vel NNR-NNT		IGSweekly CC		IGSweekly NNR-NNT	
	Horizontal	Up	Horizontal	Up	Horizontal	Up	Horizontal	Up
BRAZ	7,916	5,686	3,487	1,930	3,569	3,438	2,878	0,999
CHPI	7,470	5,531	3,862	1,757	3,704	2,711	3,402	1,054
CONZ	5,179	3,692	3,133	1,685	3,461	3,543	3,142	1,379
CRO1	5,795	4,980	5,314	2,429	4,224	3,894	4,620	1,886
GLPS	4,388	7,013	4,012	2,491	4,589	3,664	4,163	1,375
ISPA	7,694	7,064	3,643	1,848	5,105	3,904	2,696	1,344
LPGS	4,561	4,639	3,670	2,560	2,708	2,281	3,515	2,366
MANA	15,239	9,955	4,515	4,485	4,680	3,636	4,391	2,365
SANT	3,660	4,787	2,545	1,229	3,298	3,631	2,634	1,046
SCUB	6,892	4,318	4,495	1,893	3,503	3,450	4,341	1,387
UNSA	4,169	4,302	1,621	0,891	2,315	3,018	1,712	0,818
Mean value	6,633	5,633	**3,663**	**2,109**	3,741	3,379	**3,408**	**1,456**
Standard deviation	3,235	1,790	1,002	0,938	0,846	0,507	0,907	0,530

Table 2. RMS in the reference points according to strategy of adjustment.

Analyzing the average values of residuals, they confirm that the minor values obtain in both cases in which there was applied the strategy Minimum constraint (NNR-NNT). Between these two options the one that minor deformation produces is the one that uses the IGS weekly coordinates (last column, table 1 and Fig.11.). It presented an average residual of 3.40 mm in horizontal component and 1.45 mm in height.

The solutions obtained by means of coordinates constrain linking to the weekly coordinates of the IGS also they present minimal deformations. The strategy that should have discarded was CC with coordinates IGS05 updated by speeds, since in this case they were duplicating and even they were trebling the values of introduced deformation.

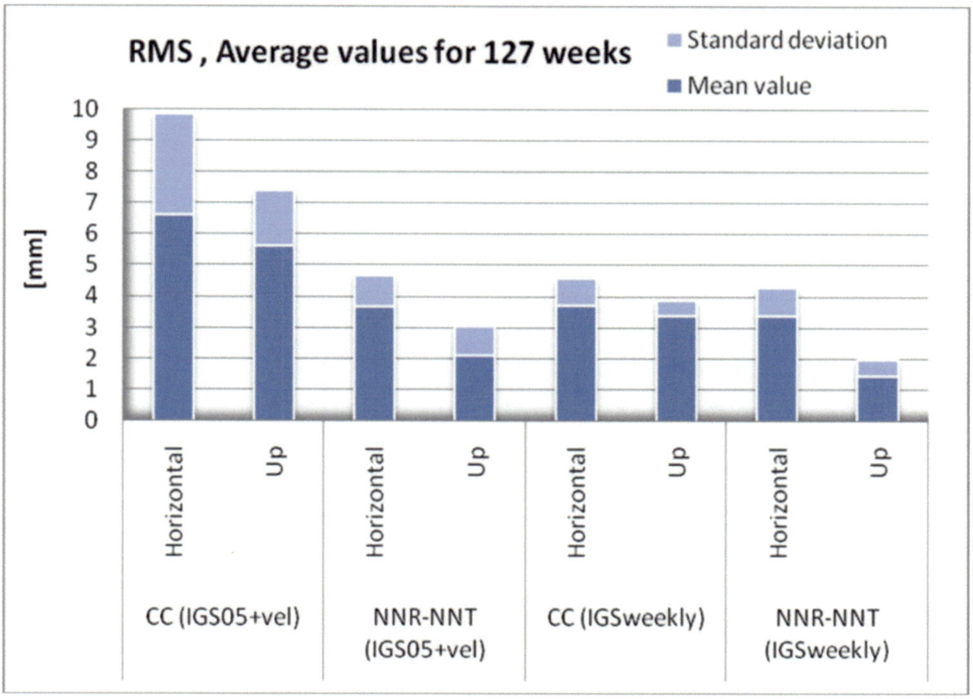

Figure 11. Deformation of the network according to the strategy of adjustment.

3.5. Frame accuracy

In the previous section we arrived at the solution that allows us to deform as little as possible the network. However, this solution will not be adequate if the accuracy of the frame is lost. It's time to reach equilibrium in the balance. Finding the point at which the chosen solution provides lower distortion and greater accuracy

In order to quantify the accuracy of the frame over the adjusted network, the extent of estimated corrections of the coordinates (new coordinates - prior coordinates) in the fiducial points were analyzed. They should not exceed the error value with which they were determined in order to preserve the precision of the reference frame that intend to densify.

The estimated corrections were analyzed for each of the four strategies of adjustment used. In each case they were compared to the a-priori error of fiducial coordinates and the optimum solution was selected together with the analysis of deformation. In the above mentioned solution 80% of the fiducial points presented accuracy to the reference frame within 0.005m., thus complying with the objective pursued. (Fig.12).

Final result was selected as better strategy of adjustment to link the network SIRGAS-CON-D-South to the strategy "Minimum constraint" adopting as reference coordinates the correspondents to the IGS weekly solution.

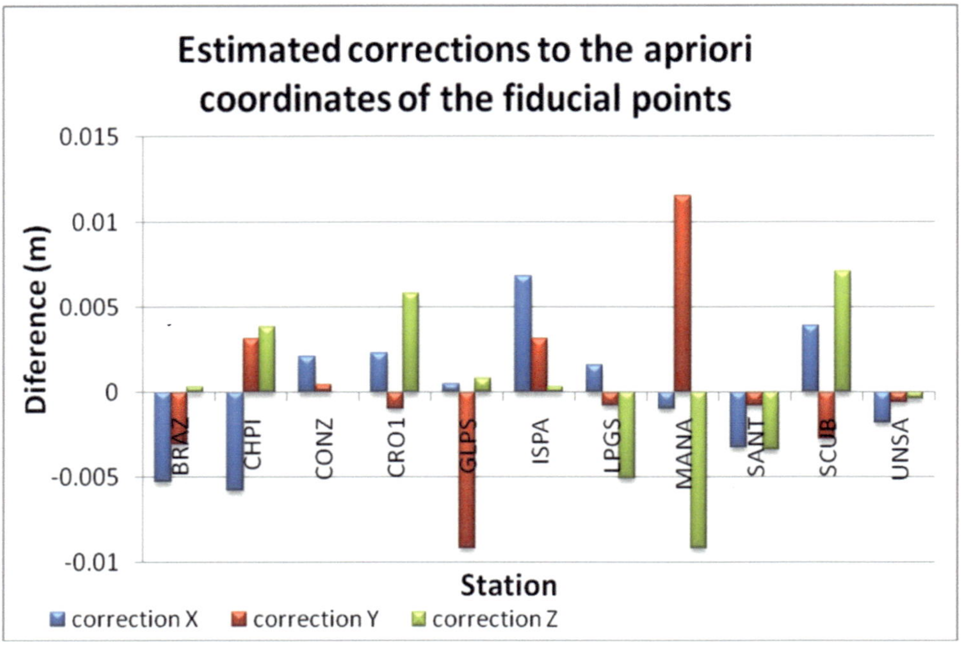

Figure 12. Frame accuracy. Estimated corrections in the fiducial points.

4. Conclusions

The absence of result analysis in the adjustment can bring about one of the following problems: the adoption of an inappropriate frame, its loss of accuracy, or an unnecessary loss of the precision if the Network to be adjusted

To ignore seasonal variations at reference stations can introduce systematic errors in the datum realization and the networks can be significantly deformed [13].

These effects are larger in regional networks than in the global net, especially in regions with strong seasonal variations such as Latinamerica [2].

The geometry of the quasi free network is always deformed, when the geodetic datum is introduced. This deformation is specifically larger at the remote sites of the network, and in fiducial points

Updating strategies based on the linear movement of the reference stations introduce important errors into the station positions, mainly at the reference stations. This is a consequence of constraining a seasonal signal to be a linear trend [13].

It is necessary that the reference frame definition include, together with the usual linear terms, seasonal variations in order to improve the modelling of the reference site motions and to make it more reliable.

In the mean time, weekly solutions of regional reference frames shall be aligned to IGS frames by constraints to the IGS weekly coordinates [5].

Author details

María Laura Mateo
Instituto Argentino de Nivología, Glaciología y Cs Ambientales – CONICET, Argentina
Facultad de Ingeniería -Universidad Nacional de Cuyo, Argentina

María Virginia Mackern
Facultad de Ingeniería -Universidad Nacional de Cuyo, Argentina
Facultad de Ingeniería -Universidad Juan Agustín Maza, Argentina

5. References

[1] Drewes, Hermann (2009). Reference Systems, Reference Frames and the Geodetic Datum – basic considerations. Sideris, M.G. (Ed.): Observing our Changing Earth, IAG Symposia, Vol. 133, 3-9, Springer 2009.

[2] Sanchez L. and Mackern M.V. (2009), Datum realization for the SIRGAS weekly coordinates, Geodesy for Planet Earth, IAG Scientific Assembly 2009; Buenos Aires, Argentina.

[3] Mackern, M. V and Brunini, C. (2004). "El ajuste de Redes en la Geodesia del siglo XXI." Tópicos de Geociencias. Un volumen de estudios Sismológicos, Geodésicos y Geológicos en homenaje al Ing. Fernando Séptimo Volponi. VOL. 1. PAG 115 – 142. EDIT. EFU. Fundación Universidad Nacional de San Juan. ISBN N° 950-605-340-5. San Juan. Argentina.

[4] Mackern, María Virginia (2003). Materialización de un Sistema de Referencia Geocéntrico de alta precisión mediante observaciones GPS. Tesis Doctoral, Universidad Nacional de Catamarca. Argentina .

[5] Mateo M.L. (2011). Tesis doctoral: Determinación precisa de velocidades en las estaciones GNSS de medición continua de América Latina, Red SIRGAS. Facultad de Ingeniería – Universidad Nacional de Cuyo, Mendoza, Argentina.

[6] Dach, R., Hugentobler, U., Fridez, P. and Meindl, M.: Bernese GPS Software version 5.0, Astronomical Institute, University of Berne, Switzerland, 2007.

[7] Brunini, C.; Sánchez, L.; Mackern, M.V.; Mateo, M.L.; Martínez, W.; Luz, R. (2010). SIRGAS: the geodetic reference frame in Latin America and the Caribbean. IAG Commission 1 Symposium 2010 Reference Frames for Applications in Geosciences. Marne la Vallee, Francia. Octubre de 2010.

[8] Mateo, M. L.; Calori, A. V.; Mackern, M. V. and Robin, A. M.. (2010). Evolución de la red SIRGAS-CON-D-Sur. Aportes del centro de procesamiento CIMA. Reunión SIRGAS (Sistema de Referencia para las Américas) 2010. Lima, Peru. Noviembre 2010.

[9] Altamimi, Z., Collilieux, X., Legrand, J., Garayt, B., Boucher, C.(2007): ITRF2005: A new release of the International Terrestrial Reference Frame based on time series of station positions and Earth Orientation Parameters, Journal of Geophysical Research, 112, B09401,x. doi:10.1029/2007JB004949.

[10] Altamimi, Z., (2010) ITRF2008 and the IGS Contribution. IGS Workshop 2010 – Newcastle, UK

[11] Ferland, R. (2006). [IGSMAIL-5447]: Proposed IGS05 Realization. http://igscb.jpl.nasa.gov/mail/igsmail/2006/msg00170.html.

[12] Drewes, Hermann (2005). Deformation of the South American Crust Estimated from Finite Element and Collocation Methods. Springer Berlin Heidelberg, Volume 128, pag. 544-549. 2005

[13] Mackern, M. V.; Mateo, M. L.; Robin, A. M. and Calori, A. V. (2009): A Terrestrial Reference Frame (TRF), coordinates and velocities for South American stations: contributions to Central Andes geodynamics. Adv. Geosci., 22, 181–184, 2009. www.adv-geosci.net/22/181/2009/

Chapter 3

GPS and the One-Way Speed of Light

Stephan J.G. Gift

Additional information is available at the end of the chapter

http://dx.doi.org/10.5772/50670

1. Introduction

The constancy of the speed of light in a vacuum is a fundamental idea in modern physics and is the basis of the standard of length in metrology since 1983. Its genesis is in the theory of special relativity introduced 100 years ago by Albert Einstein who postulated that light travels at a constant speed in all inertial frames [1-3]. There have been numerous experiments [3] over the past century that test light speed constancy under a variety of conditions and they almost all yield a value c (in vacuum). The first experiment among these that was taken as indicating light speed constancy is the celebrated Michelson-Morley experiment of 1887 that searched for ether drift based on interferometer fringe shifts [4]. This experiment involved interfering light beams that traversed orthogonal paths on a movable apparatus. It was designed to reveal the speed of the Earth's orbital motion through the hypothesized ether using the expected change in light speed arising from movement with or against the associated ether wind. The observed fringe shift was significantly less than what was expected as a result of the revolving Earth. The null result was interpreted as an indication of light speed constancy. This basic experiment was repeated many times over the years with essentially the same results. In 1925 Miller did appear to achieve positive fringe shifts [5] but it was later argued that this resulted from diurnal and seasonal variations in equipment temperature [6].

In 1964 Jaseja et al. introduced a major enhancement of this basic experiment [7]. These researchers employed laser technology to realize a sensitivity increase of 25 times the original experiment but detected no change in the system's beat frequency within its measurement accuracy. A later improved version of the Jaseja experiment by Brillet and Hall [8] searched for light speed anisotropy in the form of changes in the resonant frequency of a cavity resonator. They claimed a 4000-fold improvement over the results of Jaseja et al. and again detected no change. The conclusion from these experiments was that light speed is constant.

Modern versions of the Michelson-Morley experiment operating along the lines of the approach by Brillet and Hall use electromagnetic resonators that examine light speed

INTECH
open science | open minds

© 2012 The Author(s). Licensee InTech. This chapter is distributed under the terms of the Creative Commons Attribution License http://creativecommons.org/licenses/by/3.0), which permits unrestricted use, distribution, and reproduction in any medium, provided the original work is properly cited.

isotropy. These systems compare the resonant frequencies of two orthogonal resonators and check for changes caused by orbital or rotational motion. Several experiments of this type have been conducted including experiments by Hermann et al.[9], Muller et al. [10] and Eisele et al. [11]. These experiments have progressively lowered the limit on light speed anisotropy with the most recent measurement being $\delta c / c < 10^{-17}$ where δc is the measured change in light speed. It should be noted however that Demjanov [12] and Galeav [13] have reported positive fringe shifts in recent Michelson-Morley type experiments but these have received little or no attention.

As a result of these many negative tests the almost universal belief among physicists is that the postulate of light speed constancy has been confirmed. However Zhang [3] has shown that what these experiments have established is two-way light speed constancy but that one-way light speed constancy remains unconfirmed. A few experiments testing one-way light speed have been conducted including those by Gagnon et.al. [14] and Krisher et.al. [15] but these too are not true one-way tests because of the apparent inability to independently synchronize the clocks involved [3].

The global positioning system (GPS) utilizes advanced time-measuring technology and appears to provide the means to accurately determine one-way light speed. It is a modern navigation system that employs synchronized atomic clocks in its operation [16]. This system of synchronized clocks enables the accurate determination of elapsed time in a wide range of applications including time-stamping of financial transactions, network synchronization and the timing of object travel. Based on the IS-GPS-200E Interface Specification [17], GPS signals propagate in straight lines at the constant speed c (in vacuum) in an Earth-Centered Inertial (ECI) frame, a frame that moves with the Earth but does not share its rotation. This isotropy of the speed of light in the ECI frame is utilized in the GPS range equation to accurately determine the instantaneous position of objects which are stationary or moving on the surface of the Earth.

Using the system, Marmet [18] observed that GPS measurements show that a light signal takes about 28 nanoseconds longer traveling eastward from San Francisco to New York as compared with the signal traveling westward from New York to San Francisco. Kelly [19] also noted that measurements using the GPS reveal that a light signal takes 414.8 nanoseconds longer to circumnavigate the Earth eastward at the equator than a light signal travelling westward around the same path. Marmet and Kelly both concluded that these observed travel time differences in the synchronized clock measurements in each direction occur because light travels at speed $c - v$ eastward and $c + v$ westward relative to the surface of the earth. Here v is the speed of rotation of the Earth's surface at the particular latitude. This research by Marmet and Kelley was the precursor to a series of papers by this author on the use of GPS technology in the unambiguous demonstration of one-way light speed anisotropy. In this chapter we bring this material together in one place so that the full impact of the technology on this important issue can be better appreciated and the significant results made available to a wider audience.

2. Clock synchronization

In light speed determination, synchronized clocks are required for the timing of a light pulse as it propagates between two separated points. In this regard, the IEEE 1588 Standard for a Precision Clock Synchronization Protocol for Networked Measurement and Control Systems defines synchronized clocks as follows: "Two clocks are synchronized to a specified uncertainty if they have the same epoch and measurements of any time interval by both clocks differ by no more than the specified uncertainty. The timestamps generated by two synchronized clocks for the same event will differ by no more than the specified uncertainty." In other words clocks are synchronized if they indicate the same times for the same events and this is realised using a clock synchronization procedure. This is the logical and widely accepted meaning of synchronized clocks and is the one adopted in the chapter. Unfortunately some authors have created a degree of confusion by referring to other modes of clock operation as synchronized clocks, modes which result from what may be referred to as "clock synchronization schemes". Using such clocks, light speed measurement will show a dependence on these different clock synchronization schemes since differently synchronized clocks will measure different time intervals for the same light signal transmission. In fact virtually any value of speed can be obtained by suitably "synchronizing" the measuring clocks and according to Will [20], "a particularly perverse choice of synchronization can make the apparent speed...infinite"! These "apparent" speeds bear no relation to physical reality and are meaningless. A proper clock synchronization method is one that results in clock operation such that clocks indicate the same times for the same events and light speed can be reliably measured using such synchronized clocks.

The synchronization approach discussed by Einstein [1] involves the consideration of two clocks A and B at rest at different points in an inertial frame. Let a ray of light propagate from A directly to B and be reflected at B directly back to A. Let the start time at A as indicated on the clock at A be t_A and let the time of arrival of the light ray at B as indicated on the clock at B be t_B. Finally, let the time of arrival of the reflected ray back at clock A be t'_A. Then, Einstein declared [1] that the two clocks are synchronized if

$$t_B - t_A = t'_A - t_B \tag{1}$$

This synchronization technique demands that "the "time" required by light to travel from A to B equals the "time" it requires to travel from B to A" i.e. that light travels with the same speed in both directions which Einstein "established *by definition*". However it is precisely the light speed in the different directions that we wish to measure and therefore it would be circular logic to assume a priori that light speed in both directions is the same in order to synchronize the measuring clocks.

The GPS utilizes a clock-synchronization procedure that has been exhaustively tested and rigorously verified [16, 17] and now forms part of the specification for the GPS. This procedure for the synchronization of clock stations is also contained in standards published by the CCIR, a committee of the International Telecommunications Union in 1990 and 1997 [19]. Similar rules were established in 1980 by the Consultative Committee for the Definition of the Second (now the Consultative Committee for Time and Frequency (CCTF)) [19]. In these synchronization procedures, the synchronization of two clocks fixed on the moving Earth is accomplished by transmitting an electromagnetic signal from one clock to the other assuming the postulate of the constancy of the speed of light c then applying a correction to the elapsed time that is said to arise because of the rotating Earth. This adjustment is called the "Sagnac correction" and is today automatically applied to all electromagnetic signals transmitted around the Earth in order to achieve clock synchronization.

This synchronization algorithm has been tested and confirmed in numerous experiments. While there is disagreement about the underlying theory, the procedure works. The resulting GPS clocks are synchronized according to the IEEE definition and enable the determination of one-way light speed relative to observers situated on the rotating Earth. The simple exercise is the transmission of a light or electromagnetic signal between separated GPS clocks fixed on the surface of the Earth and the division of the fixed distance between the clocks by the measured time interval between transmission and reception of the light signal. Since light was used to synchronize the clocks the objection to light speed measurement using these same clocks might be raised. This objection can however be answered by observing that the synchronized clocks have been rigorously and exhaustively tested and verified. In any event such a measurement will serve as a check on the requirement that the measured light speed be consistent with the assumed constant light speed c involved in the synchronization process that follows from the application of the postulate of light speed constancy.

3. One-way light speed test using the GPS clocks

In this section the synchronized clocks of the GPS are used in the one-way determination of the speed of light by timing the transmission of a light signal travelling between two fixed points on Earth. In order to exclude issues associated with the curvature of the Earth's surface and non-inertial frames, we consider a clock A located in a building in one city say Scarborough in the Republic of Trinidad and Tobago and another clock B located at the same latitude in the same building and a short distance l away from clock A. This is shown in figure 1.

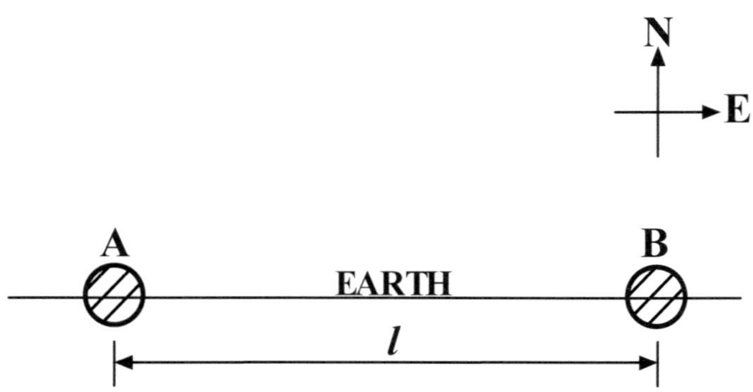

Figure 1. GPS Clocks A and B at fixed positions on Earth

Under such circumstances light travels between the two clocks in a straight line in an approximately inertial frame which is the same as that used in the performance of the many light speed measurements in which c is the reported value [7-11]. It is particularly noteworthy that the one-way experiment by Krisher et al. [15] which searched for light speed changes resulting from rotation of the Earth extended 21 km across the surface of the Earth and appears to violate the inertial requirement. Yet these authors who claimed $\delta c / c < 3.5 \times 10^{-7}$ gave no consideration to any non-inertial effects.

3.1. Eastward transmission

We establish synchronization of the GPS clocks by transmitting a light signal from clock A to clock B. Using the CCIR synchronization rules involving the assumed constancy of the speed of light c along with the so-called Sagnac adjustment, the total time Δt for light to travel the path from clock A to clock B is given by [16, 18]

$$\Delta t = \int_{path} \frac{d\sigma'}{c} + \frac{2\omega_E}{c^2} \int_{path} dA_z' \tag{2}$$

where $d\sigma'$ is infinitesimal distance in the moving frame, ω_E is the angular velocity of the rotating Earth and dA_z' is the infinitesimal area in the rotating coordinate system swept out by a vector from the rotation axis to the light pulse and projected onto a plane parallel to the equatorial plane. Carrying out the integration associated with (2) yields

$$\Delta t = \frac{l}{c} + 2A_z' \frac{\omega_E}{c^2} \tag{3}$$

where l is the distance between the two stations both moving at speed v the speed of the Earth's surface at that latitude. Let the circumference of the Earth at that latitude be l_C and let the corresponding radius be r. Then the area A_z' is given by

$$A_z' = \frac{l}{l_C} \pi r^2 \tag{4}$$

Since $\omega_E = v / r$ and $l_C = 2\pi r$, substituting equation (4) in (3) gives

$$\Delta t = \frac{l}{c} + \frac{lv}{c^2} \tag{5}$$

The first term in (5) corresponds to the light travel time under the assumption of constant light speed c and the second term is the so-called "Sagnac correction" that is said to be required because of the Earth's rotation. This total elapsed time must now be added to clock B such that if at the instant of light transmission the time on clock A is t_A, then at the instant of reception the time t_B on clock B is set to $t_B = t_A + \Delta t$. After this procedure Clocks A and B are synchronized.

Following the synchronization verification of the GPS clocks A and B, we use them to measure one-way light speed. Thus at a specified time, a light signal is transmitted eastward from clock A directly to clock B. Because the clocks have been synchronized using (5), the time interval $\Delta t = t_B - t_A$ between the transmission and reception of the signal is exactly that given in (5) as

$$\Delta t = t_B - t_A = \frac{l}{c} + \frac{lv}{c^2} \tag{6}$$

Equation (6) is like a law of nature as it indicates the time for light to travel eastward between two points at the same latitude fixed on the surface of the Earth. It means therefore that an actual clock measurement for the time of transmission is not required since clock behavior for eastward travel is fully represented by equation (6). This equation therefore makes available the full precision of the GPS clocks anywhere in the world without the need for actual clocks and is therefore a very useful result.

Using this elapsed time in speed determination, since the distance between the two clocks is l, it follows that the one-way speed of light c_{AB} traveling eastward between the two clocks is given by

$$c_{AB} = \frac{l}{\Delta t} = \frac{l}{\frac{l}{c} + \frac{lv}{c^2}} = c(1+\frac{v}{c})^{-1} = c(1-\frac{v}{c}+o(\frac{v}{c})) = c - v, v << c \tag{7}$$

Thus the synchronized clocks of the GPS give a one-way eastward light speed measurement of $c_{AB} = c - v$ relative to the surface of the Earth and not light speed $c_{AB} = c$ required by the postulate of the constancy of the speed of light.

3.2. Westward transmission

For westward transmission we again establish synchronization of the GPS clocks, using the rules of the CCIR, by transmitting a light signal from clock B to clock A. In this case the total time Δt for light to travel the path from clock A to clock B is given by [16, 18]

$$\Delta t = \frac{l}{c} - 2A'_z \frac{\omega_E}{c^2} \tag{8}$$

which reduces to

$$\Delta t = \frac{l}{c} - \frac{lv}{c^2} \tag{9}$$

Again, the first term in (9) is the elapsed time assuming constant light speed c and the second term is the so-called "Sagnac correction" said to be required because of the Earth's rotation. This total elapsed time must be added to clock A such that if at the instant of light transmission the time on clock B is t_B, then at the instant of reception the time t_A on clock A is set to $t_A = t_B + \Delta t$. After this procedure clocks B and A are synchronized.

Using these clocks to conduct a one-way light speed test, at a specified time, an observer at clock B sends a light signal westward to an observer at clock A. Because the clocks are synchronized using (9), the time interval $\Delta t = t_A - t_B$ between the transmission and reception of the signal is given in (9) as

$$\Delta t = t_B - t_A = \frac{l}{c} - \frac{lv}{c^2} \tag{10}$$

Equation (10) is essentially a law of nature as it provides the time for light to travel westward between two points at the same latitude fixed on the surface of the Earth. It means therefore that an actual clock measurement for the westward time of transmission is not required since clock behavior is fully represented by equation (10). Equation (10) therefore brings the full precision of the GPS clocks to everyone anywhere in the world without the need for actual clocks! This constitutes another very useful result.

Using the time found in (10) for one-way light speed in the westward direction, since the distance between the two clocks is l, it follows that the one-way speed of light c_{BA} traveling westward between the two clocks is given by

$$c_{BA} = \frac{l}{\Delta t} = \frac{l}{\frac{l}{c} - \frac{lv}{c^2}} = c(1 - \frac{v}{c})^{-1} = c(1 + \frac{v}{c} + o(\frac{v}{c})) = c + v, v << c \tag{11}$$

Thus the synchronized clocks of the GPS give a one-way westward light speed measurement of $c_{BA} = c + v$ relative to the surface of the Earth and not light speed $c_{BA} = c$ required by the postulate of light speed constancy.

The results in equations (7) and (11) confirm the independent claims of Marmet and Kelley: Light travels faster westward than eastward relative to the surface of the Earth. Specifically the one-way measurement of light speed using GPS data in (6) clearly indicates that a signal sent eastward travels at speed c minus the rotational speed of the Earth v at that latitude giving $c - v$. The GPS data available in (10) also shows that a signal sent westward travels at speed c plus the rotational speed of the Earth v at that latitude giving $c + v$. These generalized results were first reported by Gift [21].

We are now better able to understand why the times Δt in (6) and (10) enable the synchronization of two clocks in GPS: Time interval (6) is the time for light to travel from clock A to clock B at the actual speed $c_{AB} = c - v$ relative to both clocks and time interval (10) is the time for light to travel from clock B to clock A at the actual speed $c_{BA} = c + v$ relative to both clocks. At the constant speed c required by the postulate of light speed constancy the associated time l/c for light to travel from one clock to the other when added to the receiving clock does not result in synchronization because this is not the true travel time. In the CCIR synchronization rules the time l/c is adjusted by $\pm lv/c^2$ in order to compensate for the real changes in light speed $c \pm v$ that occur relative to the clocks. In view of these results, the interpretation that the time Δt is the time required for light to travel between clocks at constant speed c, with a correction added because of the rotating Earth is now known to be invalid.

4. Michelson-Morley experiment using the GPS clocks

With the availability of synchronized clocks, the Michelson-Morley experiment can be conducted with direct timing of the signals traversing the orthogonal arms of the apparatus. Such an approach was previously considered but never executed because of insufficient timing resolution. The approach proposed here does not encounter this problem since the novel feature of the method is that the light travel time is directly available from the GPS clock synchronization algorithm adopted by the CCIR. This renders actual signal timing with physical clocks completely unnecessary [22].

The basic configuration of the original Michelson-Morley experiment [4] is shown in figure 2 where the apparatus is moving with velocity v through the hypothesized ether in direction PM1. Light from a source S splits into two beams at beam-splitter P. One beam travels from P to mirror M1 and back and is reflected at P into the interferometer I. The second beam is reflected at P to mirror M2 and back and passes through P into the interferometer I where both beams form an interference pattern.

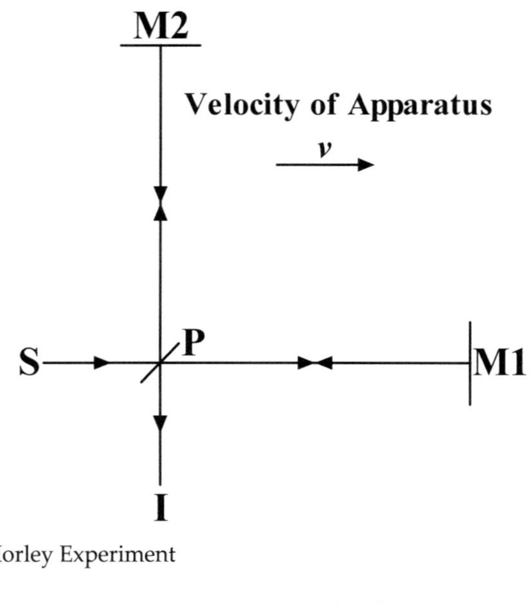

Figure 2. Michelson-Morley Experiment

In the frame of the moving apparatus as a result of ether drift, the resultant light speed between P and M1 would be $c - v$ toward M1 and $c + v$ toward P while the resultant light speed between P and M2 would be $(c^2 - v^2)^{1/2}$ in both directions. For optical path lengths $PM1 = l_1$ and $PM2 = l_2$ the time $t_1(a)$ for the light to travel from P to M1 is given by

$$t_1(a) = \frac{l_1}{c - v} \tag{12}$$

and the time $t_1(b)$ for the light to travel from M1 to P is given by

$$t_1(b) = \frac{l_1}{c + v} \tag{13}$$

Therefore the round-trip time along PM1 is given by

$$T_1 = t_1(a) + t_1(b) = \frac{2l_1}{c(1 - v^2 / c^2)} \tag{14}$$

The time $t_2(a)$ for the light to travel from P to M2 is given by

$$t_2(a) = \frac{l_2}{\sqrt{c^2 - v^2}} \tag{15}$$

and the time $t_2(b)$ for the light to travel from M2 to P is given by

$$t_2(b) = \frac{l_2}{\sqrt{c^2 - v^2}} \tag{16}$$

Therefore the round-trip time along PM2 is given by

$$T_2 = t_2(a) + t_2(b) = \frac{2l_2}{c(1 - v^2 / c^2)^{1/2}} \tag{17}$$

The time difference $\Delta T = T_1 - T_2$ is given by

$$\Delta T = T_1 - T_2 \approx \frac{2(l_1 - l_2)}{c} + \frac{2l_1 v^2}{c^3} - \frac{l_2 v^2}{c^3} \tag{18}$$

If the apparatus is turned through 90° so that PM2 is in the direction of motion, the time difference becomes

$$\Delta T' = T_1' - T_2' \approx \frac{2(l_1 - l_2)}{c} + \frac{l_1 v^2}{c^3} - \frac{2l_2 v^2}{c^3} \tag{19}$$

The change in these time differences is

$$\Delta = \Delta T - \Delta T' = (l_1 + l_2)\frac{v^2}{c^3} \tag{20}$$

If $l_1 = l_2 = l$ then this reduces to

$$\Delta = \frac{2l}{c}\frac{v^2}{c^2} \tag{21}$$

A fringe shift proportional to this value (given by $\delta = \frac{c}{\lambda}\Delta$) is expected to appear in the interferometer. The time difference $\Delta = \frac{2l}{c}\frac{v^2}{c^2}$ is second-order and is significantly reduced by length contraction arising from motion through the ether [12]. This is why Michelson-Morley type experiments have been largely unsuccessful in detecting ether drift.

The accurate synchronized clocks in the GPS are now used to directly determine one-way light travel time. Thus in a modification of the original Michelson-Morley apparatus GPS clocks are placed at P, M1 and M2 in fig.2. Additionally the arm PM1 is oriented along a line of latitude and the arm PM2 is positioned along a line of longitude. As a result of the rotation of the Earth there is movement of the apparatus at velocity $v = w$ in the direction PM1 towards the East where w is the rotational speed of the surface of the Earth at the particular latitude.

4.1. Time measurement along PM1

The time $t_1(a)_{GPS}$ measured by the GPS clocks at P and M1 for the light to travel from P to M1 is [16, 18, 21]

$$t_1(a)_{GPS} = \frac{l_1}{c} + \frac{l_1 w}{c^2} \tag{22}$$

while from equation (12) of ether theory

$$t_1(a) = \frac{l_1}{c - w} \approx \frac{l_1}{c} + \frac{l_1 w}{c^2}, w << c \tag{23}$$

Hence $t_1(a)_{GPS} = t_1(a)$ and ether drift arising from the rotation of the Earth is detected. The time $t_1(b)_{GPS}$ measured by the GPS clocks for the light to travel from M1 to P is [16, 18, 21]

$$t_1(b)_{GPS} = \frac{l_1}{c} - \frac{l_1 w}{c^2} \tag{24}$$

while from equation (13) of ether theory

$$t_1(b) = \frac{l_1}{c + w} \approx \frac{l_1}{c} - \frac{l_1 w}{c^2}, w << c \tag{25}$$

Hence $t_1(b)_{GPS} = t_1(b)$ and ether drift arising from the rotation of the Earth is again detected. From ether theory as well as clock measurement, the difference in the out and back times along PM1 is given by

$$\Delta t_1 = t_1(a) - t_1(b) = \frac{2l_1}{c}\frac{w}{c} \tag{26}$$

Result (26) is first-order and therefore not affected by second-order length contraction as is the second-order result (21) in the conventional Michelson-Morley type experiments. Equation (26) has been extensively verified in GPS operation.

4.2. Time measurement along PM2

The time $t_2(a)_{GPS}$ for the light to travel from P to M2 measured by the GPS clocks at P and M2 is [16, 18]

$$t_2(a)_{GPS} = \frac{l_2}{c} \tag{27}$$

while from equation (15) of ether theory

$$t_2(a) = \frac{l_2}{\sqrt{c^2 - w^2}} \approx \frac{l_2}{c}, w << c \tag{28}$$

Hence $t_2(a)_{GPS} = t_2(a)$ and ether theory is confirmed by GPS measurement. The time $t_2(b)_{GPS}$ for the light to travel from M2 to P measured by the GPS clocks is [16, 18]

$$t_2(b)_{GPS} = \frac{l_2}{c} \tag{29}$$

while from equation (16) of ether theory

$$t_2(b) = \frac{l_2}{\sqrt{c^2 - w^2}} \approx \frac{l_2}{c}, w << c \tag{30}$$

Hence $t_2(b)_{GPS} = t_2(b)$ and ether theory is again confirmed by GPS measurement. From ether theory as well as GPS clock measurement, the difference in the out and back times along PM2 is given by

$$\Delta t_2 = t_2(a) - t_2(b) = 0 \tag{31}$$

This has been confirmed by actual GPS measurements which have shown that unlike East-West travel, there is no time difference between light travelling North and light travelling South.

The modified Michelson-Morley experiment using synchronized GPS clocks to measure light travel times out and back along the arms of the apparatus has detected ether drift resulting from the rotation of the Earth. The clocks have directly confirmed the light travel times for changed light speeds $c \pm w$ in the East-West direction arising from the drift of the ether as the apparatus moves through the medium at speed w corresponding to the speed of rotation of the Earth's surface at the particular latitude. The experiment is operated within the dimensions of the original Michelson-Morley apparatus where the frame is considered to be approximately inertial and where special relativity has been universally applied [2]. This negates any objections about rotating coordinates and non-inertial frames which are never raised in the original Michelson-Morley experiment or in any of the several modern versions of the experiment [7-11].

5. One-way light speed using the range equation

In section 3 in the determination of one-way light speed, the CCIR algorithm was used to determine flight time for light transmission between two fixed points on the same latitude. In this section the range equation used in the GPS to evaluate distance and determine position is employed in the determination of flight time. Specifically by substituting known spatial positions in the range equation, light travel times can be determined without the direct use of the GPS clocks. These times are then used to determine one-way light speed in the East-West direction.

The range equation is central to the operation of the GPS. It holds in an ECI frame which is a frame that moves with the Earth as it revolves around the Sun but does not share its rotation. It is given by [16]

$$\left|\overline{r}_r(t_r) - \overline{r}_s(t_s)\right| = c(t_r - t_s) \tag{32}$$

where t_s is the time of transmission of an electromagnetic signal from a source, t_r is the time of reception of the electromagnetic signal by a receiver, $\overline{r}_s(t_s)$ is the position of the source at the time of transmission of the signal and $\overline{r}_r(t_r)$ is the position of the receiver at the time of reception of the signal. Using elapsed time measurements determined by the GPS clocks and the light speed value c in this equation, position on the surface of the Earth can be accurately determined. It has been exhaustively tested and rigorously verified and has resulted in the worldwide proliferation of the GPS.

Wang [23] has used the range equation operating in the ECI frame to show that the speed of light is dependent on the observer's uniform motion relative to the ECI frame. He did this by using the range equation to determine elapsed time and concluded that the successful application of the range measurement equation in GPS operation is inconsistent with the principle of the constancy of the speed of light. Instead of using the synchronized clocks of the GPS, we use the range equation (32) of the GPS to determine elapsed time for light traveling between two known adjacent points at the same latitude fixed on the surface of the rotating Earth. We then use this time and the known distance between the two fixed points to calculate the one-way speed of light.

Consider figure 3 in which two adjacent GPS stations A and B are at the same latitude and fixed on the surface of the Earth a distance l apart with B East of A. Since the Earth is rotating, the stations are moving eastward at speed v the Earth's surface speed at that latitude. Let l be sufficiently small by for example having stations A and B in the same building such that the stations are moving uniformly in the same direction at speed v relative to the ECI frame. In such circumstances stations A and B constitute an approximately inertial frame moving at speed v relative to the ECI frame, again similar to the many light speed experiments conducted to verify light speed constancy.

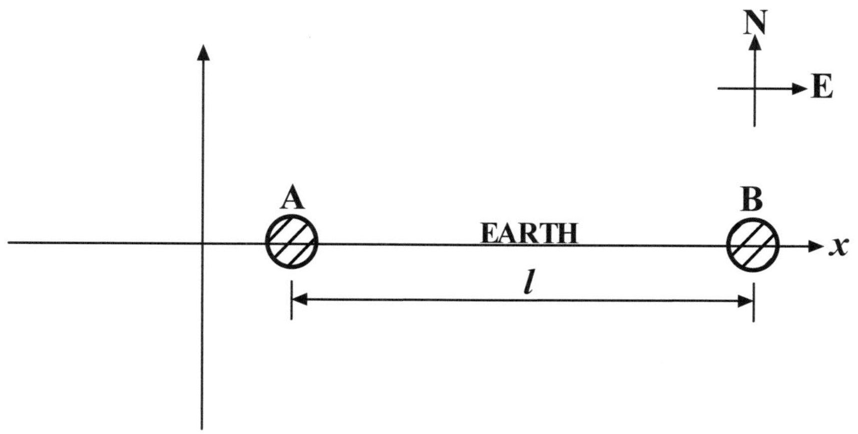

Figure 3. GPS Stations A and B at fixed positions on Earth

5.1. Eastward transmission

Let station A transmit a signal eastward at time t_I to station B which receives it at time t_F. On an axis fixed in the ECI frame along the line joining the two stations with the origin west of station A, let $x_A(t)$ be the position of station A at time t and $x_B(t)$ be the position of station B at time t. Then from the range equation (32),

$$x_B(t_F) - x_A(t_I) = c(t_F - t_I) \tag{33}$$

where $x_B(t_F)$ is the position of station B at time t_F and $x_A(t_I)$ is the position of station A at time t_I. Since the stations are moving uniformly in the same direction at speed v relative to the ECI frame, it follows that the relation between the position $x_B(t_F)$ of station B at the time of reception of the signal and its position $x_B(t_I)$ at the time of emission of the signal is given by

$$x_B(t_F) = x_B(t_I) + v(t_F - t_I) \tag{34}$$

Substituting for $x_B(t_F)$ from (34) in (33) yields

$$x_B(t_I) - x_A(t_I) + v(t_F - t_I) = c(t_F - t_I) \tag{35}$$

This gives the elapsed time as

$$(t_F - t_I) = \frac{l}{c - v} \tag{36}$$

Therefore the speed c_{AB} of the light traveling from station A to station B is given by separation l divided by elapsed time $(t_F - t_I)$ which using (36) is

$$c_{AB} = \frac{l}{(t_F - t_I)} = \frac{l}{l/(c-v)} = c - v \tag{37}$$

5.2. Westward transmission

Let station B transmit a signal westward at time t_I to station A which receives it at time t_F. Then using the range equation (32) and noting that $x_B(t_I) > x_A(t_F)$,

$$x_B(t_I) - x_A(t_F) = c(t_F - t_I) \tag{38}$$

where $x_B(t_I)$ is the position of station B at time t_I and $x_A(t_F)$ is the position of station A at time t_F. Since the stations are moving uniformly in the same direction at speed v relative to the ECI frame, the relation between the position $x_A(t_F)$ of station A at the time of reception of the signal and its position $x_A(t_I)$ at the time of emission of the signal is given by

$$x_A(t_F) = x_A(t_I) + v(t_F - t_I) \tag{39}$$

Substituting for $x_A(t_F)$ from (39) in (38) yields

$$x_B(t_I) - x_A(t_I) - v(t_F - t_I) = c(t_F - t_I) \tag{40}$$

This yields the elapsed time as

$$(t_F - t_I) = \frac{l}{c+v} \tag{41}$$

Therefore the speed c_{BA} of the light traveling from station B to station A is given by separation l divided by elapsed time $(t_F - t_I)$ which using (41) is

$$c_{BA} = \frac{l}{(t_F - t_I)} = \frac{l}{l/(c+v)} = c + v \tag{42}$$

The results in equations (37) and (42) first reported in [24] indicate that light travels faster westward than eastward relative to the surface of the Earth. In particular the one-way determination of light speed using the range equation of the GPS establishes in (37) that a signal sent eastward travels at speed c minus the rotational speed of the Earth v at that latitude giving $c - v$. The range equation data also shows in (42) that a signal sent westward travels at speed c plus the rotational speed of the Earth v at that latitude giving $c + v$. This is true for the short-distance travel in the approximately inertial frame considered here as well as long-distance circumnavigation of the Earth [19] and fully corroborates the light speed determined in section 3 using the synchronized GPS clocks [21].

6. Test of the light speed invariance postulate using the range equation

The direct one-way tests above reveal that light travels faster West than East and does so for short-distance travel which approximates an inertial frame or large-distance travel such as circumnavigating the Earth. This finding contradicts the light speed invariance postulate of special relativity according to which the speed of light is constant in all inertial frames [1-3]. A particularly interesting interpretation of this postulate was presented by Tolman in 1910 [25]. Referring to a similar figure 4 he said the following:

> "A simple example will make the extraordinary nature of the second postulate evident. *S* is a source of light and *A* and *B* two moving systems. *A* is moving towards the source *S*, and *B* away from it. Observers on the systems mark off equal distances *aa′* and *bb′* along the path of the light and determine the time taken for light to pass from *a* to *a′* and *b* to *b′* respectively. Contrary to what seem the simple conclusions of common sense, the second postulate requires that the time taken for the light to pass from *a* to *a′* shall measure the same as the time for the light to go from *b* to *b′*."

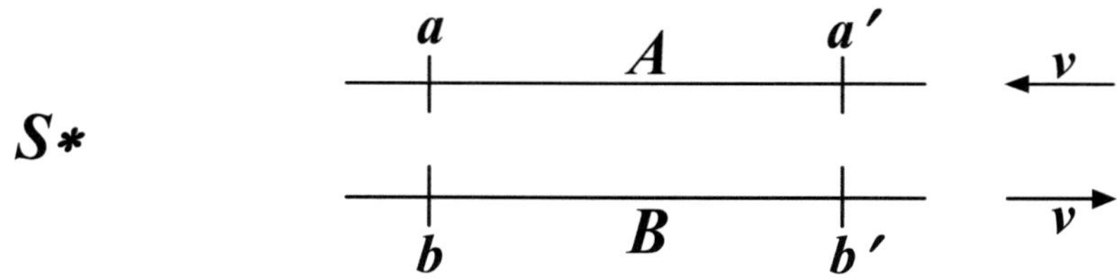

Figure 4. Test of Light Speed Invariance

The range equation of the GPS now makes it possible to test this prediction of the light speed invariance postulate of special relativity [26].

In figure 4 consider a light source S fixed in the ECI frame and two systems A and B moving uniformly in the ECI frame. A is moving towards the source at a constant speed v relative to the ECI frame and B is moving away from the source at a constant speed v relative to the ECI frame. Observers on the two systems mark off equal distances aa' and bb' equal to D along the light path and determine the time taken for the light to pass from a to a' and b to b' respectively. As Tolman has observed, "the second postulate [of special relativity] requires that the time taken for the light to pass from a to a' shall measure the same as the time for the light to go from b to b'." According to the second postulate this measured time must be D/c on both systems. This prediction will now be tested using the range equation of the GPS.

6.1. Analysis on system A

Let GPS stations be placed at a and a' respectively. On an axis fixed in the ECI frame along the line aa' joining the two stations with station a closer to the origin O than station a' and taking positive values, let $x_a(t)$ be the position of station a at time t and $x_{a'}(t)$ be the position of station a' at time t. Let light from S arrive at station a at time t_I and later at station a' at time t_F. Then using the range equation (32) and noting that $x_{a'}(t_F) > x_a(t_I)$,

$$x_{a'}(t_F) - x_a(t_I) = c(t_F - t_I) \tag{43}$$

Since both stations are moving uniformly toward S at speed v relative to the ECI frame, it follows that the relation between the position $x_{a'}(t_F)$ of station a' at the time of its reception of the signal and its earlier position $x_{a'}(t_I)$ is given by

$$x_{a'}(t_F) = x_{a'}(t_I) - v(t_F - t_I) \tag{44}$$

Substituting $x_{a'}(t_F)$ from (44) in (43) yields

$$x_{a'}(t_I) - x_a(t_I) - v(t_F - t_I) = c(t_F - t_I) \tag{45}$$

Since $x_{a'}(t_I) - x_a(t_I) = D$ then (45) becomes

$$x_{a'}(t_I) - x_a(t_I) = D = (c + v)(t_F - t_I) \tag{46}$$

Hence for an observer on system A the range equation gives the time for light to travel between a and a' as

$$t_F - t_I = \frac{D}{c + v} \tag{47}$$

Thus the light travel time measured by an observer on A is $D/(c+v)$ and not D/c as required by the light speed invariance postulate. Therefore the light speed $c_{aa'}$ detected on system A for the light traveling from station a to station a' is given by the fixed length D divided by elapsed time $(t_F - t_I)$ which using (47) is given by

$$c_{aa'} = \frac{D}{(t_F - t_I)} = \frac{D}{D/(c+v)} = c + v \tag{48}$$

6.2. Analysis on system B

Let GPS stations be placed at b and b' respectively. On an axis fixed in the ECI frame along the line bb′ joining the two stations with station b closer to the origin O than station b′ and taking positive values, let $x_b(t)$ be the position of station b at time t and $x_{b'}(t)$ be the position of station b′ at time t. Let light from S arrive at station b at time t_I and later at station b′ at time t_F. Then using the range equation (32),

$$x_{b'}(t_F) - x_b(t_I) = c(t_F - t_I) \tag{49}$$

Since both stations are moving uniformly away from S at speed v relative to the ECI frame, it follows that the relation between the position $x_{b'}(t_F)$ of station b′ at the time of its reception of the signal and its earlier position $x_{b'}(t_I)$ is given by

$$x_{b'}(t_F) = x_{b'}(t_I) + v(t_F - t_I) \tag{50}$$

Substituting $x_{b'}(t_F)$ from (50) in (49) yields

$$x_{b'}(t_I) - x_b(t_I) + v(t_F - t_I) = c(t_F - t_I) \tag{51}$$

Since $x_{b'}(t_I) - x_b(t_I) = D$ then (51) becomes

$$x_{b'}(t_I) - x_b(t_I) = D = (c - v)(t_F - t_I) \tag{52}$$

Hence for an observer on system B the range equation gives the time for light to travel between b and b' as

$$t_F - t_I = \frac{D}{c - v} \tag{53}$$

Thus the light travel time measured by an observer on B is $D/(c-v)$ and not D/c as required by the light speed invariance postulate. Therefore the light speed $c_{bb'}$ observed on system B for the light traveling from station b to station b' is given by the fixed length D divided by elapsed time $(t_F - t_I)$ which using (53) is given by

$$c_{bb'} = \frac{D}{(t_F - t_I)} = \frac{D}{D/(c - v)} = c - v \tag{54}$$

Equations (47) and (53) indicate that the light travel times over the distance D on systems A and B are $D/(c+v)$ and $D/(c-v)$ respectively and not the value D/c required by the second postulate of special relativity. This is because the light speeds observed on systems A and B in equations (48) and (54) are $c+v$ and $c-v$ respectively and are different from the value c required by the second postulate. The light speed variation demonstrated here in the ECI is exactly what is observed using the synchronized clocks of the GPS in section 3 and the range equation of the GPS in section 5 in the frame of the surface of the rotating Earth.

7. One-way light speed in the sun-centered inertial frame

The demonstration of light speed anisotropy has thus far been confined to a space on or close to the surface of the Earth. It is possible to broaden the scope of the discussion beyond the terrestrial frame to the region encompassed by the solar system. In this regard Wallace [27] used published interplanetary data to present evidence of light speed $c+v$ relative to the moving Earth where v is the Earth's orbital speed. This light speed variation for light travelling through space is exactly what has been observed on the orbiting Earth for light from planetary satellites in the Roemer experiment [28] and for light from stars on the ecliptic in the Doppler experiment [29]. Light speed anisotropy arising from the orbital motion of the Earth has also been reported for light propagation over cosmological distances [30] and in the Shtyrkov experiment involving the tracking of a geostationary satellite [31].

This phenomenon of light speed anisotropy arising from light transmission through space has recently been investigated by this author using planet and spacecraft tracking technology [32]. Specifically range equations operating in the solar system barycentric or sun-centered inertial (SCI) frame used in tracking planets and spacecrafts were used to determine the one-way speed of light reflected from a planet or spacecraft and observed from the orbiting Earth moving in the solar barycentric frame. Time measurement was effected using atomic clocks based on Coordinated Universal Time (UTC) and the spatial coordinates were taken relative to the SCI frame. The equations are given by [33]

$$c\tau_u = |r_B(t_R - \tau_d) - r_A(t_R - \tau_d - \tau_u)| \tag{55}$$

$$c\tau_d = |r_A(t_R) - r_B(t_R - \tau_d)| \tag{56}$$

where t_R is the time of reception of the signal, τ_u and τ_d are the up-leg and down-leg times respectively, r_A is the solar-system barycentric position of the receiving antenna on the Earth's surface, r_B is the solar-system barycentric position of the reflector which is either a responding spacecraft or the reflection point on the planet's surface and c is the speed of light in the SCI frame. In practice in order to obtain values for τ_u and τ_d, the two equations must be solved iteratively.

Using these equations in a novel approach it was found [32] that light travels from the reflector to Earth at a speed $c+v$ relative to the Earth for the Earth moving toward the reflector at orbital speed v and light travels at speed $c-v$ relative to the Earth for the Earth moving away from the reflector at orbital speed v. This light speed variation for light traveling in the SCI frame confirms the earlier finding of Wallace [27] for light travel through space and is consistent with light speed changes observed on the orbiting Earth for light from planetary satellites in the Roemer experiment [28] and for light from stars on the ecliptic in the Doppler experiment [29].

8. Conclusion

Measuring the speed of light has for many years been a major activity in science. Following the introduction of special relativity theory in 1905 in which light speed invariance was postulated, light speed tests assumed even greater significance. Numerous experiments have been conducted over the past century the vast majority of which appear to confirm the postulate. A careful examination by Zhang [3] however revealed that while two-way light speed constancy has been confirmed, one-way light speed constancy has not. Indeed these experiments by their very nature seem unable to detect one-way light speed variation and therefore cannot be used to fully test the postulate. The main contribution of this chapter is the use of GPS technology in testing one-way light speed and the demonstration that light speed does in fact vary contrary to the postulate of light speed invariance. This technology includes the synchronized clocks of the GPS, the range equation of the GPS and the UTC clocks and range equations used in tracking of planets and space missions.

The light speeds $c \pm v$ in the East-West direction determined using both the GPS clocks and range equation in the frame of the surface of the Earth are different from the results of the many light speed experiments [7-11, 14, 15] conducted in the same frame which all give c. These non-constant light speed values $c \pm v$ induced by the rotation of the Earth contradict the postulate of light speed constancy since the postulate requires constant light speed c for light traveling eastward or westward between the two clocks. In his consideration of light travel between San Francisco and New York, Marmet [18] has remarked that "Unless we accept the absurd solution that the distance between [New York] to [San Francisco] is smaller than the distance between [San Francisco] and [New York], we have to accept that in a moving frame, the velocity of light is different in each direction" a difference that "is even programmed in the GPS computer in order to get the correct Global Positioning." Wang [23] has also argued that the successful application of the range equation in GPS operation is inconsistent with the postulate of the constancy of the speed of light.

Apart from this demonstration of the postulate's inapplicability in the frame of the rotating Earth, the range equation was used to directly test this postulate in a form expressed by Tolman and again showed that the postulate of light speed constancy is not valid even in the ECI frame. Light speed changes were also observed for light travelling through space. Based on range equations employing UTC measurements and spatial coordinates relative to the solar-system barycentric frame, the speed of light reflected from a body such as a planet or a space vehicle and travelling to Earth was found to be $c \pm v$ where v is the orbital speed of the Earth toward or away from the reflecting surface at the time of reflection of the signal.

It is clear therefore that GPS technology very easily demonstrates that light speed is not constant and hence that the light speed invariance postulate which leads to the Lorentz Transformation and special relativity is invalid. This significant finding has profound implications for modern physics and metrology where light speed constancy is a foundation tenet. Moreover this light speed variability indicates the existence of a preferred frame, the search for which interestingly was the original objective of the Michelson-Morley experiment.

In order to confirm this preferred frame detection, the GPS clocks were utilized in a modified Michelson-Morley experiment where the clocks replaced the interferometer. The clocks measured light travel times along the arms of the apparatus and revealed ether drift arising from the Earth's rotation. This direct determination of the light travel times rendered the measurement essentially immune to the second-order length contraction phenomenon which negates the fringe shift in the conventional Michelson-Morley experiments. The GPS technique did not require actual time measurement but utilized light travel time that is directly available from the CCIR clock synchronization algorithm. The modified experiment succeeded in detecting ether drift for rotational motion while the majority of other Michelson-Morley-type experiments are considered to have produced null results. In the approximately inertial frame of the experiment, special relativity is directly applicable and predicts a zero time-of-flight difference between equal orthogonal arms and hence a null result [2].

Contrary to this prediction of special relativity, the modified Michelson-Morley experiment detects non-zero time-of-flight differences corresponding to ether drift and thereby reveals a preferred frame as previously reported by Gift [34] and Shtyrkov[31] and also by Demjanov [12] and Galaev [13]. This is consistent with the preferred frame associated with the set of "equivalent" transformations identified by Selleri [35]. This set contains all possible transformations that connect two inertial frames under a set of reasonable assumptions and which differ only by a clock synchronization parameter. This includes the Lorentz Transformation of special relativity and the Inertial Transformation which yields a modern ether theory [35, 36]. Using this "equivalent" set Selleri [36] and Gift [37] have identified the Inertial Transformation and the associated modern ether theory as the space-time theory that best accords with the physical world. Light speed variation, so easily demonstrated by GPS technology and which invalidates the Lorentz Transformation, decidedly confirms the Inertial Transformation that predicts it.

Thus the modern ether theory based on the inertial transformation is a robust replacement for special relativity [35, 36] and the transition to this new theory is facilitated by the similarity of the structure of the members of the set of "equivalent" transformations. Such a transition can usher in a period of renewed scientific discovery as areas that are now prohibited can legitimately be explored. A good example of this is the case where Lorentz covariance imposed by relativistic considerations was relaxed as a result of which a new quantum theory of magnetism emerged that for the first time provided convincing explanations for the chemical reactivity of free radicals, the covalent bonds underpinning organic chemistry and the celebrated Pauli Exclusion Principle [38].

In view of the incontrovertible demonstrations of light speed variation using GPS technology presented in this chapter, investigation into the properties of the Inertial Transformation and the nature of the associated Modern Ether Theory should be the main focus of space-time research in the twenty-first century.

Author details

Stephan J.G. Gift
Department of Electrical and Computer Engineering, Faculty of Engineering,
The University of the West Indies, St.Augustine, Trinidad and Tobago, West Indies

9. References

[1] Einstein, A., On the Electrodynamics of Moving Bodies, p 37-65 in The Principle of Relativity edited by H.A. Lorentz et al., Dover Publications, New York, 1952.

[2] Rindler, W., Relativity Special, General and Cosmological, 2nd edition, Oxford University Press, New York, 2006.

[3] Zhang, Y.Z., Special Relativity and its Experimental Foundations, World Scientific, Singapore, 1997.

[4] Michelson, A.A. and E.W. Morley, The relative motion of the Earth and the luminiferous aether, American Journal of Science, ser.3, v.34, 333-345, 1887.

[5] Miller, D. C., The ether-drift experiment and the determination of the absolute motion of the earth, Review of Modern Physics, 5, 203, 1933.

[6] Shankland, R. S., Mc Cuskey, S. W., Leone, F. C., & Kuerti, G., New analysis of the interferometer observations of Dayton C. Miller, Review of Modern Physics, 27, 167, 1955.

[7] Jaseja, T.S., A. Javan, J. Murray and C.H. Townes, Test of Special Relativity or of the Isotropy of Space by Use of Infrared Masers, Physical Review 133, A1221-A1225, March 1964.

[8] Brillet, A. and J.L. Hall, Improved Laser Test of the Isotropy of Space, Physical Review Letters, 42, 549-552, February 1979.

[9] Hermann, S., A. Senger, E. Kovalchuk, H. Müller and A. Peters, Test of the Isotropy of the Speed of Light Using a Continuously Rotating Optical Resonator, Physical Review Letters, 95, 150401, 2005.

[10] Müller, H., P.L. Stanwix, M.E. Tobar, E. Ivanov, P. Wolf, S. Herrmann, A. Senger, E. Kovalchuk and A. Peters, "Relativity tests by complementary rotating Michelson-Morley experiments". Physical Review Letters 99 (5): 050401, 2007.

[11] Eisele, C, A. Nevsky and S. Schiller, Laboratory Test of the Isotropy of Light Propagation at the 10^{-17} Level, Physical Review Letters, 103, 090401, 2009.

[12] Demjanov, V.V., Physical interpretation of the fringe shift measured on Michelson interferometer in optical media, Physics Letters A 374, 1110-1112, 2010.

[13] Galaev, Y.M., The Measuring of Ether-Drift Velocity and Kinematic Ether Viscosity Within Optical Wavebands, Spacetime & Substance 3 (5,15) (2002) p207-224

[14] Gagnon, D.R., Torr, D.G., Kolen, P.T. and Chang, T., Guided-Wave Measurement of the one-way Speed of Light, Physical Review A, 38, 1767, 1988.

[15] Krisher, T.P., Maleki, L., Lutes, G.F., Primas, L.E., Logan, R.T., Anderson, J.D. and Will, C.M., Test of the Isotropy of the One-way Speed of Light Using Hydrogen-Maser Frequency Standards, Phys. Rev. D 42, 731, 1990.

[16] Ashby, N., Relativity in the Global Positioning System, Living Reviews in Relativity, 6, 1, 2003.

[17] IS-GPS-200E 8 June 2010, http://www.gps.gov/technical/icwg/IS-GPS-200E.pdf accessed August 21, 2011.

[18] Marmet, P., The GPS and the Constant Velocity of Light, Acta Scientiarum, 22, 1269, 2000.

[19] Kelly, A., Challenging Modern Physics, BrownWalker Press, Florida, 2005.

[20] Will, C.M., Clock synchronization and isotropy of the one-way speed of light, Physical Review D, 45, 403, 1992.

[21] Gift, S.J.G., One-Way Light Speed Measurement Using the Synchronized Clocks of the Global Positioning System (GPS), Physics Essays, Vol.23, No2, pp 271-275, June 2010.

[22] Gift, S.J.G., Successful Search for Ether Drift in a Modified Michelson-Morley Experiment Using the GPS, Applied Physics Research, Vol. 4, No. 1, pp 185-192, February 2012.

[23] Wang, R., Successful GPS Operations Contradict the Two Principles of Special Relativity and Imply a New Way for Inertial Navigation-Measuring Speed Directly, Proceedings of the IAN World Congress in Association with the U.S. ION Annual Meeting, 26-28 June 2000, San Diego, CA.
[24] Gift, S.J.G., One-Way Light Speed Determination Using the Range Measurement Equation of the GPS, Applied Physics Research, 3, 110, 2011.
[25] Tolman, R., The Second Postulate of Relativity, Physical Review, 31, 26, 1910.
[26] S.J.G. Gift, Another Test of the Light Speed Invariance Postulate, Modern Applied Science, 5, 152, 2011.
[27] Wallace, B. G., Radar Testing of the Relative Velocity of Light in Space, Spectroscopy Letters 2, 361, 1969.
[28] Gift, S.J.G., Light Speed Invariance is a Remarkable Illusion, Physics Essays, 23, 1, 2010.
[29] Gift, S.J.G., Doppler Shift Reveals Light Speed Variation, Apeiron, 17, 13, 2010.
[30] Nodland, B. and Ralston, J. P., Indication of Anisotropy in Electromagnetic Propagation over Cosmological Distances, Physical Review Letters, 78, 3043, 1997.
[31] Shtyrkov, E.I., Observation of Ether Drift in Experiments with Geostationary Satellites, Proceedings of the Natural Philosophy Alliance, pp201-205, 12th Annual Conference, Storrs CT, 23-27, May 2005.
[32] Gift, S.J.G., One-way Speed of Light Using Interplanetary Tracking Technology, submitted for publication.
[33] Myles Standish, E. and Williams, J. G., Orbital Ephemerides of the Sun, Moon and Planets, p16, http://iau-comm4.jpl.nasa.gov/XSChap8.pdf accessed May 17, 2012.
[34] Gift, S.J.G., The Relative Motion of the Earth and the Ether Detected, Journal of Scientific Exploration, 20, 201, 2006.
[35] Selleri, F., Recovering the Lorentz Ether, Apeiron, 11, 246, 2004.
[36] Selleri, F., La Relativita Debole, Edizioni Melquiades, Milano, 2011.
[37] Gift, S.J.G., Separating Equivalent Space-Time Theories, Apeiron, 16, 408, 2009.
[38] Gift, S.J.G., A Quantum Theory of Magnetism, Progress in Physics, 1, 12, 2009.

Section 2

Wireless Sensor Network Localization Techniques

Chapter 4

Indoor Positioning System Based on the Ultra Wide Band for Transport Applications

F. Elbahhar, B. Fall, A. Rivenq, M. Heddebaut and R. Elassali

Additional information is available at the end of the chapter

http://dx.doi.org/10.5772/50017

1. Introduction

Outdoor positioning can be improved with the start-up of Galileo. So, the accuracy will usually be in the order of a few metres. With Galileo services, the users needs require the same performances in outdoor and indoor applications. This is more obviously true in the construction, hypermarket, museums, where location awareness can become a crucial parameter for value-added services, especially for Galileo-GPS services. Positioning in difficult environments, especially in indoor, represents a current limitation for localisation systems (GPS/Galileo). In fact, indoor positioning faces additional difficulties as compared to outdoor positioning. Attenuation and multipath reflections of the line-of-sight (LOS) signal (or direct path) by the walls, floors, and ceiling of a tunnel are the main factors preventing typical GPS receivers from functioning indoors. Most of the time, the sum of multipath signals is stronger than the direct path signal, thereby preventing the receiver from accurately calculating the time of arrival [1]. The multipath signal distorts the cross correlation function peak, as detected by a receiver. The scientific and industrial community, especially in transport applications, considers that it is important to provide a positioning function in indoor (tunnel, station...) with a good performance of about a few centimetre. So, in order to reach this performance level, different techniques are under development, such as Ultra Wide Band technology. The UWB promises to overcome the power consumption and accuracy limitations of both the GPS and WLAN, and is more suitable for indoor location-based applications. In fact, Ultra Wide Band (UWB) technology provides high accuracy positioning in the multipath and confined environments typically found inside buildings. Integration of UWB with GPS or Galileo can provide a seamless transition from outdoor to indoor position and vice-versa. The ranging accuracy expected from UWB systems should be better than 0.5m in severe multipath environments [2, 3]. This chapter focuses on the indoor positioning system using the Ultra Wide Band, especially for transport applications. The first section is dedicated to introducing the indoor positioning application [3]. After this introduction, we present a brief review of some relevant work [7–9]. In indoor positioning system especially for transport applications two scenarios are considered: the self localisation and server-based localisation

INTECH
open science | open minds

© 2012 The Author(s). Licensee InTech. This chapter is distributed under the terms of the Creative Commons Attribution License http://creativecommons.org/licenses/by/3.0), which permits unrestricted use, distribution, and reproduction in any medium, provided the original work is properly cited.

([4–6]). In the self localisation, called the positioning system, the mobile receives signals from the bases station (the transmitter) and interprets the received-signals as ambient information for localising its position in a local coordinate system. The server-based localisation (fixed points) is based on the measure of the signals radiated from a mobile. Then, using the signals measured at distributed sensors, the server estimates the position of the mobile. In the second section of this chapter, we present two approaches for UWB positioning. The first one is based on the Direct Sequence Spread Spectrum UWB technique. The second one is based on the new waveforms, using the orthogonal waveforms, called the Modified Gegenbauer Functions. These functions allow specific waveforms for each transmitter with a good orthogonal propriety [3]. Theses approaches are studied and evaluated in terms of localisation errors for different multipath channels. The multipath channel effect in the UWB environment is introduced and analysed in this section for indoor positioning applications. In the third section, the test results using laboratory instruments are presented to validate the simulation and analytical results, given in the second section. In the fourth section, the first results concerning a new positioning approach for railway application, using UWB radio and Time Reversal techniques, are given. The last section of this chapter is dedicated to proposing conclusions and recommendations for future research.

2. Introduction of the indoor positioning system

In indoor positioning system two scenarios are considered: the self localisation and server-based localisation. In the self localisation, the mobile receives signals from the Base Stations BS (the transmitter) and interprets the received-signals as ambient information for localising its position in a local coordinate system. The block schema of this scenario is illustrated in figure 1. Each station (access point) sends the radio frequency signal and the processing, used to calculate of the mobile position, is realised in the mobile. The server-based localisation (fixed points) is based on the measure of the signals radiated from a mobile. Then, using the signals measured at distributed sensors, the server estimates the position of the mobile. The block schema of the second scenario is illustrated in figure 2.

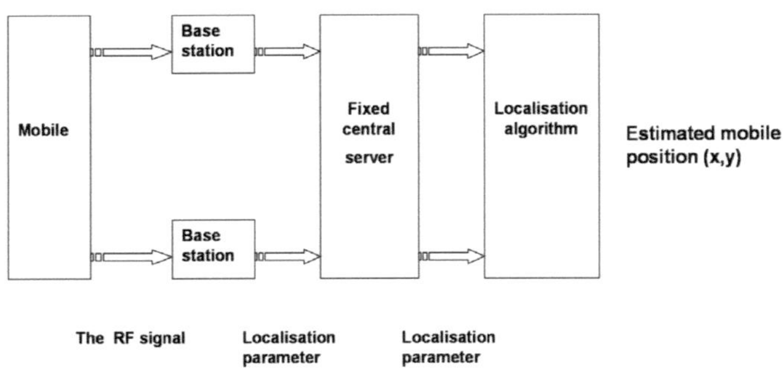

Figure 1. The self localisation: the positioning system approach.

Positioning techniques exploit one or more characteristics of radio signals to estimate the position of their source or the self position. Depending on accuracy requirements and constraints on transceiver design, various signal parameters can be employed. Commonly, a single parameter is estimated for each received signal. However, it is also possible to estimate multiple signal parameters in order to improve positioning accuracy. Some of the parameters that have been used for positioning are the Received Signal Strength Intensity (RSSI), the

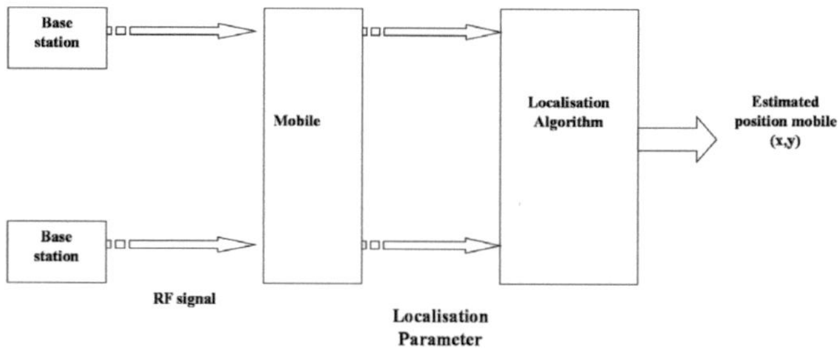

Figure 2. The server-based localisation approach.

Angle Of Arrival (AOA), the Time Of Arrival (TOA) and the Difference of Time Of Arrival (TDOA).

- Received Signal Strength Intensity (RSSI): provides information about the distance between two points. The main idea behind the RSS approach is that if the relation between distance and power loss is known, the RSS measurement at a point can be used to estimate the distance between that point and the transmitting point, assuming that the transmitted power is known.
- The Angle Of Arrival (AOA): this technique is based on determining the direction of the arrival signal. A stationary device measures the angle of arrival of the signal sent by a mobile device. Location can be estimated through triangulation if at least two stationary devices perform measurement. Measuring angles requires a specific antenna array. The angle information is obtained at an antenna array by measuring the difference in arrival times of an incoming signal at different antenna elements.
- Time Of Arrival (TOA): In TOA positioning, a mobile device sends a signal to a stationary device, which sends it back to the mobile device. The mobile device measures the round-trip time (RTT) of the signal. This leads to a circle, whose radius corresponds to half of the RTT and whose centre is on the location of the stationary device. Location of the mobile device can be approximated to be at the intersection of at least three measured circles. This technique requires accurate clocks because a $1\mu sec$ error in timing equals to a 300 m error in distance estimate. Thus, the accuracy is too low for TOA positioning.
- Time Difference Of Arrival (TDOA): TDOA positioning is developed to eliminate the tight synchronization requirement of TOA. In fact TOA range measurements require synchronisation among the mobile station (MS) and the Base Station BS (transmitter). However, TDOA measurement can be obtained even in the absence of synchronisation between MS and BS, if there is synchronisation among the base stations. In this case, the difference between the arrival time of two signals travelling between the MS and BS is estimated. This locates the MS on a hyperbola with foci at the BS.

Instead of performing a single measurement such as RSSI, AOA or TOA, a point can be estimated using a combination of position-related parameters. Such hybrid approaches can provide more accurate information about the position of the mobile than the approaches that estimate a single position parameter. Various combinations on measurement, such as TOA/AOA, TDOA/AOA and TOA/RSSI, are possible depending on accuracy requirement, complexity constraints, propagation environment and necessary processing delay [13, 18].

3. Indoor positioning application

Indoor localisation is the determination of the position of a device or a person in an indoor environment. While the localisation in outdoor environments can in most cases, efficiently and accurately use the GPS, or some more accurate variants like D-GPS, these systems are usually not efficient indoors. Typically, the transit satellite signal is highly damped when traversing the building and is not sufficient for a receiver to be detected. Satellite signals which are reflected by surrounding buildings may be detected through windows, but the path length for the satellites will be altered differently each from another and the position determination will be inaccurate to some 10..1000 meters.Therefore, the investigation of position detection technologies suitable for indoor environments is a current research and development topic.

3.1. Large public indoor positioning application

In recent years the applications for indoor localisation information have been developed and their use has increased. The new technologies continue to improve with better and more accurate positioning information. So, the new applications have been developed for the mass consumer markets. Different indoor location services and applications are offered in various domains. Location-based advertising has an application in a scenario in which the cell phone of a visitor to a mall can be located and used to display an advertisement of a shop very near to where its holder is located. Person and asset tracking applications in schools: surveillance of the children, criminals can be monitored using the positioning system, items can be tracked in warehouses, materials and equipment in manufacturing areas.

3.2. Indoor positioning application for transport applications

For transport applications, especially in subway transportation areas, it is very important to ensure the positioning service. In fact, the subway line is divided in parts called sections of about 500 m in length. When a train is in a section, it is declared to be engaged and no coach can go in until the train leaves it. This is the safety system adopted in most of the current networks. In this scope, only a few trains (say from 2 to 4, due to limited emitted power) will be allowed to receive or transmit data or video in any given area, including messages broadcast to passengers or security information sent to or from the control center, such as train status or problems encountered on the track or onboard. For this application, the distance between a train and a preceding one must be known to a precision in the sub-meter range over distances higher than one hundred meters in the train location application. So, for this application, it is important to use the positioning system able to operate with good performance in terms of the precision, positioning error and the processing delays. Another example can be cited concerning the public transport. In fact, knowing the position of our train or bus allows us to have pertinent information such as the lists of closest hotels, spectales and restuarants.These examples demonstrate that localisation may be needed as a key component for numerous domains.

3.3. Existing technical solutions for indoor positioning

Several techniques and commercial systems are used for indoor localisation. For exemple, a WLAN-based system is used to calculate positions by measuring the received signal strength (RSS). RADAR [16] (Microsoft) and Place Lab (Intel) are WLAN-based positioning systems.

The infrared-based method is used with the sensors that recognize the unique ID codes of infrared devices. Although the structure of this system is simple and the cost is low, a limited visibility range and line-of-sight (LOS) obstructions are its weak points. Ultrasonic-based system uses the difference in the transfer speed between RF and ultrasonic signals. This system has the advantages of 3D position recognition, low-power, and low cost. The Cricket system (MIT) and the Active bat system (AT&T Lab) [17] both use ultrasonic technology. Another radio frequency technique identification RFID system utilises a tag-and-reader scheme. Such tags contain circuitry that gain power from radio waves emitted by readers in their vicinity. They use this power to reply their unique identifier to the reader. This technique is very attractive because of the reasonable system price, and reader reliability but the major disadvantage of this technique is the very limited range, lower than $10m$.

Using current generation non-dedicated narrow band WLAN/WPAN derived technologies, a few meter localisation accuracy is achievable in indoor environments. However, this level of performance appears insufficient for some specific applications and services for example transport applications. These applications usually necessitate high performance in terms of precision and processing delay. The UWB technique is an excellent signalling choice for high accuracy localisation in short to medium distance, due to its high time resolution and inexpensive circuit.

Commonly, a UWB signal is defined to be a signal with a fractional bandwidth B_{fr} of more than 20% or an absolute bandwidth B of at least 500 MHz. The absolute bandwidth is calculated as the difference between the upper frequency f_h and the lower frequency f_l of the -10 dB emission point (equation 1).

$$B = f_h - f_l \tag{1}$$

The fractional bandwidth is defined as:

$$B_{fr} = \frac{B}{f_c} \tag{2}$$

where f_c is the center frequency ans is given by equation 3:

$$f_c = \frac{f_h + f_l}{2} \tag{3}$$

Due to their large bandwidth, UWB signal is characterized by very short duration waveforms. A UWB system transmits ultra short pulses with a low duty cycle. So, the ratio between the pulse transmission instant and the average time between two consecutive transmissions is usually kept small. There are two competing technologies for the UWB Wireless communication systems, namely: Impulse Radio (IR) and Multi-band OFDM (MB-OFDM). The IR-UWB technique is based on the transmission of very short pulses with relatively low energy. The IR-UWB provides lower data rates (a few Mbits/s) at higher ranges (tens of meters) with the possibility to have the positioning function. The MB-OFDM system gives potentially very high rates (in the order of 500 Mbits/s) with very short ranges (a few meters), the very wide frequency band used being divided into 14 sub-bands (500 MHz each). Several proposals based on these two technologies have been submitted to the IEEE 802.15 [19]. Both technologies are valid and credible. In addition to high rate Wireless communication systems, UWB signals have also been considered for low rate Wireless communication that focus on low power and low complexity devices. The IEEE formed the task group 4a standard for alternative PHY. The IEEE 802.15.4a provides high precision ranging /positioning and ultra power consumption. This standard is considered in this chapter. Two positioning systems are studied, DS-CDMA/UWB, and an original solution based on the orthogonal waveforms, and compared in terms of positioning errors.

4. The UWB proposed positioning system

4.1. The UWB waveforms

Some common UWB waveforms include Gaussian, monocycle pulse; pulse based on the modified Hermite polynomials. For example, the Gaussian and monocycle pulses (figure 3), given, respectively, the equation 4:

$$\begin{aligned} w(t) &= \exp\left(-\tfrac{t}{\tau}\right)^2 \\ v(t) &= \tfrac{t}{\tau}\exp\left(-\tfrac{t}{\tau}\right)^2 \end{aligned} \tag{4}$$

with τ the pulse duration.

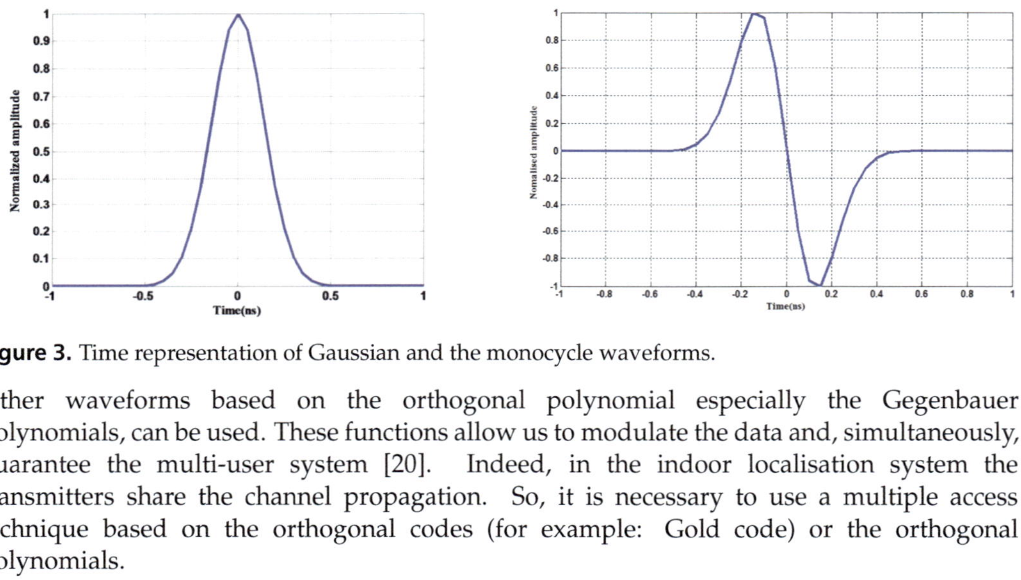

Figure 3. Time representation of Gaussian and the monocycle waveforms.

Other waveforms based on the orthogonal polynomial especially the Gegenbauer polynomials, can be used. These functions allow us to modulate the data and, simultaneously, guarantee the multi-user system [20]. Indeed, in the indoor localisation system the transmitters share the channel propagation. So, it is necessary to use a multiple access technique based on the orthogonal codes (for example: Gold code) or the orthogonal polynomials.

The MGF $G_n(\beta, x)$ uses the weight function $W(x) = (1 - x^2)^{\beta - 1/2}$ where $\beta > -1/2$ is a shape parameter, n is the degree of the function and x is the variable. These functions are orthogonal on the interval $[-1, 1]$ for $m \neq n$:

$$\int_{-1}^{1} w(x) G_n(\beta, x) G_m(\beta, x) dx = 0 \tag{5}$$

These functions can be defined by a recurrence relation. Furthermore, they satisfy the differential equation 6.

$$G_n(\beta, x) = 2\frac{n + \beta - 1}{n} x G_{n-1}(\beta, x) - \frac{n + 2\beta - 2}{n} G_{n-2}(\beta, x) \tag{6}$$

Their expressions for the first few orders are given by the following equations:

$$\begin{aligned}
G_0(\beta, x) &= (1-x^2)^{\beta-1/2} \\
G_1(\beta, x) &= 2\beta x(1-x^2)^{\beta-1/2} \\
G_2(\beta, x) &= \beta[-1+2(1+\beta)x^2](1-x^2)^{\beta-1/2} \\
G_3(\beta, x) &= \beta(1+\beta)[-2x+(2+\beta)\tfrac{4x^3}{3}](1-x^2)^{\beta-1/2} \\
G_4(\beta, x) &= \beta(1+\beta)\left[\tfrac{1}{2}-2(2+\beta)x^2+(2+\beta)(3+\beta)\tfrac{2x^4}{3}\right](1-x^2)^{\beta-1/2}
\end{aligned} \quad (7)$$

The waveforms of the MGF G_n are shown in figure 4, for $n = 1$ to 4 and $\beta = 1$ versus time normalized to waveform duration T. They are normalized here so as to have an energy of unity. Using Gegenbauer waveforms, a positioning system may be built, which requests up to 4 transmitters.

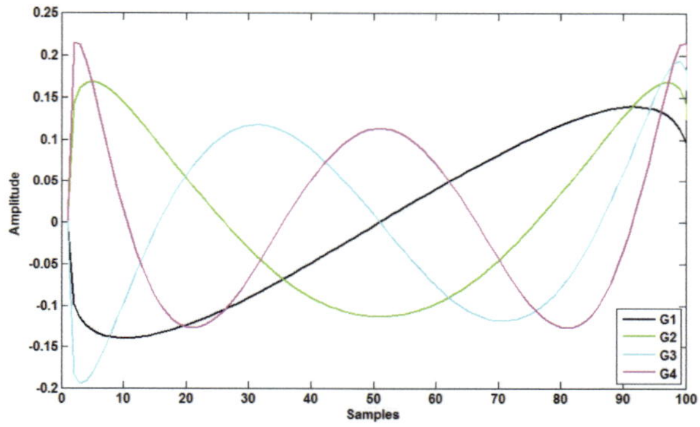

Figure 4. Time representation for modified Gegenbauer functions of orders $n = 1$ to 4.

4.2. The Positioning techniques used

In order to lessen synchronisation effects, we use the TDOA technique. This technique makes use of the difference of TOA's at the participating transmitters. In TDOA approach, the difference between the arrival times of two signals travelling between the mobile and the two reference points is estimated. This locates the mobile on a hyperbola, with foci at the two reference nodes. Making use of triangulation, an exact positioning of mobile can be found. The relationship between the range difference and the TDOA measurement between the receivers is given by equation 8.

$$\hat{d}_{TDOA} = R_{i,j} = c\tau_{i,j} = c(\tau_i - \tau_j) = R_i - R_j \quad (8)$$

where $\tau_{i,j}$ is TDOA between receiver i and j, $R_{i,j}$ is the range difference, τ_i and τ_j are TOA arrival estimates at transmitter (point) i and j, while R_i and R_j are range estimates at transmitter i and j. Figure 5 illustrates the principle of th TDOA technique in 2D.

There are 2 ways of obtaining TDOA estimates [8]. The first way which makes use of the cross correlation estimation technique of the received signal $r_1(t)$ and $r_2(t)$ to calculate the delay corresponding to the largest cross-correlation value.

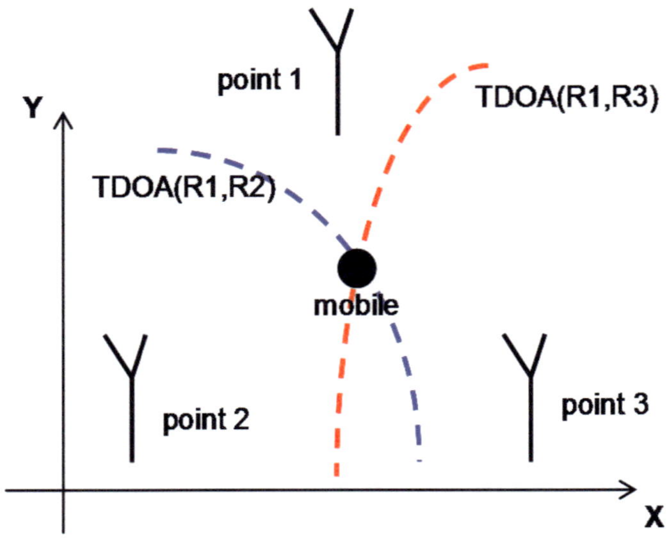

Figure 5. 2D TDOA Positioning technique.

For example, in the case of the server-based localisation, consider a signal $x(t)$ being radiated by a mobile and being received by two points (stations).

$$\begin{aligned} r_1(t) &= A_1 x(t - y_1) + n_1(t) \\ r_2(t) &= A_2 x(t - y_2) + n_2(t) \end{aligned} \tag{9}$$

Equation 9 represents the received signals at the two points (Stations). A_1 and A_2 are the amplitudes of the received signals with delays y_1 and y_2, corrupted with noise $n_1(t)$ and $n_2(t)$. It is assumed that $x(t)$, $n_1(t)$ and $n_2(t)$ are real and jointly stationary, zero mean random processes and that $s(t)$, $n_1(t)$ and $n_2(t)$ are uncorrelated.

$$\begin{aligned} r_1(t) &= x(t) + n_1(t) \\ r_2(t) &= Ax(t - Y) + n_2(t) \end{aligned} \tag{10}$$

where A is the amplitude ratio between the two received signals and $Y = y_2 - y_1$. TDOA estimation requires estimation of values of Y. A simple cross correlation technique is illustrated in figure 6.

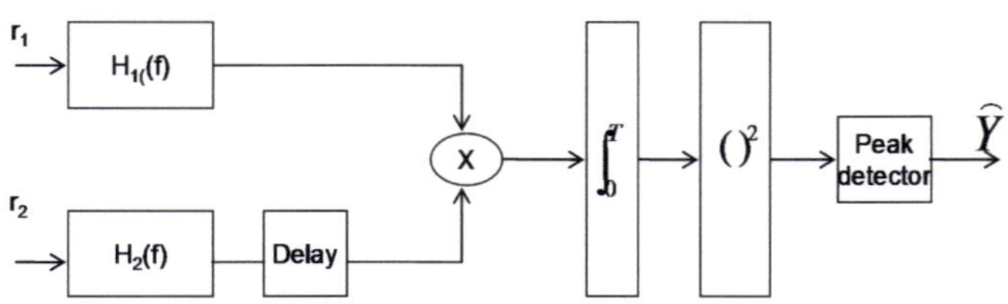

Figure 6. The Cross Correlation Method for TDOA estimation.

The cross correlation function of these two received is given by equation 11.

$$c(t) = \frac{1}{T} \int_0^T r_1(t) r_2(t + \tau) dt \tag{11}$$

with T is the period of observation.

TDOA estimate $\hat{Y}$ is the value of τ that maximizes the cross correlation is given by equation 12.

$$\hat{Y} = \arg(\max |c(t)|) \tag{12}$$

The second method uses the substraction at the TOA estimates from two transmitters (points) to produce a relative TDOA estimate (equation 8). This requires a knowledge of timing at the two transmitters and thus requires a strict clock synchronization between the two transmitters. Also, this method has an advantage of eliminating the errors in TOA estimates common to all the transmitters. After the TDOA estimate step, a hyperbolic position location algorithm is used to produce an accurate and unambiguous solution to the position location algorithm.

Once the TDOA estimates have been obtained, they are converted into range difference measurements. Thereafter, these measurements are converted into nonlinear hyperbolic equations [11]. Several algorithms have been proposed for this purpose having different complexities and accuracies [12]. Here, we will focus on the mathematical model for hyperbolic TDOA equations based on the Chan technique. In fact, a non-iterative solution capable of achieving optimum performance in terms of positioning error was proposed by Chan [13]. This solution is a non-iterative and has a higher noise threshold than the others methods [12]. Furthermore, it provides an explicit solution that is not available for example in the Taylor-series method [13]. This method is used in the simulations given in this chapter, because essentially it is less sensitive to the channel propagation noise.

Let (x, y) be the source location (mobile), and (Xi, Yi) be the known location of ith Base Station BS or transmitter, where $i = 2, 3...M$, M being the total number of BSs taking part in the position location process. Moreover, assume that $BS = 1$ is the controlling BS. The range difference between source and the ith BS is given by equation 13.

$$R_i = \sqrt{(X_i - x)^2 - (X_i - y)^2} \tag{13}$$

Now, the range difference between base stations with respect to $BS = 1$ is given by equation 14.

$$R_{i,1} = cd_{i,1} = R_i - R_1 \tag{14}$$

where c is the signal propagation speed ($3.10^8 m/s$) and $d_{i,1}$ is the range difference distance between i^{Th}BS and $BS = 1$. In order to find the x and y values, Chan method is used, producing two TDOAs, for the three base stations. So, the solution for x and y in terms of $R_1$is written by equation 15:

$$\begin{pmatrix} x \\ y \end{pmatrix} = \begin{pmatrix} X_{2,1} & Y_{2,1} \\ X_{3,1} & Y_{3,1} \end{pmatrix}^{-1} \left\{ \begin{pmatrix} R_{2,1} \\ R_{3,1} \end{pmatrix} R_1 + \frac{1}{2} \begin{pmatrix} R_{2,1}^2 - K_2 + K_1 \\ R_{3,1}^2 - K_3 + K_1 \end{pmatrix} \right\} \tag{15}$$

with

$$\begin{cases} K_1 = X_1^2 + Y_1^2 \\ K_2 = X_2^2 + Y_2^2 \\ K_3 = X_3^2 + Y_3^2 \\ \\ R_{2,1} = cd_{2,1} \\ R_{3,1} = cd_{3,1} \end{cases}$$

On the right side of the above equation, all the quantities are known quantities, except R_1. Therefore, the solution of x and y will be in terms of R_1. When these values of x and y are substituted into the equation $R_{2,1}$, a quadratic equation in terms of $R_1$is produced. Once the roots for R_1 are known, values of x and y can be determined. It should be noted that only the positive R_1 root must be considered. One of the roots of the quadratic equation is, in fact, either negative or too large to be within the cell radius.

4.3. Positioning system based on the DS-CDMA technique

The first presented system is based on the DS-CDMA technique. So, for each emitter (Base Station) a pseudo code is attributed. In this study, the gold code is chosen due to its good orthogonality propriety. The bloc diagram for each transmitter (BS) is described in figure 7. The transmitter is composed of the coded and the modulated (antipodal modulation) operation using Gaussian waveforms. The antipodal modulation (analog to binary phase shift keying) is used for all data bits from each transmitter. The receiver unit, figure 8, consists of the demodulated and decoded function in order to retrieve the signal of each BS. Finally, the localisation technique based on Time Difference of Arrival is used to calculate the estimated position of the mobile.

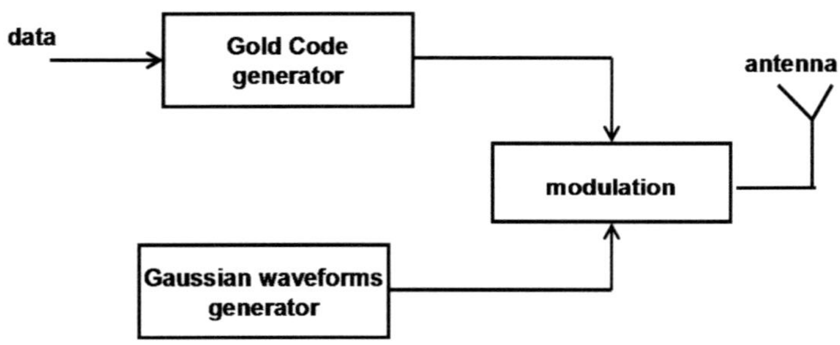

Figure 7. Block diagram of transmitter (BS).

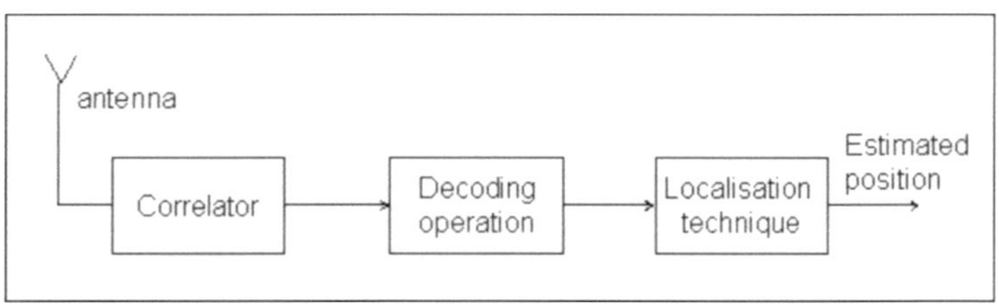

Figure 8. Block diagram of the receiver (MS).

4.4. System based on the Modified Gegenbauer Functions

The second system is based on the Modified Gegenbauer Function MGF. In this case, we attributed one MGF order for each BS. For example, order 1 (G1) is for SB1, order 2 (G2) for BS2 and order 3 (G3) for BS3. The Block diagram of the transmitter is given by figure 9. The receiver unit consists in demodulating and calculating the MS position using the TDOA technique figure 10.

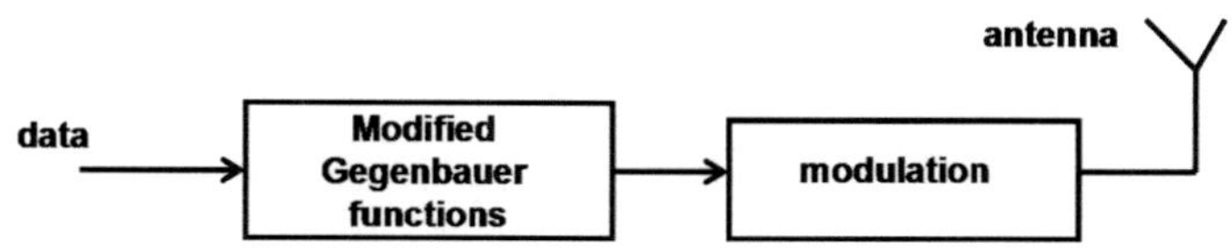

Figure 9. Block diagram of the transmitter (BS).

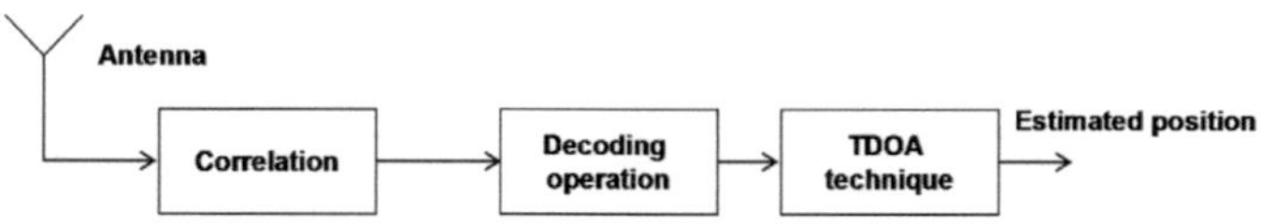

Figure 10. Block diagram of the receiver (Mobile).

5. Simulation results

In order to compare two systems DS-CDMA /UWB and the MGF/UWB, we calculate the positioning error considering different channel propagation models. So, in order to send information from one point to another, the transmitted signal must travel though the propagation channel to reach the receiver (mobile station). In this chapter, two channels are used: the Additive White Gaussian Noise AWGN channel with a uniform spectral power density and the IEEE 802.15.3a channel. The proposed system was considered to deliver positioning in tunnels, especially subway transportation areas; hence the choice of using a IEEE 802.15.3a channel in order to evaluate, in simulation, the proposed solutions. The IEEE 802.15.3a model was developed from around 10 contributions, all referring to distinct experimental measurements, performed in indoor residential or office environments [5]. The IEEE 802.15.3a model is based on the Saleh Valenzuela formalism. Parameters are provided to characterize the clusters and ray arrival rates (Λ and λ), as well as the inter and intra-cluster exponential decay constants (Γ and γ). Four sets of parameters are provided to model the four following channel types (figure 11):

- The channel model CM 1 corresponds to a distance of 0-4 m in a residential Light Of Sight LOS situation;
- The channel model CM 2 corresponds to a distance of 0-4 m in an residential Non Light Of Sight NLOS situation;
- The channel model CM 3 corresponds to a distance of 4-10 m in an office NLOS situation;
- The channel model CM 4 corresponds to an Office NLOS situation with a large delay spread $\tau_{rms} = 25$ ns.

The key parameter of the UWB signal is the choice of waveforms. Two waveforms are used, in this section, and evaluated in terms of the positioning errors in IEEE 802.15.3a cases. The first one is the Gaussian pulse, the second waveform is the Modified Gegenbauer pulse. Figures 12 and 13 illustrate the results of the simulations realised in Matlab software. These figures show that the MGF wavefroms give better results than the monocyle waveforms in terms of positioning error. Especially, in the case of a very noisy channel SNR> -9 dB. These performances decrease in the absence of line of sight. The MGFs are less sensitive to the propagation channel effects than the monocycle waveforms.

In figure 14, the positioning errors are evaluated for different SNR values and different waveforms numbers in two cases: the proposed system based on the CDMA-UWB with gold

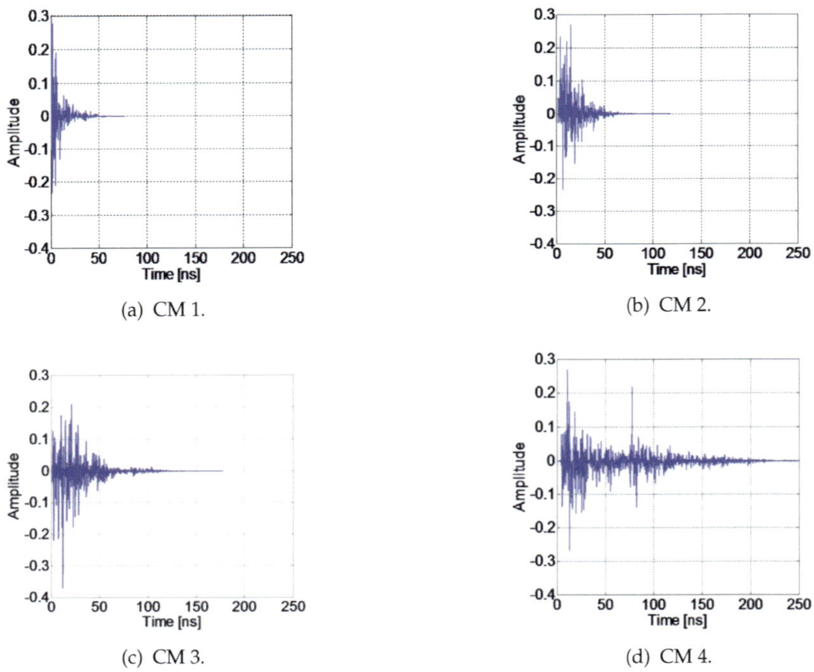

(a) CM 1. (b) CM 2.

(c) CM 3. (d) CM 4.

Figure 11. Channel impulse responses for CM 1, CM 2, CM 3 and CM 4.

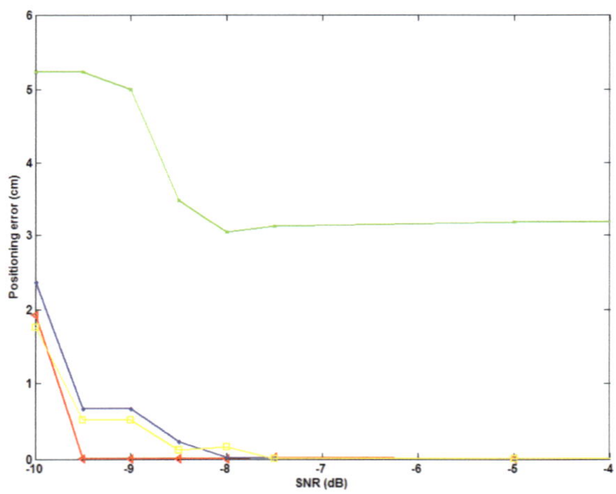

Figure 12. The positioning error in case of MGF in IEEE 803.15.3a channel.

code (7 chip) and the second proposed system based on the Modified Gegenbaeur functions. In this case, the channel effect is a simple AWGN channel. We show that, when increasing the number of MGF, the positioning error decreases. For example, the positioning error is less than 1.5 cm for SNR > -10 dB when we attribute seven Gegenbauer pulses per base station. For DS-CDMA solution, using code Gold length N = 7 chip, the positioning error is higher than 1.5 cm for SNR > -6 dB. We conclude that MGFs give a better performance than the

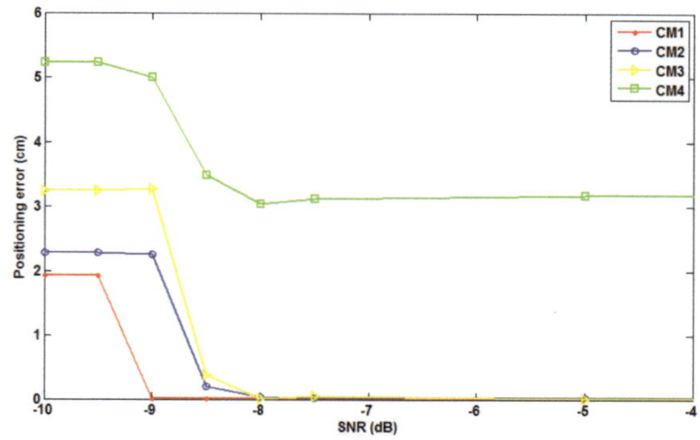

Figure 13. The positioning error in case of monocyle waveforms in IEEE 803.15.3a channel.

DS-CDMA, even if we use one order for each transmitter system. Another adavantge of the MGF positioning system is the processing delay lower than the DS-CDMA solution. In fact, in the MGF positioning system the modulation and the multiple access technique are realised simulatneously. These performances increase if we attribute more than one Modified Gegenbauer pulse per transmitter.

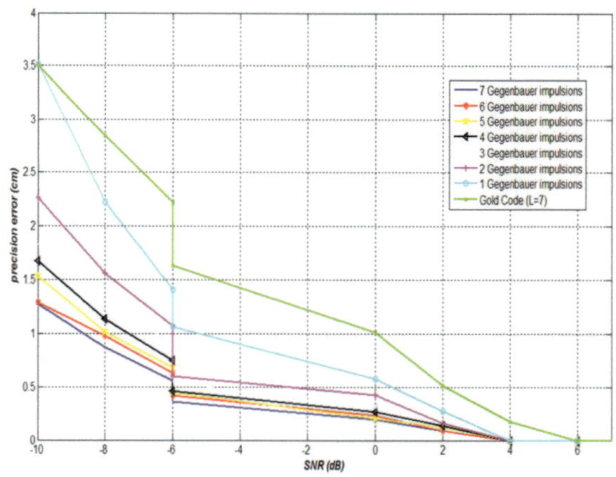

Figure 14. The positioning error for different SNR values.

6. Measurements results

An experimental setup was established to validate the proposed indoor positioning system. the tested system is based on the gegenbauer waveforms. The measurement setup is illustrated in figures 15 and 17. S_1, S_2 and S_3 are the position of transmitting antennas. The transmitter signal is generated using Arbitrary Waveform Generator 10GHz and the monopole antenna omni-directional adapted to the 800 MHz - 19 GHz band (figure16). The received signal is measured by oscilloscope. The mobile position is calculated using the TDOA technique. The test results, realised using the configuration 17, are illustrated in figure 18 and in table 1. These results show that the positioning error is repetitively about $18cm$.

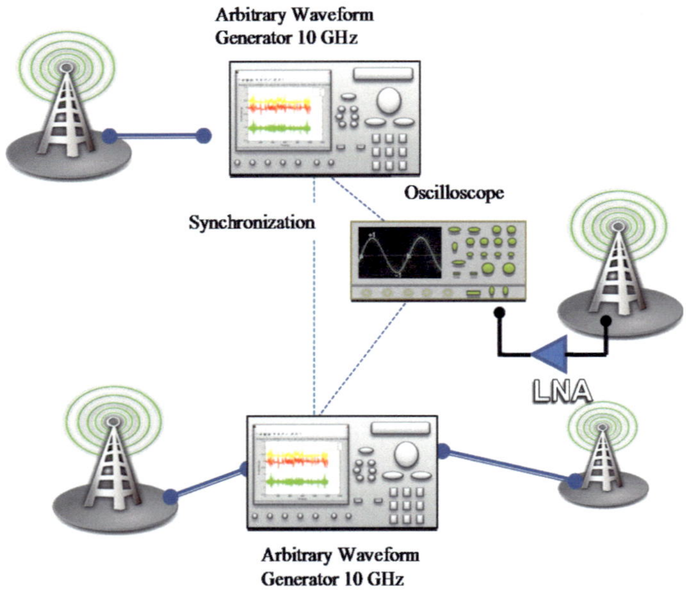

Figure 15. Experimental setup for indoor positioning system.

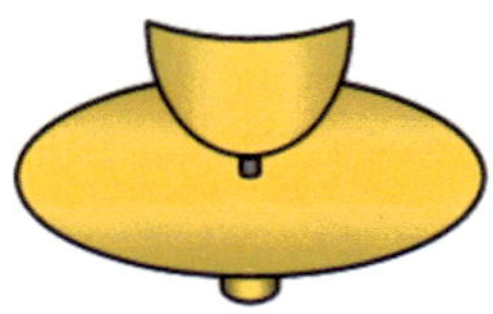

Figure 16. Monopole antenna.

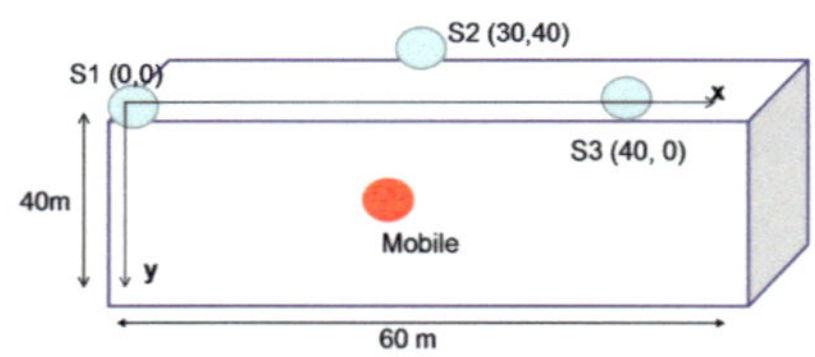

Figure 17. Configuration used in the measurements.

Acquisition	1	2	3	4	5
Positioning error (cm)	17,5	12	19	18,5	20,03
processing delay (s)	0,87	0,79	0,67	0,80	0,82

Table 1. Positioning error results

7. Future work

In the future work, localisation systems for railway transport using UWB radio and Time Reversal (TR) techniques will be studied and evaluated. In fact, Time Reversal channel pre-filtering facilitates signal detection and helps increasing the received energy in a targeted area. In this context, a proposed UWB-TR techniques for the precise location of trains is

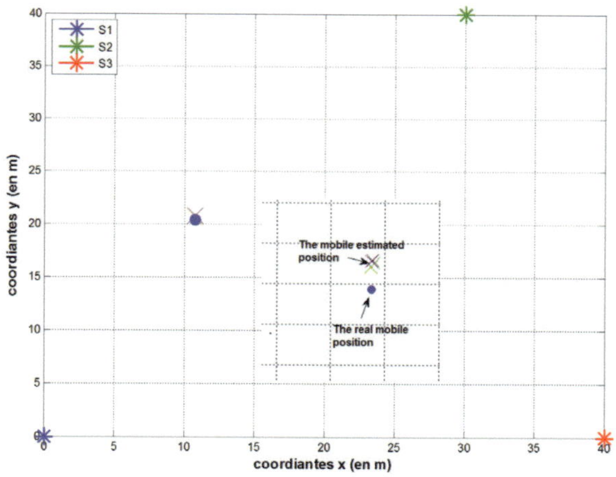

Figure 18. The test results.

illustrated in figure 19. It shows the particular case of a railway tunnel. The balises are geo-referenced. UWB/TR balises will be installed on the side of the track. The UWB/TR balises are kilometer markers. On arrival in the range of the UWB communication, the train computes its absolute localisation to the balises using time of arrival information. Moreover, several simple UWB transmitters are located in the balises to enable focalization. The local Channel State Information (CSI) between any balise transmitter and virtual optimal balise localisation along the track is identified a single time during the initial installation. This information is then introduced as pre-filtering data in the different UWB transmitters. Therefore focalization is obtained in this virtual localisation area whenever the train passes, potentially improving the absolute localisation process.

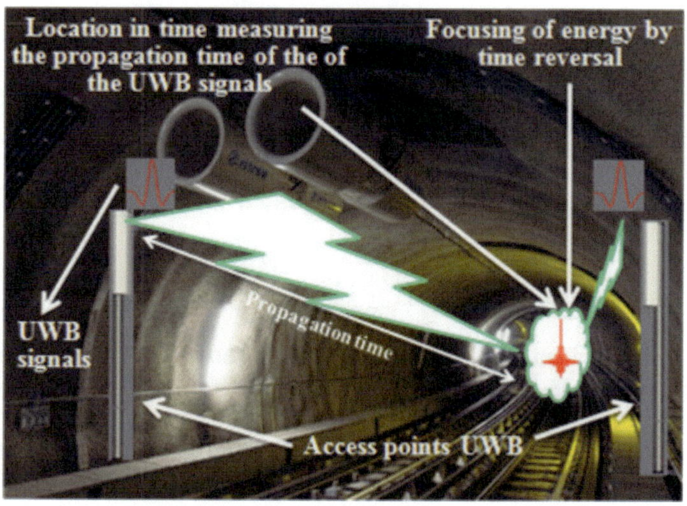

Figure 19. TR-UWB localisation system proposed.

The principle of the proposed TR-UWB system uses three stages: first the channel impulse response is measured and recorded at the transmitter Tx. Then, this impulse response is reversed in time and transmitted in the propagation channel to the receiver Rx. The original signal we have chosen to be transmitted, associated with the impulse response, is the second derivative of Gaussian function; it is an ultra short pulse with duration of 500 picoseconds. This principle can then be described by noting s(t) the transmitted pulse, $h(t)$ the complex impulse response of the channel and $h^*(-t)$ the complex conjugate of the time reversed version of $h(t)$. We note by $y(t)$ the received signal without TR and $y_{rt}(t)$, the received signal with TR at the receiver. Their expressions are given by equations 16 and 17:

$$y(t) = s(t) \otimes h(t). \quad (16)$$

$$y(t) = s(t) \otimes h^*(-t) \otimes h(t). \quad (17)$$

where $\otimes$ represents the convolution operation. From equation17, we deduce the equivalent impulse response $h_{eq}(t)$ which corresponds to the autocorrelation function of the channel equation 18:

$$h_{eq}(t) = h^*(-t) \otimes h(t). \quad (18)$$

The autocorrelation function is used to evaluate temporal focusing. This characteristic is very beneficial for the application to the UWB system [14] [15].To study the temporal focusing, we evaluate the focusing gain (FG), by considering the impulse response channel $h(t)$ and the equivalent impulse response channel $h_{eq}(t)$. FG is then defined as equation 19:

$$FG_{dB} = 10log_{10}\left(\frac{max(|h_{eq}(t)|^2}{max(|h(t)|^2)}\right) \quad (19)$$

For performance evaluation in terms of temporal focusing, we calculated the focusing gain (FG), for different channel models. FG is obtained after determining the Power Delay Profile in the case of UWB-IR is denoted PDP and PDP_{TR} in the case of TR-UWB. PDP determines average power of scattering components occurring with propagation delay, it gives the intensity of a signal received through a multipath channel as a function of time delay. Expression of PDP and PDP_{TR} in both cases is then given by equations 20 and 21.

$$PDP = |h(t)|^2 \quad (20)$$

$$PDP_{TR} = \left|h_{eq}(t)\right|^2 \quad (21)$$

The equation giving the expression of FG can be written equation 22:

$$FG_{dB} = 10log_{10}\left(\frac{max(PDP)}{max(PDP_{RT}}\right) \quad (22)$$

Figures 20 to 23 show a comparison between PDP and PDP_{TR}, and, then, we can find a temporal focusing and increase the amplitude of the power from PDP to PDP_{TR}. This translates into results on the evaluation of FG in table 2.Thus we find, using successively $CM1$ to $CM4$, the FG increases due, in particular, to many multipaths. Indeed, time reversal take advantage of the complexity of the channel, which would be very beneficial for the purpose of locating in confined environments, such as tunnels.

Simulations were performed using the channel models $CM1$, $CM2$, $CM3$ and $CM4$.Table 3 presents a comparative study between the UWB-IR system and TR-UWB system, in terms of temporal focusing and Root Mean Square Error RMSE of localisation, treated in a particular

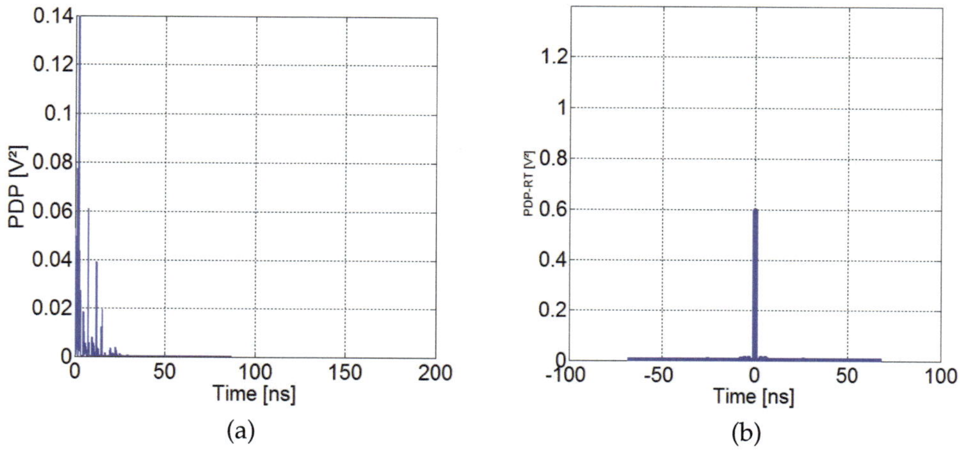

Figure 20. $(a)PDPand(b)PDP_{TR}forCM1$.

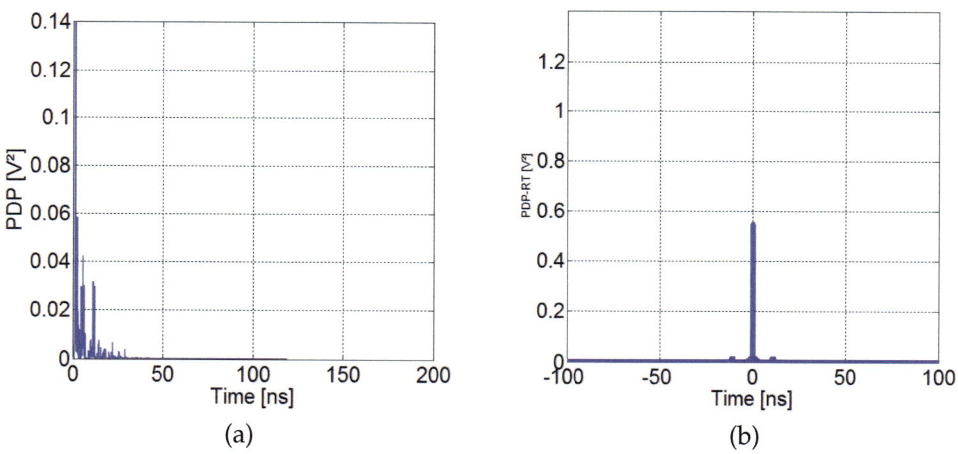

Figure 21. $(a)PDPand(b)PDP_{TR}forCM2$.

Channel model	CM 1	CM 2	CM 3	CM 4
GF [dB]	7.27	8.03	9.94	11.69

Table 2. Focusing gain FG for different channel model

case. This study shows that, with the combination of UWB in TR, we get better information in terms of localisation error. Indeed, in the particular case treated, the RMSE is 7.45 cm in the case of $CM3$ for the UWB-IR system, whereas it is only 1.12 cm for the TR-UWB system. This remark also applies to $CM1, CM2$ and $CM4$.

These first results show that the proposed solution allow us to significantly reduce the localisation errors, especially of the channel propagation environments complex. The next step will be dedicated to validating these results through experimentations in real environment.

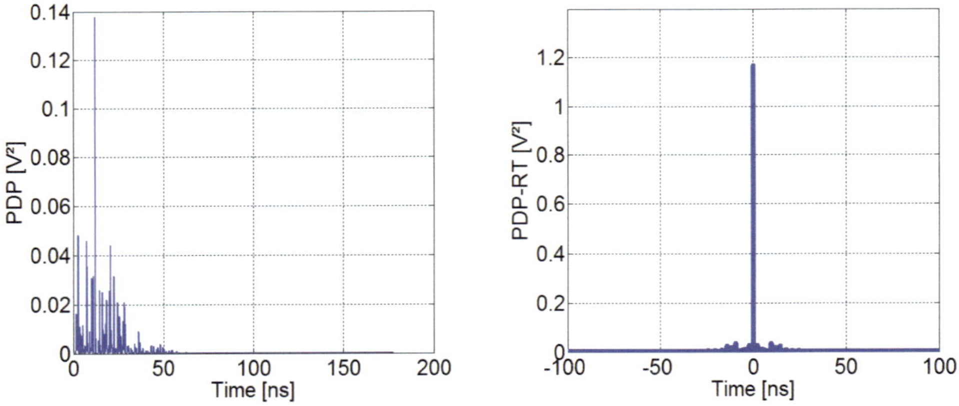

(a) (b)

Figure 22. $(a)PDPand(b)PDP_{TR}forCM3$.

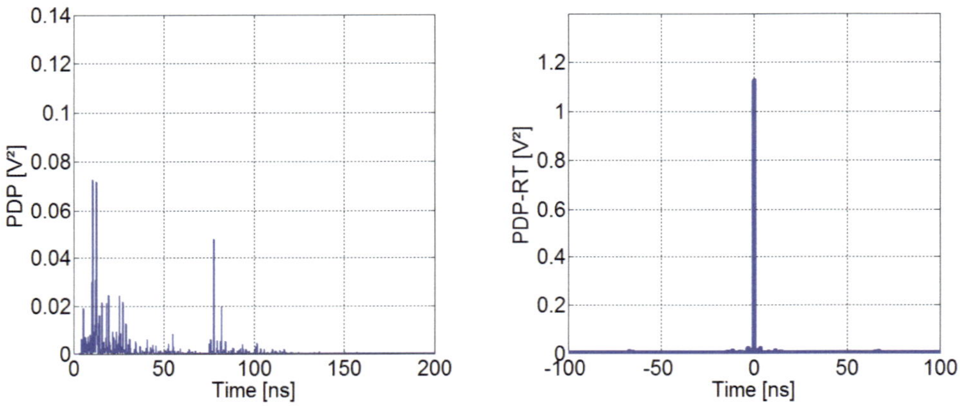

(a) (b)

Figure 23. $(a)PDPand(b)PDP_{TR}forCM4$.

Channel model	FG[dB]	RMSE[cm] for UWB	RMSE[cm] for UWB-TR
CM1	7.27	2.08	$= 0$
CM2	8.03	2.66	$= 0$
CM3	9.94	7.45	1.12
CM4	11.69	17.82	12.78

Table 3. Comparison between UWB-IR and UWB-TR in terms of FG and positionnig error in CM1, CM 2, CM3 and CM4, SNR=8 dB

8. Conclusion

In this chapter, the indoor positioning system based on the UWB technique is presented. The indoor localisation application is given, especially for railway transport. The existing indoor positioning technique is presented. The UWB technique chosen to establish positioning system is introduced. Two proposed systems are presented and evaluated in terms of localisation error using AWGN and IEEE 802.15.3a channels. The first one is based on the

Direct Sequence Spread Spectrum UWB technique. The second one is based on the Modified Gegenbauer orthogonal waveforms. The simulation results show that the second proposed system is less sensitive to noise. This is due to the good orthogonal propriety of the MGF functions. These results are validated by tests using laboratory instruments. Thereafter, the new positioning system for railway application using UWB radio and Time Reversal techniques is presented. The first results show that the combination of UWB and Time Reversal TR can reduce the localisation error thanks to the characteristic of TR technique, including the temporal focusing. In the future work, these simulations results will be validated by tests in real environments.

Author details

Fouzia Elbahhar, B. Fall and Marc Heddebaut
Univ Lille Nord de France, F-59000 Lille, France and IFSTTAR/LEOST, F-59650 Villeneuve dŠAscq - France

Atika Rivenq
Univ Lille Nord de France, F-59000 Lille, France and IEMN (UMR 8520 CNRS) Dept. OAE, UVHC F-59313 Valenciennes - France

Raja Elassali
TIM/ ENSA, UCAM, Avenue Prince Moulay Abdellah, B.P511-40000 Marrakech - Morocco

9. References

[1] Ingram.S.J., Harmer.D., Quinlan. M "UltraWideBand indoor positioning systems and their use in emergencies", Monterey, Position Location and Navigation Symposium, (PLANS) April 2004, pp. 706 - 715.

[2] Elbahhar.F, Lamari. A, Rivenq. A, Rouvaen. J. M, Heddebaut.M, Boukour.T, Sakkila. L "Novel approach of UWB multi-band system based on orthogonal function for transports applications", The European Physical Journal Applied Physics, March 2011, 53,

[3] Lamari.A, Elbahhar.F, Rivenq.A, Rouvaen J. M, Heddebaut.M "Performance Evaluation of a Multi-band UWB localisation and Communication System based on Modified Gegenbauer Functions", Wireless personal communications (Kluwer), Vol. 34 , Issue 3, pp. 255 - 277, Aug. 2008.

[4] Chen.P.C, "A non line of sight error mitigation algorithm in location estimation". In proc. IEEE int. conf.wireless commun, networking (WCNC vol.1, New Orleans, LA, sep.1999, pp. 316 - 320.

[5] Pahlavan.K, Krishnamurthy. P, Beneat.A ;Worcester Polytech. Inst., MA "Wideband radio propagation modeling for indoor geolocation applications", Communications Magazine, IEEE Volume: 36 Issue:4 , pp. 60 - 65, Apr 1998.

[6] Bhargava. V, Sichitiu. M. L, "Physical authentication through localisation in wireless local area networks", Proceedings of the IEEE Global Telecommunications Conference, pp. 2658 - 2662, 2005

[7] Hatami. A, Pahlavan.K "Performance comparison of RSS and TOA indoor geolocation based on UWB measurement of channel characteristics", Proceedings of the IEEE 17th

International Symposium on Personal, Indoor and Mobile Radio Communications, pp. 1 - 6, Helsinki, Finland.
[8] Yu.K, Montillet.J.-P, Rabbachin.A "UWB location and tracking for wireless embedded networks". Signal Processing, Vol. 86, No., (2006), pp. 2153 - 2171.
[9] Zhang.M, Zhang. S, Cao. J, Mei. H "A novel indoor localisation method based on received signal strength using discrete Fourier transform", Proceedings of the First International Conference on Communications and Networking in China, pp. 1 - 5, Beijing, China.
[10] Joseph C. Liberti, Jr., Theodore S. Rappaport, "Smart Antennas for Wireless Communications: IS 95 and Third Generation CDMA Applications", Prentice Hall communications engineering and emerging technologies series, 1999
[11] B.friedlander," A passive localisation algorithm and its accuracy analysis", IEEE Journal of oceanic engineering vol.OE-12, no.1, pp.234-244, January 1987
[12] Y.T.Chan and K.C.Ho "A simple and efficient estimator for hyperbolic location", IEE Transactions on signal processing, vol, 42, no.8, pp.1905-1915, August 1994.
[13] M.Laoufi "Localisation dŠusagers de la route en détresse par réseau de radiocommunications cellulaire dŠappel dŠurgence dédié" Ph.D. thesis, University of Valenciennes, France, 2003.
[14] H. Saghir, M. Heddebaut, F. Elbahhar, J.M. Rouvaen, A. Rivenq, "Train-to-wayside wireless communication in tunnel using ultra-wideband and time reversal", Ed. Elsevier, Transportation Research Part C: Emerging Technologies vol. 17, Issue 1, pp. 81-97, Feb 2009.
[15] D. Abassi-moghadam, D. Tabataba Vakili, "Channel characterization of time reversal UWB communication systems", Wiley International Journal of Communication Systems, pp 601-614, published online 18 July 2010.
[16] P. Bahl and V. Padmanabhan, "RADAR: an in-building RF-based user location and tracking system", Proceedings of the Nineteenth Annual Joint Conference of the IEEE Computer and Communications Societies, Vol. 2, 2000.
[17] M. T. Shiraji, S. Yamamoto, "Human Tracking Devices: the Active Badge/Bat and Digital Angel / Verichip systems", Project paper 1, ECE 399, Oregon State University, USA,Fall Term 2003.
[18] N.Obeid, M.Heddebaut, F.Elbahhar, C.Loyez, N.Rolland, "Millimeter Wave Ultra Wide Band Short Range Radar localisation Accuracy", VTC Spring 2009.
[19] Z.Ahmadian,"Performance Analysis of the IEEE 802.15.4a UWB System", Communications", IEEE Transactions on, Volume: 57, Issue: 5 1474 - 1485,May 2009.
[20] F. Elbahhar, A. Rivenq-Menha j, J.M. Rouvaen, M. Heddebaut and T. Boukour, "Comparison between DS-CDMA and Modified Gegenbauer Functions for a multi-user communication Ultra Wide Band system", IEE. Proceedings Communications, 152, 1021 - 1028, 2005.

Chapter 5

Ultra Wide Band Positioning Systems for Advanced Construction Site Management

Alberto Giretti, Alessandro Carbonari and Massimo Vaccarini

Additional information is available at the end of the chapter

http://dx.doi.org/10.5772/48260

1. Introduction

Phenomena of scarce performances and rapid obsolescence of buildings, entailing high maintenance costs, are very often the consequence of a loose integration between the operational phases of the building process. The separation of technical and management competences, the lack of coordination between high level strategic and procurement decisions and on-site construction management, make the achievement of designed performances a particularly fragile process and it is one of the side effects of a number of the singular complexities of the building construction process. Nearly every facility is, in fact, custom designed and nomadic. Construction sites are very dynamic working environments, often changing their layout on a daily or weekly basis. They are affected by unpredictable and uncontrollable external events such as weather, the availability of local resources, etc. (Behzadan et al. 2008). Furthermore, large building projects involve thousands of parts and components while changes to design plans at construction time are not uncommon. Building parts and components are mostly made or assembled on-site, standardization is rather low. Adjustments are also made on site, at times without even updating original building plans. Consequently, the management of building construction is a rather complex task. The high investments normally required for construction facilities call for the optimization of workflow efficiency and of the related financial effort. On the Health & Safety side, workers' safeguard in construction sites is always among project managers' major concerns. The complexity of on-site working conditions requires the careful planning and coordination of several crews and equipment in order to ensure safety. The optimization of safe working conditions places many constraints on: mobilization, transportation, collaboration of equipment, work interference, tight schedules and spatial constraints. The traditional programmatic approach, based on a careful but a-priori planning of the working conditions and on the preliminary identification of the critical conditions, is not capable of coping effectively with the high dynamic nature of the

INTECH
open science | open minds

© 2012 The Author(s). Licensee InTech. This chapter is distributed under the terms of the Creative Commons Attribution License http://creativecommons.org/licenses/by/3.0), which permits unrestricted use, distribution, and reproduction in any medium, provided the original work is properly cited.

construction site, affected by frequent rescheduling and repositioning of the working activities. In this operational context, support to the management of critical performance factors such as production progress, production quality and workers' safety requires the implementation of very flexible and effective real-time monitoring systems: systems capable of accurately tracking workers, goods and the position of equipment so that workflow information can be provided to the programming/control department in real-time and critical, hazardous situations can be predicted. The harsh, highly dynamic nature of the construction environment calls for position tracking systems capable of working without line of sight, systems accurate enough to track assets in real-time with an error comparable with a worker's footprint (about 0.60 m) with a signal ranges of hundreds meters, and whose deployment does not cause high impact and costs to the layout management of construction sites, while requiring low maintenance.

Recently, a set of new low cost and highly standardized technologies, like Global Positioning Systems (GPS), Wireless Networks, Radio Frequency Identification (RFID), the IEEE 802.15.4 (ZigbBee) communication protocol, etc. provided the technological basis for the dynamic position tracking of workers and materials, allowing the development of innovative real-time operational and management services for highly automated nomadic construction facilities (Fontana & Gunderson 2002), (Khoury & Kamat 2009). Position tracking is achieved by establishing a referenced radio frequency environment, made up of a series of transmitters/receivers nodes, and by assigning a sensor to each tracked entity. The sensor is capable of receiving/transmitting signals from/to the reference environment. Locating a node in these wireless environments therefore involves the collection of location information from radio signals traveling between the target and the reference nodes. Radio signal's information regarding the angle of arrival (AOA), the signal strength (SS), or time of arrival (TOA) have been used to determine the location of the target node (Caffery 2000). The AOA-based positioning technique involves measuring angles of the target node seen by the reference nodes, which is done by means of antenna arrays. To determine the location of a node in a two-dimensional (2-D) space, it is sufficient to measure the angles of the straight lines connecting the node and two reference nodes. The AOA approach is not well suited for position tracking in construction sites because AOA information in the non-line-of-sight situation rapidly becomes a major error source in positioning and tracking. SS relies on a path-loss model. The distance between two nodes can be calculated by measuring the energy of the received signal at one node. This distance-based technique requires at least three reference nodes to determine the 2-D location of a given node, using triangulation (Caffery 2000). Unfortunately, SS-based positioning algorithms are very sensitive to the characteristics of the channel. The variance of a distance estimation obtained from SS measurements is lower bounded by the Cramér-Rao lower bound (CRLB) (Qi & Kobayashi 2003):

$$\sqrt{Var(\hat{d})} \geq \frac{\ln 10}{10} \cdot \frac{\sigma_{sh}}{n_p} \cdot \hat{d} \tag{1}$$

where $\hat{d}$ is the distance between the two nodes, n_p is the path loss factor, and σ_{sh} is the standard deviation of the zero mean Gaussian random variable representing the log-normal channel shadowing effect (where a large obstruction obscures the main signal path between transmitter and receiver). In general, since the best achievable limit depends on the channel parameters and the distance between the two nodes, the highly dynamic nature of constructions sites hinders precise control of the positioning accuracy. Nevertheless, in cases where the target node can be kept close to one or more reference nodes, the SS measurements may provide enough precision for practical purposes (Sahinoglu & Catovic 2004). In fact some applications of the SS method based on ZigBee have been developed for real-time construction management (Carbonari et al. 2010). In these cases the system is designed in such a way that the reference radio frequency frame overlays a rather tight zoning over the working environment, so that, wherever the target node is positioned, it is in proximity of one or more reference nodes. SS localization is performed through the application of the "Weighted Centroid Localization" (Blumenthal et al. 2007), according to which the position of the target node is computed as a weighted average value of the positions of the routers closest to the tag. Technically, a mobile node who wants to evaluate its position sends a broadcast message to all the reference nodes in the network in its radio range. Each router sends its PAN (Personal Area Network) coordinator a short message whose payload contains its fixed coordinates and the message's RSSI (Received Signal Strength Indication), measured in dBm. The RSSI is then used by a central application server, directly connected to the PAN coordinator via TCP/IP, to estimate the weighting coefficient associated to each router's coordinates for implementing the Weighted Centroid Localization algorithm. TOA positioning techniques rely on measurements of the travel times of signals between nodes. If two nodes have a common clock, the node receiving the signal can determine the time of arrival (TOA) of the incoming signal that is time-stamped by the reference node. The Global Positioning System (GPS) (Ergen et al. 2007) was one of the early systems providing TOA based position tracking support to construction management applications. GPS is a space-based satellite navigation system that provides location and time information in all weather, anywhere on or near the Earth, where there is an unobstructed line of sight to four or more GPS satellites. A GPS receiver calculates its position by precisely timing the signals sent by GPS satellites high above the Earth. Each satellite continually transmits messages that include the time the message was transmitted and the satellite position at the time of the message transmission. The receiver uses the messages it receives to determine the transit time of each message and computes the distance to each satellite. The application of the GPS technology for position tracking on construction sites is severely limited by the non-line-of-sight (NLOS) occurring in indoor environments. Nevertheless, within this limit, a number of construction management applications, concerning outdoor activities, built by integrating GPS position data with the information embedded into Radio Frequency Identifiers (RFID) have been proposed in the last decade. Integrated RFID and GPS technology have been developed for the purpose of: tracking highly customized prefabricated components, avoiding delays in construction (Ergen et al. 2007); embedding RFID tags in building components to store design data, which can be passed to the people in charge of maintenance during the operational phase (Cheng et al. 2007) and improving the efficiency of tool tracking and availability by using RFID tags (Gajamani & Varghese 2007). The FutureHome EU funded project

(Abderrahim et al. 2005) has developed systems for product and process analysis suited to manufactured and prefabricated construction solutions. The advantages conveyed by mobile computing in construction management have also been described in (Rebolj et al. 2001), the most relevant ones being: information embedding; supply delivery records and progress updates directly at the jobsite; access to as-built and up-to-date documents; rapid communication and collaboration throughout the entire project life cycle. On the Health & Safety management side, (Abderrahim et al. 2005) propose a mechatronic helmet, using a GPS antenna and a bidirectional communication system, for workers' safety control. The helmet is capable of position tracking and is integrated in a computer based control system that implements a rather articulated safety control policy. (Wang et al. 2004) developed a policy for collision detection among construction equipment. The role of handheld and wearable computing has also been widely investigated (Fuller et al. 2002).

2. Ultra Wide Band geolocation systems

Ultra Wide Band (UWB) is a mature TOA geolocation technology that has undergone considerable development in the last decades.

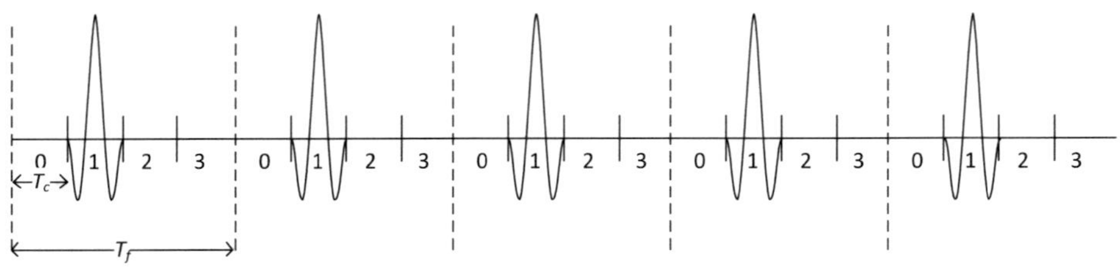

Figure 1. A sample transmitted signal from a time-hopping impulse radio UWB system. T_f is the frame time and T_c is the chip interval.

The majority of the initial concepts and patents for ultra-wideband (UWB) technology originated in the late 1960s at the Sperry Research Center (Sudbury, MA), then part of the Sperry Rand Corporation, under the direction of Dr. Gerald F. Ross. UWB is a technology for transmitting information spread over large bandwidth (>500MHz) while allowing spectrum sharing with other users. It is based on the transmission of a baseband impulse, made of a few cycles of an RF carrier, with low duty cycles resulting in a very low power spectral density, as typically produced by impulse or step-excited antennas (Figure 1). The main advantages in using UWB for geolocation are the absence of steady waves generated by obstacles and the immunity w.r.t frequency selective effects. Denoting with *c* the speed of light, with SNR the signal-to-noise ratio, with *S(f)* the Fourier transform of the transmitted signal and with β the effective (or root mean square) signal bandwidth defined by:

$$\beta \equiv \left[\frac{\int_{-\infty}^{\infty} f^2 \left| S(f) \right|^2 df}{\int_{-\infty}^{\infty} \left| S(f) \right|^2 df} \right]^{\frac{1}{2}} \tag{2}$$

a distance $\hat{d}$ between two nodes obtained from TOA estimation when a single-path additive white Gaussian noise (AWGN) channel is assumed, is subject to the following lower limit for the achievable accuracy (Poor 1994), (Cook & Bernefeld 1970):

$$\sqrt{Var(\hat{d})} \geq \frac{c}{2\sqrt{2}\sqrt{SNR}\beta'} \tag{3}$$

Equation (3) suggests to increase SNR or the effective signal bandwidth for improving accuracy of a TOA based approach: this implies that the use of very large bandwidths (such as in UWB radios) allows, at least in principle, extremely accurate location estimates. For example, for a received UWB pulse of 1.5 GHz bandwidth with SNR=0dB the lower limit for accuracy falls below one inch. In order to achieve these theoretical estimation accuracies, clock synchronization and clock jitter between the nodes become key factors (Shimizu & Sanada 2003). Time-difference-of-arrival (TDOA) technique (Caffery 2000) can be employed in order to relax the need for synchronization between a given node and the reference nodes, but synchronization among the reference nodes is always required. By estimating the TDOA of two signals traveling between the given node and two reference nodes, the actual location of the node is restricted on a hyperbola, with foci at the two reference nodes. Therefore, a third reference node is needed for localization on a two dimensional space.

A typical UWB geolocation system is made of one or more tags and of multiple UWB daisy-chained (usually with CAT-5 cables) beacons, that relay processed time-of-arrival data. A UWB Tag sends out sequence of packet bursts. The frequency-locked UWB receivers measure times-of-arrival and a central processor accurately estimates tag position from the set of differential times-of-arrival by minimizing an error functional. Given N reference nodes with locations θ_i, i=1,…,N, a node located in unknown location θ and a number of N TOA measurements τ_i, i=1,…,N with reliability described by scalar weighting factors w_i, i=1,…,N and defining the distance $d_i(\theta) = \theta - \theta_i$ between the given node and the ith reference node, the following low-complexity least squares (LS) approach can be used to formulate the error functional whose minima is the node location estimate (Caffery 2000):

$$\hat{\vartheta} = \arg\min_{\vartheta} \sum_{i=1}^{N} \omega_i \left[\tau_i - d_i(\vartheta) / c \right]^2 \tag{4}$$

In practical UWB systems, the TOA is usually detected using a correlation-based estimation algorithm that calculates the time at which a matched filter output peaks. In principle, especially in indoor installations, the detected TOA may not be the true TOA since multiple replicas of the transmitted signal, due to multipath propagation, may partially overlap and shift the position of the correlation peak. Due to the large bandwidth of the UWB signal, multipath components are usually resolvable without the use of complex algorithms, since the detection of the first arriving signal path is usually possible with the necessary accuracy (Gezici et al. 2005).

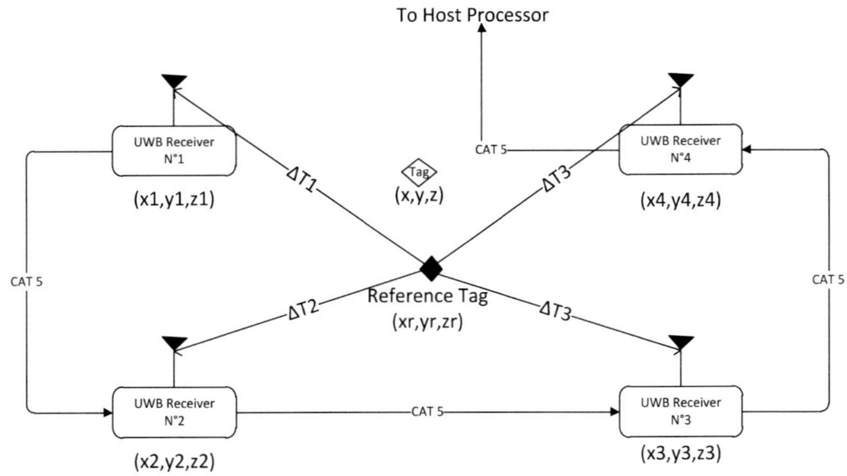

Figure 2. A typical UWB configuration

A second source of position estimation error is non-line-of-sight (NLOS) occurrence between the transmitter and the receiver which occurs when the direct line-of-sight (LOS) between two nodes is completely blocked. In these cases only reflections of the UWB pulse reach the receiving node and the delay of the first arriving pulse is always bigger than the true TOA: the extra distance traveled by the pulse generates a positive bias in the measured time delay called NLOS error. Since, LS technique in (4) gives the optimal results in terms of Maximum Likelihood Estimate (MLE) provided that the measurement error is a zero mean Gaussian random variable with known variance, a positive bias would produce large errors in the location estimation. Additional positive bias is introduced when (especially in NLOS propagation) the first arriving pulse is not the strongest pulse and conventional TOA estimation method select the strongest path for estimating TOA. A unified analysis of the NLOS location estimation problem is introduced in (Gezici et al. 2005) where asymptotically optimal estimators (even in presence of statistical NLOS information) are presented. For UWB systems it is shown that the first arriving signals from the LOS nodes can be used to get an asymptotically optimal receiver performance. Figure 3a shows a simple location estimation scenario, where six reference nodes are trying to locate the target node in the middle, and figure 3b shows the minimum positioning error versus bandwidth for different number of NLOS nodes (Gezici et al. 2005).

In the optimal case, the high bandwidth allows UWB to theoretically provide position accuracy below 30cm. Nevertheless, the required presence of at least one LOS receiver and the fact that, for a given installation, the number of NLOS receiver significantly affects the system position estimation accuracy, opens the floor to further investigations concerning the performance of UWB position tracking systems in real working environments. At present only few analyses have been carried out concerning the application of UWB to the construction sector. The following sections report on practical applications of UWB in real world construction environments and further discusses their implications for the development of work progress and health & safety management.

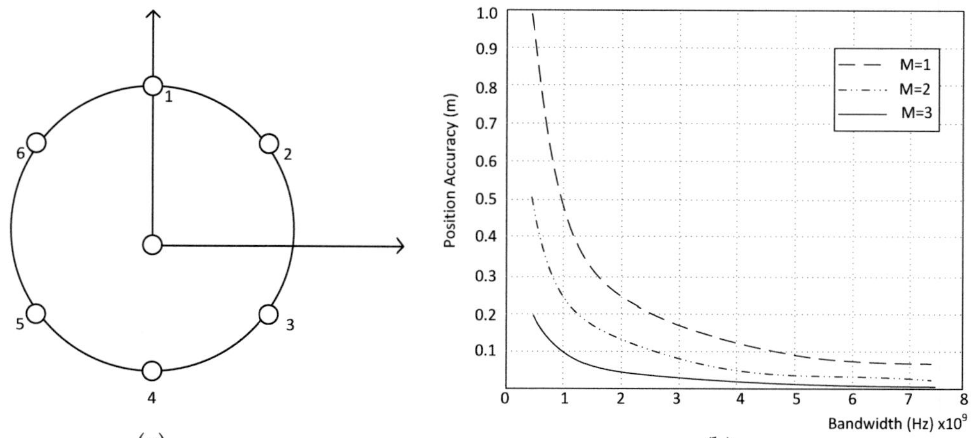

(a) (b)

Figure 3. (a) A simple location estimation scenario where the target node is in the middle of six reference nodes located uniformly around a circle. (b) The minimum positioning error plotted against the effective bandwidth for different numbers of NLOS nodes, M, at SNR = 0 dB. For M = 1, node 1; for M = 2, node 1 and 2; and for M = 3, node 1, 2, and 3 are the NLOS nodes.

3. UWB position tracking in building construction sites

The application of UWB technology to construction sites has both practical and technological implications. From the practical standpoint the main issue is the operational impact on site layout. UWB needs CAT5 cables to link receivers to central units in order to transmit data and to keep them in sync. Laying CAT5 cables on a construction site is usually a tricky task because the harsh and continuously varying environment imposes having to place receivers along boundaries, minimizing crossing paths as much as possible. This can be a severe constraint on the design of UWB tracking systems concerning the minimization of NLOS paths between receivers and tags. Furthermore, TDOA systems require one or more reference tags, used only for sync purposes, to be placed in locations with at least 3 LOS paths to as many receivers. The more the construction progresses, the more likely it is that this condition will be violated, requiring a reconfiguration of the UWB system. On the technological side, the main issues concerning the application of UWB position tracking to construction sites are the operational range, the accuracy and the reliability of the position tracking of material/workers proceeding at usual transport/walking speed through eventually obstructed NLOS conditions. An operational range between 50-70m is usually required to cover a typical medium dimensioned construction site with a reasonably sized receiver set-up where, for example, typical medium sized site corresponds to blocks of flats or medium public building. The operational range of an UWB system is related to signal attenuation depending on the distance from the receivers, on the number of interfering obstacles, and on the power of the tags. Commercial systems offer tags with power ranging from 300mW to 1W. On-field experiments have demonstrated that high power tags are required when signal attenuation occurs; like those caused by building envelopes and internal wall partitions. On the other hand, low power tags can be used in the early phases of building construction, when work is typically carried out in large open areas. Since these two distinct operating conditions substantially affect, as we will later demonstrate, both operational range and tracking accuracy, a fundamental guideline for UWB system design in construction site emerges. The UWB set-up should be arranged in two different installation phases. In an early construction phase (which typically encompasses the site set-up, the excavation, the foundation and the elevation of the frame structure) a boundary receiver lay-out is sufficient to accurately trace workers, equipment and material position. A second phase, starting from the construction of the envelope walls, requires one or more complementary UWB systems to be laid down in the scaffolding boundary, i.e. possibly using the scaffold structure to hang receivers. The scaffold structure can be used to trace workers on different floors as well, implementing a 3D receiver configuration. Multi-floor slab paths should be minimized in this case. Figure 4 illustrates an example of the two UWB set-up phases proposed for a typical construction made up of two blocks of flats. Concerning the early construction phases, (Cheng et al. 2011) reports about an accurate assessment of UWB error ranges tracking worker positions in a 2400m² construction pit scenario, and in a large 64000m² lay down yard. In the construction pit case, the receiver configuration diameter was 65m and in the lay down yard, the installation diameter was 270m. The registered errors are approximately 0.4m and approximately 1.5m for the construction pit and the lay down yard set-ups, respectively. These results show quite clearly that in the initial excavation, foundation and frame elevation phases UWB technology can be successfully applied for real-time tracking of workers and material. To extend these results to the following phases of the building construction, when envelope and partition walls are elevated, the degradation of the tracking accuracy must be assessed, since in these cases the minimum number of receivers (i.e. three for 2D tracking, four for 3D tracking) with LOS transmission path may fail to occur.

(a) (b)

Figure 4. Typical UWB plant layouts in different construction phases. (a) Boundary layout is sufficient to trace workers and materials during the early phases. (b) Complementary UWB systems should be laid down to track works in advanced construction phases. Black triangles represent receivers.

We will further detail this noteworthy operational condition later in this section. Figure 5 shows a typical construction site in an advanced construction phase, immediately following the completion of the reinforced concrete frame and floor slabs (a), and after the completion of the envelope walls and internal partition walls (b). During these phases the work frequently moves from buildings' exterior to their interior, with a great deal of activities performed on scaffolding. Hence, the signal quality provided from inside a framed building and from workers moving on scaffolding to a boundary positioned receiver must be assessed. The authors carried out a number of experiments to test these specific operational conditions. The experiments were conducted on a block of flats (Figure 5) built with a reinforced concrete frame structure and light masonry walls. External hollow walls including 0.05 m polystyrene insulation, as in Figure 6a, had the external wall layer of solid bricks and the internal one of cellular 0.08 m blocks. Partitioning walls between apartments were made of 0.12 m thick concrete cellular blocks (Figure 6b) while the walls between rooms of the same apartment were made of 0.08 m thick cellular blocks (Figure 6c). In order to capture the effects of the envelopes and those of the partition walls, tests were performed just after the completion of the concrete frame structure (Figure 5a) and after the completion of the walls (Figure 5b). The first set of measurements was conducted just after the completion of the reinforced concrete frame and floor slabs. The 0.3W tag was subject to blinking in this configuration and therefore considered quite unreliable while the 1W power tag behaved consistently.

(a) (b)

Figure 5. Typical conditions of a construction site in an advanced construction phase, (a) just after the structural frame completion and (b) after the construction of the envelope and of the internal partitions.

Figure 6. Building technologies used for the envelope and the external partitions in the test building.

Figure 7 shows the scheme of the UWB setup used for these tests: four receivers were placed at the corners of the site (4 m high) and one reference tag in the center of the building's floor at a height of 1.20m. It is worth noticing that a metallic scaffold was installed along the building's perimeter. The entire measurement area was approximately rectangular and measured 38x35 m.

A worker moving about on the building ground slab and a worker moving on the scaffold were tracked on predetermined paths, and the error was calculated by comparing the workers' actual routes with the ones detected by the system. In general, the UWB system was able to track the workers' position both on the buildings' ground floor and when moving along the front scaffold, at different heights, with an average position error of about 0.4m, accordingly to the findings, in the early phases, reported by (Cheng et al. 2011). Analogous 2D measures were made on the first and second floor paths resulting that the receivers were able to track the worker up to the moment when the path height did not overcome the receiver height (second floor), being the receivers pointed slightly downward.

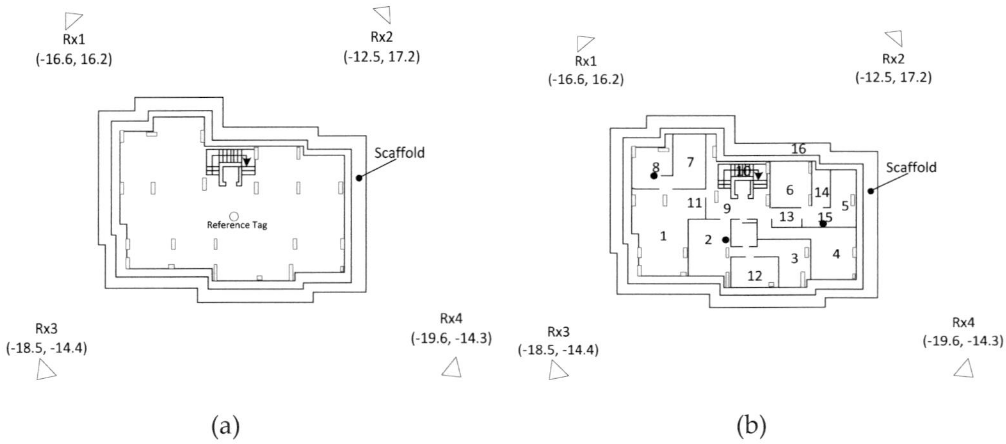

Figure 7. Lay-out of the UWB setup used for testing advanced construction phases. In both cases four receivers were placed at the corners of the construction site (4 m high). (a) Configuration after the structural frame completion. One reference tag was placed in the center of the building's floor at a height of 1.20m. (b) Configuration after the completion of the envelope and of the internal partition walls. Two more reference tags were used to enhance tracking reliability.

However, only the 1W tag worked well, the other signal (0.3W) demonstrated that it was not strong enough. Once the envelopes and internal partition walls were completed, a second measurement set was carried out. In this case the UWB system boundary set-up, as shown in Figure 7b, did not allow for real-time tracking of workers' positions. In fact, one or more receivers failed to obtain the signals from some locations inside the building from Tag 1W, hence indicating that walls act as obstacles. A number of systematic field tests were performed, encompassing 16 different positions on the first floor, as shown in Figure 7b. The results are listed in Table 1, indicating the receivers capable of getting the signal and the tracking quality for each position. For the positions identified by the "M0" type error, the tag was incapable of pointing 3 receivers at the same time. Given that three receivers are always visible, we can then infer that the frequency of the received signal was too low (low quality), due to obstacles. By comparing Table 1 with Figure 7b, relative to the walls' completion stage at the site, it is possible to infer general rules about the behavior of UWB systems in typical reinforced concrete frame constructions. Single layer walls made up of 0.08 m cellular blocks are quite transparent to UWB. Hollow walls having a double layer of blocks and internal insulation, and concrete cellular blocks both weaken UWB signals. For example, signals travel from position no. 9 to receiver 1, but not from position no. 6 to receiver 1: signal is blocked by the hollow wall between positions no. 9 and 6. A similar statement holds true for the cellular concrete block wall between positions no. 11 and 2. When three receivers are in the line-of-sight, then localization works properly (e.g. in positions no. 4, 10, 11). Instead, it could not work, even if three receivers are read, in the event that the quality of the signal received is low (such as in cases no. 5, 6, 16, 17).

Position	Receivers	Result	Errors
1	1,4	bad	M0
2	1	-	-
3	1,2	-	-
4	1,2,3	good	Ref750A
5	2,3,4	bad	M0
6	2,3,4	bad	M0, R0
7	1,3,4	blinking	M0
8	1,3,4	discrete	M0 (rare)
9	1,3,4	discrete	M0 (rare)
10	1,3,4	good	No
11	1,3,4	good	No
12	1,2	bad	M0
13	2,3	bad	M0
14	3,4	bad	M0
15	2,3	bad	M0
16	1,3,4	bad	M0

Table 1. Tracking quality in different locations.

In this case, the number of receivers used in relation to the complexity of the environment is too low. In general, UWB demonstrated to be able to pass through a maximum of two heavy walls. Therefore, in the later building construction stages, denser and closer receiver configuration, such as those depicted in figure 4b, are required. Further tests were carried out in a smaller area in order to simulate the improved configuration. The tests were done using the system setup of figure 7b but with receivers no. 2 and no. 3 moved to the right, up to positions (-2.1, -14.3) and (-2.1, -17.2) respectively. This configuration significantly reduced the receivers' distance and the tracking area, resembling the scaffold position of the receiver recommended at the beginning of this section. The resulting tracked area was limited to positions 1, 2, 7, 8 and 11. The results reported in Table 2, show that, with the exception of position no.1 (that is in contrast with Table 1, hence considered as an anomaly), the tag's signal was always received by 3 receivers, providing a significant increase in the quality of tracking. Summarizing, different results were obtained for the two construction phases identified. During the site set-up, the excavation, right through to the elevation of the frame, an accuracy of about 0.3-0.4m can easily be obtained. Therefore, up until the reinforced concrete frame structure is constructed, four receivers proved to be sufficient for monitoring the workers' movement over the entire site on the ground floor and on the scaffold. In the later construction progress phase, after the elevation of the walls, the situation changed significantly.

Position	Receivers	Results	Errors
1	1,2	bad	M0
2	1,2,3	Discrete	M0 (rare)
7	1,3,4	Discrete	M0 (frequent)
8	1,3,4	Discrete	M0 (frequent)
11	1,3,4	Discrete	M0 (frequent)

Table 2. Tracking quality after the relocation of the receivers.

In this case, the UWB system should be reconfigured to ensure sufficient accuracy for tracking workers reliably. Preliminary results demonstrated that a complementary scaffolding set-up can be used to enhance position tracking to support practical applications. In the following two sections we will point out how the results discussed in this section may be used to develop supporting systems in two main operational applications of construction management: health & safety and work progress tracking.

4. Proactive hazard detection

Notwithstanding the fact that fatalities in construction sites have recorded a decreasing trend in the European area, the construction industry still holds the poorest Health and Safety (H&S) record of any major industry, where the probability of construction workers being killed is higher than the average for all industries in Europe (European Commission 2008). Such statistics show that the major types of accidents are the result of fatal falls to a lower level, the falling of objects and collisions with means of transport and mobile plants. Traditional practices such as rewarding, training or feedback communication have been shown to be only partially effective on construction sites, as they are unstructured work places. Indeed, they are rendered more complex by the presence of different technologies, requiring the use of several types of equipment and resources at the same time. For this reason, the occurrence of accidents can be only predicted with difficulty and each site has its own peculiarities. As a consequence, much research has been devoted to the development of intelligent control systems which adopt advanced communication technologies as the means for establishing new automated control systems. The approach to intelligent support for health & safety management presently follows two main guidelines. On one hand, automation is applied to support the procedures that current European legislation and many other member countries impose to plan health & safety measures at design time, in order to discern all the procedures that need to be adopted in order to avoid accidents, according to each site's expected work schedule. Since real work schedule in the execution phase is generally different from the planned one, and since plan updates, collected during regular inspections, require a certain time to be implemented, work monitoring facilities capable of collecting sufficient identity and position related information to assess health & safety plans in near real-time have been proposed by many researchers. (Riaz at al. 2006) suggest a combination of GPS technologies and MEMS sensors while the authors in (Carbonari et al. 2010) propose an ultra-low power sensor network connected to a central system. In both cases the positional information used are rather coarse, essentially implementing a zoning framework where the presence of workers, equipment and materials in relatively wide areas, usually ranging tenths of square meters, is monitored. On the other hand, complementary approaches, see for example (Bowden et al. 2006), propose a vision where autonomous intelligent systems are able to perform more accurate real-time support to health and safety tasks and hazard prevention by issuing pre-alerts when fatalities are about to occur, by automating dangerous operations and by decreasing human interventions. In these cases, see for example (Teizer et al. 2008), workers' behavioral models are at the basis of intelligent software systems capable of recognizing dangerous situations in real-time and implementing mitigating actions. Of course, this kinds of application require far more accurate tracking information than the zoning framework required by the plan assessment approach, and UWB technology is a good candidate to provide enough accuracy and reliability to support real-time hazard detection. Since the application of UWB to hazard detection systems is still in its early phases, in the rest of this section we will describe the main issues through the discussion of an example of great generality, concerning the implementation of a hardware/software system implementing a virtual fencing facility in construction sites (Carbonari et al. 2011).

In the previous section we have demonstrated that UWB tracking systems can perform very accurate location tracking, with a 30cm accuracy, in real construction sites up until the erection of reinforced concrete structures. Instead, a different and more intense configuration for UWB receivers will be needed to perform monitoring when masonry buildings are monitored. In any case, it emerged that UWB tracking is very good for tracking from the beginning of the construction progress at least until structure frames are built and, therefore, that it would be able to support several automated hazard detection tasks like collision avoidance and virtual fencing. It is generally stated (Wang et al. 2004) that a predictive approach is required to properly manage risky situations, since hazard detection should anticipate the real occurrence of an eventual threat by a time span long enough to allow the mitigating action to be carried out. In principle, the high polling frequency of the UWB technology (from 1 to 60 Hz) becomes critical should data be provided at a speed rate that allows algorithms to properly manage risky situations by providing alerting signals in time. Proactive risk management systems are usually arranged in two levels. At the higher level, see for example (Howden et al. 2003), behavioural models are used to estimate the route that every monitored worker is going to follow and whether this path is expected to lead to risky situations. In this case, the predicting horizon is of the order of minutes. (Teizer at al. 2008) describes a new framework and its algorithms about using frequent positioning data to elicit rapid learning among construction site managers and workers in relation to safety. If accurately collected and in real-time, the location, speed and trajectory of construction resources (e.g. workers, equipment, materials) can lead to important information regarding travel patterns. This information can then be shared among project stakeholders to improve work practices, e.g. the preconditions for safe construction operations. At the lower level, within a predicting horizon of tenths of seconds, the positions of workers are continuously monitored and last minute warnings are sent if actual behaviour may lead to a possible risky situation. (Cheng et al. 2011) and (Teizer et al. 2008) give initial evidence about the potentials of UWB systems in hazard detection, concerning the movement of material and workers' interference applied to this level. Therefore, a more in depth assessment of the UWB technology in supporting the development of hazard detection algorithms is required. The virtual fencing example we are about to discuss moves in this direction. The implementation of virtual fencing logics (Figure 8) is relatively straightforward (Carbonari et al. 2011).

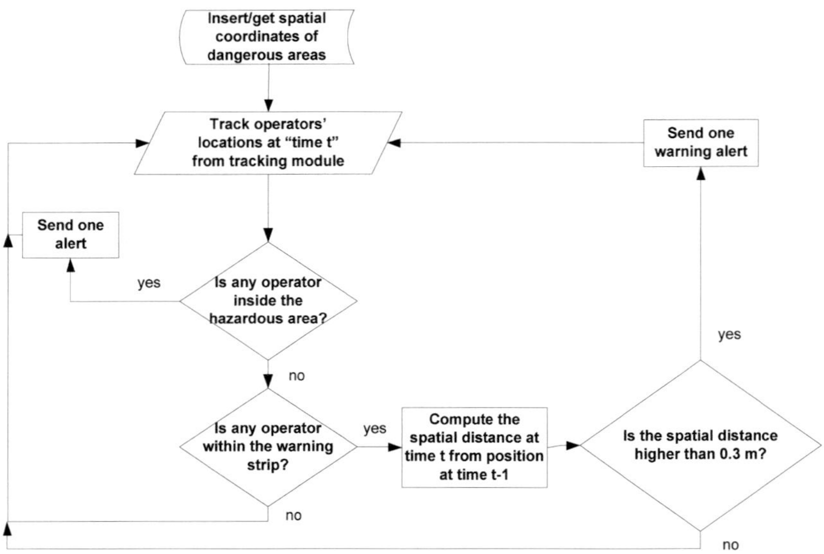

Figure 8. The flow-chart representation of virtual fencing logics.

Each forbidden area is surrounded by a "warning" strip (yellow area) as shown in Figure 9. The width of the warning strips must be determined according to tag frequency and to the speed of average operators. Workers' speed can be assumed to be fixed at 0.5 m/s, while specific testing is required to determine the most appropriate strip width. This will be further discussed later. The algorithm thus avoids the complexity of having to poll every worker on the construction facility by focusing only on the ones that have entered the yellow area. Figure 8 explains a simple virtual fence logic. The algorithm first checks whether an operator has already entered the red dangerous area; if so, a red alarm is sent. If no operator has entered the defined area, no alarms are sent, and the algorithm checks whether any operator has entered the surrounding yellow strip. If so, then the worker's actual position is compared with the previous one (recorded in the previous time step) and a warning alarm is sent only when the distance between the two positions in steps t and t–1 is greater than 0.3 m. Otherwise, the entire procedure is repeated. It is clear that tracking operators' movements within the yellow strip is critical for this application.

Some experimental campaigns, carried out by the authors, aimed at evaluating whether a real UWB position tracking system can suit a simple virtual fencing logic, like the one proposed. A number of laboratory experiments were conducted initially in order to assess the general reliability of the UWB system for this task and the width of the yellow warning strip around the red area.

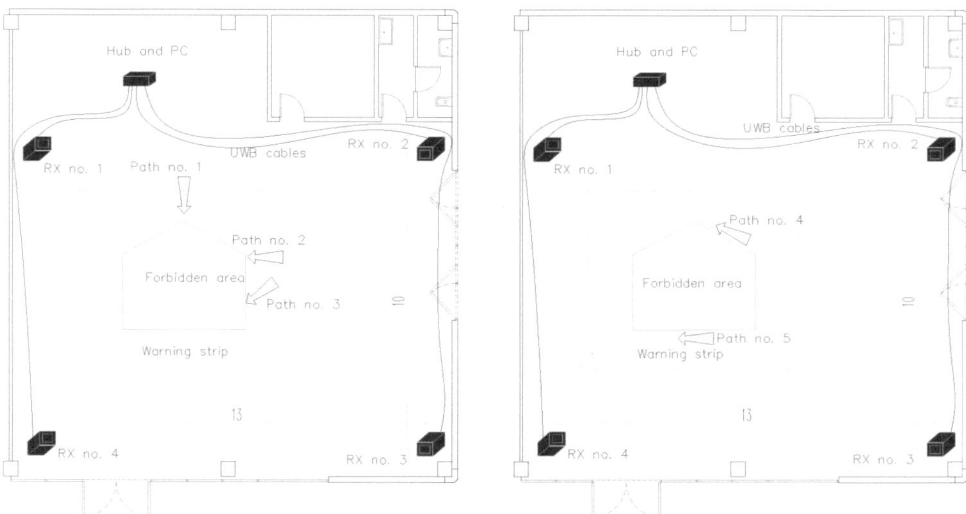

Figure 9. Plant lay-out of the laboratory experiments for the development of the UWB virtual fencing algorithm.

From empirical on-site observations, the workers' average speed was estimated to be around 0.5 m/s. Hence, in an ideal optimal localization accuracy, considering the polling frequency of the UWB system equal to 1 Hz, a yellow strip around the red area slightly wider than 1 m should be enough to obtain two resource position samples during area transit. Considering that the system usually does not localize perfectly in phase with moving workers as they cross the region boundary (i.e. in the limit case, the first localization may take place 1 s after crossing the border line), the width of the yellow strip was increased to 1.5 m. In addition, UWB has a tracking inaccuracy which, even if very low, produces a localization error fluctuating randomly around the average value, even when a worker is in fact still. The estimated position variations of the tracked resource around the mean, generally limited to a few tens of centimeters, in rare cases can reach up to 1 meter. This rare, but rather high localization error for the virtual fencing case, can cause critical inaccuracy and produce false or missed alarms. In order to verify the impact of these position errors, some tests were carried out letting one of the crew members (equipped with the 1 Hz badge tag) repeatedly cross the yellow line drawn 1.5 m from the border line of the red area and following different paths. The system was not capable of signaling the hazardous event in time in a significant number of cases (30% in the worst case). Hence, a statistical autoregressive filtering technique was applied to try to improve localization accuracy, highlighting that a system of order 3, with weighting coefficients (0.10, 0.20, 0.70), showed the best performance, providing a good trade-off between signal stability and trajectory prediction. The worst detecting condition occurs when the workers move along parallel paths (Figure 9b). In this case the number of false alarms increase unless more refined checks concerning the actual direction of the workers are carried out. This is the meaning of the dashed box section of the fencing algorithm. At the end, the UWB proved to be sufficiently reliable for further on-site testing. Table 3 reports the number of false alarms registered by the system in the laboratory experiments.

trial	Path no. 1	Path no. 2	Path no. 3	Path no. 4 (false Alarms)	Path no. 5 (false Alarms)
1	0.0	0.1	0.6	1	1
2	0.0	0.2	0.4	1	0
3	0.7	0.6	0.4	0	1
4	0.1	0.1	0.7	0	0
5	0.3	0.0	0.3	0	0
6	0.4	0.6	0.4	0	0
7	0.2	0.5	0.6	1	1
8	0.4	0.4	0.4	0	0

Table 3. Number of false alarms registered during the laboratory experimental campaign.

The field testing validation was performed on a real construction site. The system worked fairly well, reliably providing smooth paths, with a delay determined only by the time required by the UWB system to update the tag's location, and demonstrating the same precision as that shown in the laboratory tests. The assessment of overall system performance was accomplished through three trials which consisted in letting the crew member approach the predefined hazardous area. A red tape was laid out on the ground, along the border line drawn virtually in the GUI windows (Figure 10), around the hazardous zone, where warning signals were expected to be set off. The time when the signal was sent, with respect to the worker's position, was recorded. In conclusion, both the laboratory results and field testing showed that UWB position tracking technology provides the performance that may suit the requirements of real-time hazard detection algorithms. The detailed technical discussion was limited to outdoor construction sites, where UWB localization has been proven accurate enough to provide reliable data for virtual fencing.

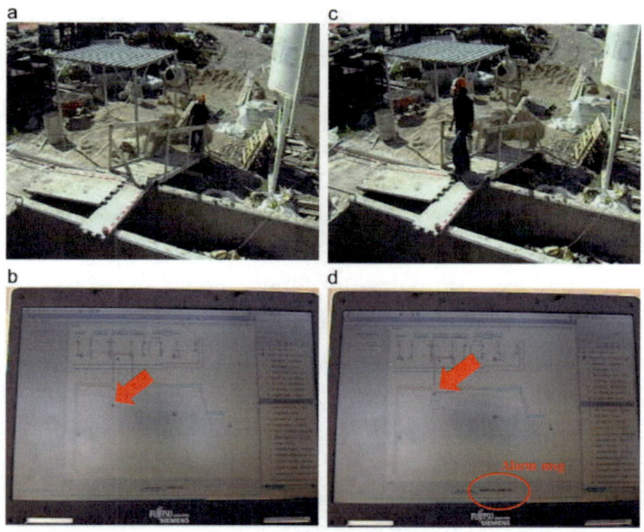

Figure 10. A snapshot of the virtual fencing experiment, conducted in a real construction site. A worker is approaching a critical area (a,b). Signaling occurs before the worker enters the area (c,d).

Despite this almost optimal operational condition, UWB technology demonstrated to be at the lowermost requirement limit (e.g. the 30cm accuracy was comparable to the 1.5m width of the yellow zone), its extension to indoor environments, which will decrease overall performance, will necessarily require a specific careful UWB lay-out design. This aspect is currently an open research point and requires further investigation.

5. Real-time work progress tracking

Building construction is an information intensive process which can easily incur in information overload of the management organization. Efficient management may be hindered by time wasted in information retrieval, by poor and complex information structuring and by delayed communications. Embedded ICT systems for improved information management and the automated control of project performances are currently the foremost frontier of construction project management (Abdelsayed & Navon 1999), (Navon & Goldschmidt 2003). By collecting low level field data (e.g. location of workers, materials and facilities) in real-time and inputting them into pattern and process recognition algorithms, these systems promise to support progress control and deviation analysis, improving the human capability of managing large and complex sets of workflow data. Past experiments demonstrated that the tracking of workers and comparisons with projects' baseline can be used to assess activities in progress and related preliminary prototypes, provided an accuracy error lower than 20% (Navon & Goldschmidt 2003). Extending this concept to automated activity progress monitoring, successful findings have been reported in the field of earthmoving control (Navon et al. 2004), supply management (Navon & Berkovich 2006), road construction (Navon & Shpatnisky 2005). The authors (Carbonari et al. 2011) argue that UWB location tracking data can be effectively applied to automated work sampling. In fact, location tracking data can be used to decompose the presence of workers and equipment in the different working area zones versus time and in relation to the trajectories of workers and material and can, in principle, be used to argue the work progress.

Robust custom designed pattern recognition algorithms are then necessary to abstract the activity type and intensity from the raw tracked data and, again, this is still an open research field. To demonstrate how UWB can be applied to work progress estimation consider the following example, excerpted by a construction work tracking campaign led by the authors during the construction of a reinforced concrete frame block of flats. Figure 11 shows some photos of the construction site and Figure 12 details the layout of the building's first floor. Site cast concrete slab erection is the monitored activity illustrated.

(a) (b) (c)

Figure 11. Pictures of the construction site set-up (a), of the receivers installation in the scaffolding (b) and of the casting phase (c) for the UWB workflow tracking experiment.

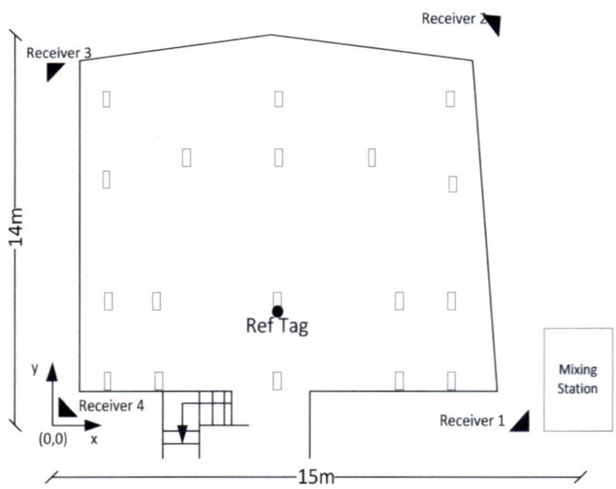

Figure 12. A photo and the plant lay-out of the on-site experiments for the UWB workflow tracking experiment.

One crew out of three was devoted to this task; two working on the site cast concrete slab and one controlling the crane trajectories. The crane moves the crane bucket from the concrete mixing station, placed approximately at referenced position (15,0) to the current concrete pouring place, where it is tended to by one of the two workers, while the other levels and vibrates the liquid concrete poured. Figures 13 shows the crane bucket X and Y coordinates time progress. It can be clearly seen that the x=15 peaks correspond to the y=0 valleys and that the pattern occurs initially about seven times close to the x=0 coordinates, with an increasing trend of the y coordinates. This is easily conceivable as representing the first of concrete casting strip, corresponding to the left most side of the floor, as actually occurred. The work then progresses with analogous trends until the crane bucket is stopped in the mixing station area of the construction site. Figure 14 shows the position of the worker on the slab. Even in this case, it can be clearly seen how the worker's position follows the crane bucket's position exactly when the latter is on the casting floor. In fact, the worker is deputed to directing the bucket locally so that the concrete is spread as uniformly as possible during the pouring phase. This is an activity constraint that is correctly and accurately represented by the UWB tracking system. Finally, the number of buckets poured can be easily counted by identifying the crane position peaks at x=15m, allowing for a very accurate work progress record. Summarizing, the proposed snapshot of a UWB tracing of a simple construction task shows that activity patterns can be clearly identified through the tracks of workers, materials and equipment. The identity information available from the unique tag identifier, and the relationships that can be drawn straightforwardly with the workflow baselines, represent a very good and reliable basis for implementing advanced work progress tracking systems in construction facilities.

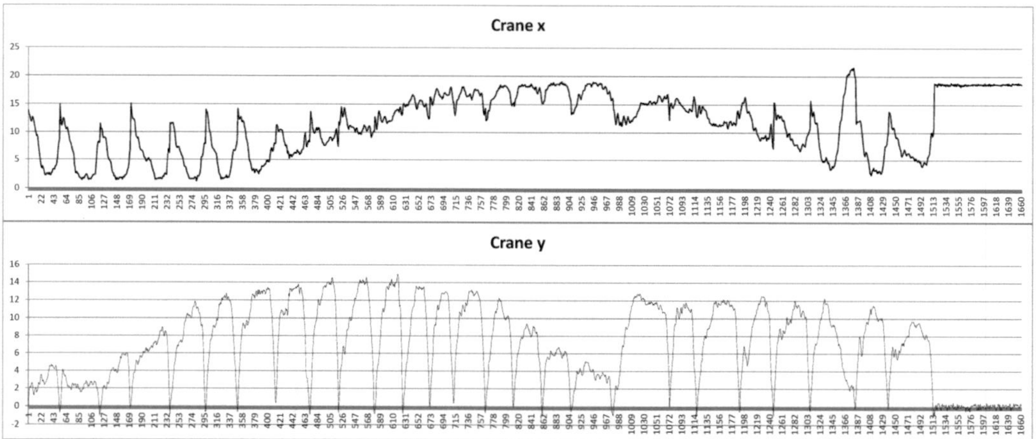

Figure 13. The time progress of the crane bucket X and Y coordinates.

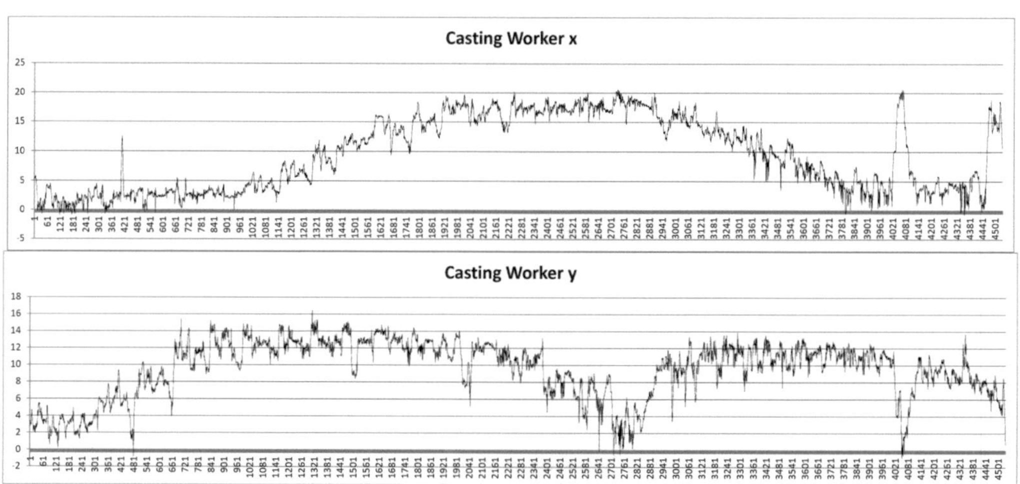

Figure 14. The time progress of the casting worker's X and Y coordinates.

6. General design guidelines

UWB position tracking has demonstrated to be completely reliable in real world construction sites up until the completion of structural frames, providing 30cm real-time accuracy. Sections four and five demonstrated that, within this operational boundary, the accuracy provided can suit both the development of advanced systems for work progress estimation and automated systems for health & safety management as well as real-time hazard risk detection. The UWB indoor tracking capabilities in a construction site boundary installation fade considerably when envelopes and partition walls are raised. In this case, two main design options are available:

- either to deploy a more fine grained UWB arrangement, with receivers installed on the scaffolds of each construction block;
- or to complement the UWB boundary installation with a flexible ultra-low power ZigBee based zoning systems (see introduction).

In the first case, the analysis carried out in section 3 pointed out that in general, UWB is able to pass through a maximum of two heavy walls. Therefore, assuming this as a general design rule for pure UWB implementation, a design procedure can be stated according to the following steps (see Fig. 15a):

1. for each indoor space draw a line of sight form its center to the outside of the building;
2. rotate the line by 360° and mark the sectors where the number of intersected heavy walls is less than two;
3. the intersections of the sectors drawn for the different indoor spaces define the candidate areas for placing UWB antennas;
4. place receivers so that each indoor space can see at least three receivers through its marked sectors.

Fig. 15a shows the optimized layout obtained with this procedure leading to the increased indoor tracking quality reported in Table 2 of section 3.

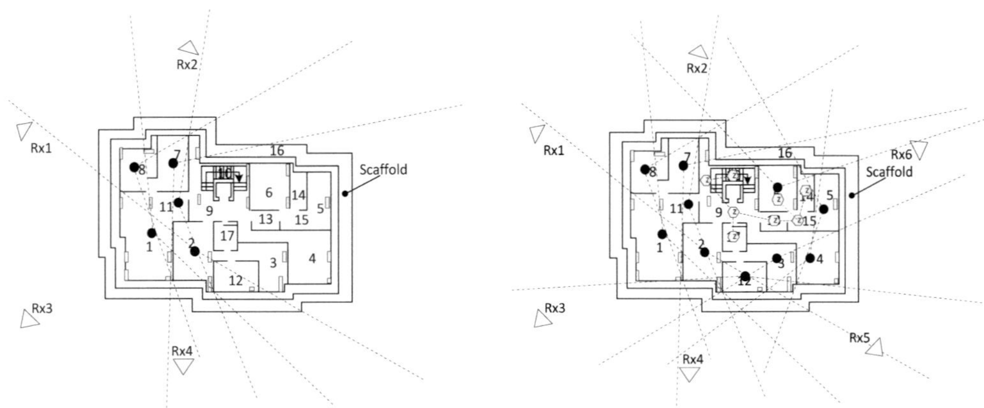

Figure 15. (a) Optimized receiver layout design for indoor spaces 1,2,3,7,8,11; (b) mixed mode UWB - ZigBee (hexagons) zoning design for mitigating loss of tracking in indoor space 9,10,13,14,15 and 17.

It may be the case that it's not possible to find a receiver layout providing three suitable receivers for all the indoor spaces of the building under construction. Fig. 15b shows that indoor spaces 9, 10, 13, 14, 15 and 17 cannot be served by any receiver boundary configuration. In this situation a hybrid mode design can be used to mitigate the localized UWB indoor loss of tracking.

The ZigBee zoning systems can be used to assess the presence of the worker in the UWB shadowed zone. Current ultra-low power ZigBee based zoning systems provide enough flexibility to be deployed without any significant impact on the construction site. Despite ZigBee zoning systems cannot provide accurate position tracking, they can still be used to support a number of high level management tasks like work progress tracking or interference monitoring among different construction tasks. Figure 15b shows the hybrid solution schema. The ZigBee network (hexagons) has been deployed to cover the most critical indoor spaces, like number 13, that cannot be easily served by boundary UWB. Some spaces like space 6 in this example can be served either by increasing the number of UWB receivers or by ZigBee zoning. It's up to the designer defining the most suitable solution. The integration of the UWB and the ZigBee systems can be either basically accomplished by switching to the second system when the first fails providing reliable data or by more complex sensor fusion approaches. However, the hybrid mode tracking position system design and its relationship with high level management tasks is still, at present, an open research issue.

7. Conclusions

In this chapter we have discussed the strength and weakness of UWB position tracking technology applied to building construction environments. UWB position tracking has demonstrated to be completely reliable in real world construction sites up until the completion of structural frames, providing support for real time management and health & safety tasks. The UWB indoor tracking capabilities in boundary installations fade considerably when envelopes and partition walls are raised. In these cases, optimized UWB design with increased number of receiver or hybrid design merging UWB with ZigBee based zoning systems are both possible. In the first case the tracking resolution of the UWB system is maintained at the expense of an increased number of receivers, rising significantly the equipment and installation costs. Anyway it is often the case that UWB boundary installation cannot suit indoor tracking in some spaces whatever number of receivers is used. In that frequent case hybrid mode position tracking design can provide support for high level management tasks. In general, hybrid mode position tracking in construction sites is still an open research issue. More specifically the open issues concern the sensor fusion approach and the relation between the tracking accuracy and the level of support of high level management tasks. Sensor fusion can be as simple as switching between systems in case of tracking failure, or much more complex as merging information at data processing level, developing hybrid TOA - SS algorithms. How optimized performance can be achieved in hybrid frameworks through careful layout design of hybrid sensor systems is a second point that requires investigation and on-site testing. In this chapter we have shown how boundary UWB set-ups can be complemented by ZigBee zoning systems deployed in the inner indoor spaces. Nevertheless in these cases some redundancy can be used to increment reliability of the UWB tracking and the robustness of the overall solution. To what extent this can be done is at present not known. Finally the localized lower resolution provided by the ZigBee zoning system may potentially downgrade the support to the high level management tasks. This depends both on the resolution and on the specific lay-out of the tracking solution with respect to the spatial arrangement of the working activities. At present there are not any insights on these advanced design features.

Author details

Alberto Giretti, Alessandro Carbonari, Massimo Vaccarini
Department of Civil and Building Engineering and Architecture,
Research Team: Building Construction and Automation, Università Politecnica delle Marche,
Ancona, Italy

8. References

Abderrahim, M.; Garcia, E.; Diez, R. & Balaguer, C. (2005). A mechatronics security system for the construction site, *Automation in Construction*, Vol. 14, No. 4, pp.460-466.

Abdelsayed, M. & Navon, R. (1999). *An information sharing, internet based, system for project control*, Civil Engineering and Environmental Systems, Vol. 16(3), pp. 211–233.

Behzadan, A., H.; Aziz, Z.; Anumba, C., J.; & Kamat, V. R. (2008). Ubiquitous location tracking for context-specific information delivery on construction sites, *Automation in Construction*, 17(6), pp. 737-748.

Blumenthal, J.; Grossmann, R.; Golatowski, F. & Timmermann, D. (2007). Weighted Centroid Localization in Zigbee-based Sensor Networks, *Proceedings of the IEEE International Symposium on Intelligent Signal Processing (WISP 2007)*, Madrid, Spain.

Bowden, S.; Dorr, A.; Thorpe, T. & Anumba, C. (2006). Mobile ICT support for construction process improvement, *Automation in Construction*, 15, pp. 664 – 676.

Caffery, J., Jr. (2000). *Wireless Location in CDMA Cellular Radio Systems*. Boston, MA: Kluwer.

Carbonari, A.; Biscotti, A.; Naticchia, B.; Robuffo, F. & De Grassi, M. (2010). A management system against major risk accidents in large construction sites, *Proceedings of the 27th ISARC*, Bratislava, Slovakia.

Carbonari, A.; Giretti, A. & Naticchia, B. (2011). A proactive system for real-time safety management in construction sites, *Automation in Construction*, Volume 20, Issue 6, pp. 686-698.

Cheng, M. Y.; Lien, L. C.; Tsai, M., H. & Chen, W., N. (2007). Open-building maintenance management using RFID technology, *Proceedings of the 24th International Symposium on Automation and Robotics in Construction – ISARC 2007*, Kochi, India.

Cheng, T.; Venugopal, M.; Teizer, J. & Vela, P., A. (2011). Performance evaluation of ultra wideband technology for construction resource location tracking in harsh environments, *Automation in Construction*, Volume 20, Issue 8, pp. 1173-1184.

Cook, C., E. & Bernfeld, M. (1970). *Radar Signals: An Introduction to Theory and Applications,* Academic Press, New York.

Ergen, E.; Akinci, B.& Sacks, R. (2007). Tracking and locating components in a precast storage yard utilizing radio frequency identification technology and GPS, *Automation in Construction,* Vol. 16, No. 3, pp.354-367.

European Commission - Directorate-General for Employment: Social Affairs and Equal Opportunities(2008). Causes and circumstances of accidents at work in the EU, Office for Official Publications of the European Communities, Luxembourg, ISBN 978-92-79-11806-7.

Fontana, R., J. & Gunderson, S., J. (2002) Ultra-wideband precision asset location system, *Proceedings of 2002 IEEE Conference on Ultra Wideband Systems and Technologies,* Baltimore, MD.

Fuller, S.; Ding, Z. & Sattineni, A. (2002). Case Study: Using the Wearable Computer in the Construction Industry, *Proceedings of the 19th ISARC, National Institute of Standards and Technology*, Gaithersburg, Maryland, pp. 551-556.

Gajamani, G., K. & Varghese, K. (2007). Automated project schedule and inventory monitoring using RFID, *Proceedings of the 24th International Symposium on Automation and Robotics in Construction – ISARC 2007,* Kochi, India.

Gezici, S.; Tian, Z.; Biannakis, G., B.; Kobayashi, H; Molisch, A., F.; Poor, H., V. & Sahinoglu Z. (2005). Localization via Ultra-Wideband Radios, *IEEE Signal Processing Magazine*, 70.

Howden, N.; Curmi, J.; Heinze, C.; Goss, S. & and Murphy, G. (2003). Operational Knowledge Representation: Behaviour Capture, Modelling and Verification, *Proceedings of the Eighth International Conference on Simulation Technology and Training (SimTecT '03),* Adelaide, Australia.

Khoury, H., M. & Kamat, V., R. (2000). Evaluation of position tracking technologies for user localization in indoor construction environments, *Automation in Construction* 18, pp. 444–457.

Navon, R. & Berkovich, O. (2006). *An automated model for materials management and control,* Construction Management and Economics, Vol. 24, pp. 635-646,.

Navon, R. & Goldschmidt, E. (2003). *Can Labor Inputs be Measured and Controlled Automatically?*, Construction Engineering and Management, July/August issue, pp. 437-445.

Navon, R.; Goldschmidt, E. & Shpatnisky, Y. (2004). *A concept proving prototype of automated earthmoving control,* Automation in Construction, Vol. 13, pp. 225-239.

Navon, R. & Shpatnisky, Y. (2005). *Filed experiments in Automated Monitoring of Road Construction,* Construction Engineering and Management, April issue, pp. 487 – 493.

Poor; H.,V. (1994). *An Introduction to Signal Detection and Estimation, 2nd ed.,* Springer-Verlag, New York.

Qi, Y. & Kobayashi, H. (2003). On relation among time delay and signal strength based geolocation methods, in *Proc. IEEE Global Telecommunications Conf. (GLOBECOM'03),* San Francisco, CA, vol. 7, pp. 4079–4083.

Rebolj, D.; Magdič, A. & Čuš Babič, N. (2001). Mobile computing in construction. *Advances in Concurrent Engineering, Proceedings of the 8th ISPE International Conference on Concurrent Engineering - Research and Applications*, California, USA.

Riaz, Z.; Edwards, D., J. & Thorpe, A. (2006). SightSafety: a hybrid information and communication technology system for reducing vehicle/pedestrian collisions, *Automation in Construction*, Vol. 15, No. 6, pp. 719-728.

Sahinoglu, Z. & Catovic, A. (2004). A hybrid location estimation scheme (H-LES) for partially synchronized wireless sensor networks, in *Proc. IEEE Int. Conf. Communications (ICC 2004)*, Paris, France, vol. 7, pp. 3797–3801.

Shimizu, Y. & Sanada, Y. (2003). Accuracy of relative distance measurement with ultra wideband system, *Proc. IEEE Conf. Ultra Wideband Systems and Technologies (UWBST'03)*, Reston, VA, pp. 374–378.

Teizer, J.; Lao, D.; Sofer, M. (2008). Rapid automated monitoring of construction site activities using Ultra-WideBand, *Proceedings of the 24th International Symposium on Automation and Robotics in Construction – ISARC 2007*, Kochi, India.

Wang, A., P.; Chen, J., C. & Hsu, P. L. (2004). Intelligent CAN-based Automotive Collision Avoidance warning system, *Proceedings of the 2004 IEEE International Conference on Networking, Sensing & Control*, Taipei, Taiwan.

Chapter 6

Capacitive User Tracking Methods for Smart Environments

Miika Valtonen and Timo Vuorela

Additional information is available at the end of the chapter

http://dx.doi.org/10.5772/50462

1. Introduction

Passive positioning for smart environments, that could enable the tracking of people unobtrusively without any conscious human interaction and without compromising the user privacy with video based solutions, has remained as a challenging problem for decades [2]. Indeed, many of the current technologies are not welcome in a domestic environment where people should be able to live in comfort, feel safe and relax. For example, people might find the positioning sensors too bulky and non-decorative, fear that their privacy is violated through them, or simply see no benefits of using them. Thus, these views pose serious challenges to the positioning technologies needed to enable the smart environments of the future.

Despite the presented problems, some specific passive user tracking technologies have greatly advanced in the last few years. Among these, capacitive user tracking techniques have established their position and have lately been shown to be usable for tracking people in three dimensions and for recognizing user contact with the objects of the environment. Therefore, this chapter will introduce four different capacitive tracking methods in sections 5–8 that can be used to realize passive and non-intrusive tracking of the human body. The presented methods, developed by Valtonen et al. in publications [47]–[51], enable together the creation of unobtrusive and user-friendly smart spaces, because they can be implemented in a non-distracting or unnoticeable way. Hence, the inhabitants of these environments can be supported in their daily tasks and concentrate on their primary goals rather than be burdened by the sensory actions of the environment.

Specifically, we at Department of Electronics, Tampere University of Technology[1] (TUT) have developed capacitive sensing methods that are able to reason human 3D position, body movement, posture and contact with the objects of the environment. The presented measurement methods and techniques are based on electric fields and on measurements of varying capacitances between the user and a set of sensing electrodes. By hiding the electrodes at least partly in the environment, the sensing can be realized in a unobtrusive way promoting

[1] Department of Electronics web page: http://www.tut.fi/en/units/departments/electronics/

INTECH
open science | open minds

© 2012 The Author(s). Licensee InTech. This chapter is distributed under the terms of the Creative Commons Attribution License http://creativecommons.org/licenses/by/3.0), which permits unrestricted use, distribution, and reproduction in any medium, provided the original work is properly cited.

the Mark Weiser's vision of calm technology [54]. Furthermore, the privacy of the observed persons is not violated by extracting visual information of the subjects. Most of the developed methods have been evaluated in multiple short and long-term tests in the *TUT Smart Home* [8, 10, 51] and have been shown to be a feasible solution for their intended tasks.

To illustrate the importance of passive user tracking techniques for smart environments, we may consider a few examples of different types of applications. First, the developed methods could be used in smart homes to control the environment proactively and autonomously. By turning off unneeded devices or appliances, and by adjusting the heating to an comfortable level based on the user position and sensed context, the energy expenditure of homes can be pushed down [2, 21, 29]. Second, in the health care sector user mental and physical activities could be remotely monitored by observing the user interactions with the home to support either the rehabilitation progress after an injury or the nurturing of the elderly [6, 20, 23, 39]. Third, in social media, user position and activity information could be automatically communicated to friends and relatives to make people more aware how their close-ones are doing [14, 31]. Finally, inhabitants of smart homes could self-evaluate and adjust their lifestyle based on the vast amount of information gathered, if it is processed to an easily digestible form [28].

2. Related work with other technologies

Video cameras have long been used for recording body movements and for positioning purposes [19, 52]. As a result, efficient algorithms for detecting human body movements from a video stream have been available for some decades. Although video-based motion capture is very efficient today, it has a major drawback with regard to user privacy. The simple fact is, many people do not want video cameras to be installed in such a personal space as their own home [46]. Even if the video stream could not be fed outside the monitored room and would only be used, for example, for context analysis, the slightest risk of hacking and intervention by an outsider prevents most people from accepting cameras in such places. That is why there has been so much research into real-time location systems (RTLS) [18] and other alternative, wireless, positioning and activity sensing methods for monitoring people in indoor environments [3, 16]. These include ultrasonic [24], radio frequency [37, 38, 44], optical [9, 22] and inertial-sensor [15, 55] based techniques that require the person to actively carry a device. However, from the occupant's point of view these techniques are neither practical nor user-friendly because the user must always carry a tag or small electronic device on their person. Moreover, visitors cannot be located in the home without first equipping them with the necessary tags.

Therefore, lately there have been several studies of passive or tagless methods for positioning the person indoors without the violation of privacy caused by visually-based solutions. One interesting approach was taken by Shwetak Patel et al. in 2008, when they published their building-ductwork-based sensing method for determining room transitions [27]. The idea in this study was to utilise the existing ductwork infrastructure of the central HVAC systems found in many homes and detect the disruptions in the airflow. Even though the idea is intriguing, the method cannot be used to accurately detect movements within a room, and even room-to-room transitions can only be detected with 75–80% accuracy.

Some years before that, in 2003, Susanna Pirttikangas et al. presented their methods for passive positioning and user footstep identification with a pressure-sensitive floor in [30].

They proposed the use of a novel Electromechanical Film (EMFi) material [26], which would produce voltage signals during pressure changes when stepped on. By placing 30-cm-wide strips of EMFi material orthogonally on the floor, they were able to calculate human X-Y position on the floor and identify human footsteps with a commendable accuracy. Later, in 2008, Jaakko Suutala and Juha Röning refined the original footstep identification methods and demonstrated a 95% classification rate with their floor system [45]. Although, these results are significant and have been proven in practice, their systems cannot recognize stationary people because of the properties of EMFi material. Thus, they cannot be used effectively for detecting, for example, a person standing still on the floor or an elderly person who may have fallen to the floor.

Meanwhile, Yoshifumi Nishida et al. published in [24] a feasible alternative for passively measuring a person's position and posture in indoor environments. Their method is based on an array of ultrasound transmitters and receivers placed in the ceiling at intervals of around 18 cm. The pilot system they developed is able to position the subject three-dimensionally within the vicinity of the sensors and can calculate the head position both horizontally and vertically with about 5 cm accuracy. Although that system gives promising results in terms of accuracy, it requires the installation of hundreds or even thousands of sensors in the ceiling, if whole apartments or houses are to be covered by the system, which is not a practical solution for tracking the users.

In 2010, Daniel Hauschildt et al. published an infrared-based positioning system [4] that operates with passive thermopile arrays placed in the corners of the monitored room. The thermopiles detect the thermal radiation of human beings and enable the tracking of either one or two persons in a 30 m^2 room with a maximum error of 26 or 68 cm, respectively. Although their system is reasonably accurate for general indoor positioning, it is prone to reflection and dynamic background radiation effects that cause the sensors to give out false readings.

3. Capacitive sensing theory

Capacitive sensing is based on measuring the capacitance between two or more electrodes or physical objects. In human tracking applications, these objects have typically been the human body and the grounded environment, but in contrast to this common measurement configuration, this chapter concentrates on methods requiring both a transmitting and receiving electrode. Independent of the used configuration, to measure the capacitance in question, an electric field is artificially created around some sensing electrodes which are then used to conduct a displacement current into or from the measuring circuitry through the electrodes. Usually, the measurement is performed with a low frequency signal (about 30–100 kHz). This ensures pure capacitive coupling between the electrodes and also prevents magnetic fields from forming in the environment [36, 51].

The electronics needed for capacitive measurement can consist of a signal generator, a phase shifter and a simple synchronous detector as shown in Figure 1. Alternatively, commercial capacitance measurement chips can be used as in publications [47]–[51], where commercial *Analog Devices AD7746* [1] sigma-delta-based capacitance-to-digital converters (CDCs) [25] were used. Whichever implementation is used, both provide comparable results (see section 9.1) and the main differences are in the ease with which the measurement circuitry can be modified and in the number of components required.

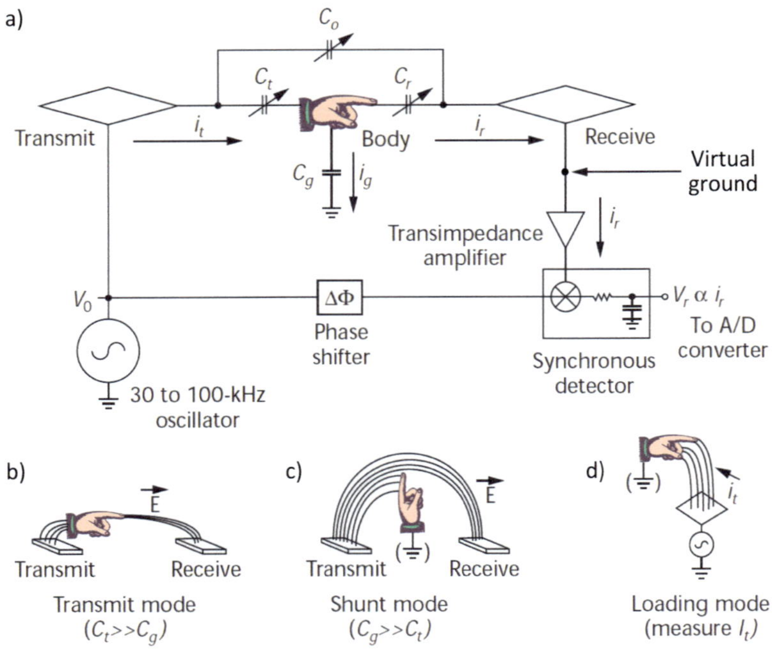

Figure 1. Capacitive sensing is typically implemented in one of its three primary sensing modes. The figure shows a) a basic measurement setup with simple measurement circuitry and the differences in b) transmit, c) shunt, and d) loading mode electrode and capacitance configurations. Although commercial capacitance measurement chips are widely available, simple measurement circuitry can be built consisting of only a signal source, a phase shifter and a synchronous detector. Modified from [42].

3.1 Sensing modes

Three primary sensing modes can be identified and used for measuring body position in relation to the electrodes: transmit mode, shunt mode, and loading mode [42]. These modes are visualized in Figure 1 with an example implementation of the measurement system. The measurement circuitry of Figure 1 always measures the serial capacitance of C_t and C_r in transmit and shunt mode by analyzing the displacement current i_r flowing from the transmitter to the receiver. In loading mode, however, only the displacement current i_t flowing from the transmitter through the body to the ground is measured and the capacitance between the electrode and the body is calculated.

Transmit mode The sensing mode used here is the transmit mode, when either the capacitance between the transmitting electrode and the human body (C_t) or the capacitance between the receiving electrode and the human body (C_r) is much greater than the capacitance between the human body and the ground (C_g). In practice, these situations occur when the measured person touches or is very close to either of the two electrodes. In the transmit mode, the body is capacitively coupled to the nearby electrode and the potential of the body changes in response to the potential of the electrode. In practice, a person usually increases the electrode area by a significant amount and reduces the observed distance between the electrodes. Therefore, the measured serial capacitance of C_t and C_r is greater when a human body is close to one of the electrodes than when there is no body nearby. This sensing mode is

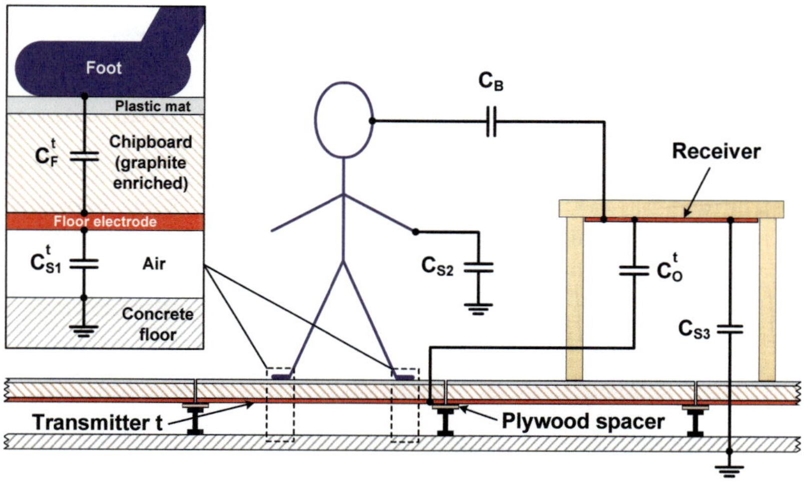

Figure 2. A simplified capacitance model of the environment with a cutaway picture of the *TUT Smart Home*. The floor is constructed from large 60×60-cm raised floor tiles and the transmitting electrodes are attached to the underside of the tiles. The system's receivers are placed in the environment, for example, under the top of the table. [51]

used in all publications [47]–[51], because it allows the electrodes to be located in a practical way that enables the distance between the user and the electrode to be measured.

Shunt mode In shunt mode, the capacitance to the ground through the human body (C_g) is considerably higher than the capacitances C_t or C_r between the measurement electrodes and the human body. Here, the human shunts the electric field to the ground and prevents the emanated displacement current from being detected by the receiver, thus decreasing the measured serial capacitance of C_t and C_r. However, the measured serial capacitance of C_t and C_r depends largely on the capacitive coupling between the measured body and the ground. Indeed, if the grounding of the body is effective and C_g is high, the measured serial capacitance of C_t and C_r is low and vice versa. Nevertheless, by measuring the received displacement current i_r at the receiver, the portion of the body between the two measurement electrodes can be deduced. If either C_t or C_r are equal to C_g, the measured serial capacitance of C_t and C_r in a measurement system is about the same as it would be without the presence of a human body.

Loading mode In loading mode, only the displacement current i_t flowing from the transmitter through the body to the ground is measured and the capacitance between the electrode and the body is calculated. This mode is often used because, apart from the surrounding ground, only a single electrode is required to make the measurement. However, as with the shunt mode, the coupling between the human body and the ground affects the resulting measurement result so it is hard to measure the absolute distance between the body and the electrode.

3.2 Capacitance model

Based on the above description of transmit mode sensing, we can develop a simplified capacitance model that describes the measured electrical properties of the test systems that were constructed in publications [47]–[51]. This simple capacitance model is shown in Figure 2 with a cutaway picture of the *TUT Smart Home* used as the test environment, with the system presented in publication [51]. The following paragraphs explain the terms used in this capacitance model.

Feet capacitance C_F^t The capacitance model incorporates two capacitances that are formed between the electrodes and the human body. First, C_F^t is the capacitance between the human feet and transmitter t. C_F^t always increases when the common area between the person and the transmitter grows, and vice versa. Likewise, C_F^t increases as the distance between the transmitter t and the body decreases. Typically, C_F^t ranges between 20–700 pF with the floor-tile types used in the test systems of publications [47]–[51] when shoes are not worn [47]. However, if shoes are worn they insulate the person from the transmitting electrodes and thus act as an additional insulator. As a result, different types of shoes significantly decrease C_F^t and thus affect the measurements. In fact, C_F^t can be up to 96% lower if shoes with 3-cm-thick soles are worn [47].

Body capacitance C_B The second of the body-related capacitances is C_B. It is formed between the body and the receiver(s) of the system, insulated by air, textiles, or wood. However, with multiple receivers, it is formed mainly between the body and the closest receiver, because the strength of the electric field is strongest near the body from which it emanates. Furthermore, because only a single receiver channel is used in all the implemented systems' hardware, the receiver signals cannot be distinguished from each other. Typically, C_B varies from around 1 to 10 pF but can be significantly higher if the tracked person is in actual contact with the receiving electrodes [47, 51].

User induced capacitance C_U^t Because the electronic circuitry which we used measures the total capacitance between the transmitting and receiving channels, C_F^t and C_B cannot be distinguished from each other in the acquired data, and so they must be combined into a single term. Also, because only a single receiver channel is used in our hardware implementations, these two capacitances are simplified in our model into one single capacitance C_U^t for any given user. Because C_F^t and C_B are connected in series, C_U^t can be defined with the equation

$$C_U^t = \frac{C_F^t C_B}{C_F^t + C_B} \tag{1}$$

assuming that any stray capacitances in the environment are discounted.

Here, C_F^t increases as the area of the feet touching the floor increases and decreases as this area decreases. Likewise, C_B increases when the person's body is brought closer to a receiver and decreases when the person recedes from the receiver.

Offset capacitance C_O^t When nobody is in the apartment, there is an offset capacitance C_O^t between each transmitter t and the receivers. Because C_O^t is in parallel with C_U^t in the capacitance model, it affects the measured total capacitance C_{TOT}^t between a transmitter t and the receivers as in the equation

$$C_{TOT}^t = C_U^t + C_O^t. \tag{2}$$

To get a reference level with which all future measurements can be compared and to obviate the effect of C_O^t, this only needs to be measured once before the person enters the measurement space. Thereafter, this measurement can be subtracted from all subsequent measurement results to yield C_U^t on its own.

Stray capacitance C_{S1}^{t} Capacitance C_{S1}^{t} occurs between each transmitter t and the ground, depending on which parts of the environmental structure are acting as an insulator. With the transmitters shown in Figure 2 in the *TUT Smart Home*, air acts as the insulator between the large bottom surface of the transmitting electrode and the ground. However, there is a significant capacitive coupling between the metal pillars and the ground, so the plywood spacers on top of the pillars are part of the insulating elements. In all of our test systems, because C_{S1}^{t} between transmitter t and a receiver is only measured by gauging the received displacement current at the receiving electrode, the amount of displacement current flowing into the environment from the transmitting electrode does not affect the result as long as the signal source is able to drive the desired waveform into the transmitter [25]. Thus, C_{S1}^{t} does not affect the measured capacitance value with a buffered output present in our test systems' hardware, and so its effect on the measurement can be neglected.

Stray capacitance C_{S2} Capacitance C_{S2} is formed between the human body and grounded objects in the environment. An increased C_{S2} can cause the received displacement current to decrease, because C_{S2} conducts a small part of the displacement current emanated by the person to the ground and thus decreases the capacitance reading C_{U}^{t}. However, C_{S2} remains fairly constant even when large and grounded conductive objects are near the measured person [47]. At most, these types of conductive objects increase C_{S2} only by some tens of percent. Because the absolute value of C_{U}^{t} is not of major interest and the error reflected in it is fairly constant, unless the person is leaning against a wall, C_{S2} can be neglected for position measurements. Likewise, because the appropriate levels for C_{U}^{t} must be calibrated to recognize user contact with certain household objects, C_{S2} does not have a significant effect on the activity measurements since its effects are calibrated at the same time.

Stray capacitance C_{S3} Capacitance C_{S3} occurs between the receivers and the ground. With a static receiver position C_{S3} remains almost constant at all times [47]. Because of the internal implementation of the *AD7746* CDC, it is theoretically insensitive to leakage current errors and parasitic capacitance to ground [25]. However, due to the inevitable imperfections in the manufacture of the CDC, the receiver should always be positioned so that C_{S3} remains below 300 pF (see figures 9–10 in [1]) to prevent large errors in the results with the measurement configurations of publications [47]–[51]. Because, when positioned correctly, C_{S3} has only a slight effect on the measured value of C_{U}^{t}, and also because the constant error caused by it is, after all, calibrated during the system initialization, it can be neglected during the practical measurements.

4. State of the art in capacitive user tracking

It is capacitive methods for human tracking applications that are attracting the most interest because 1) they provide a way of preserving personal privacy, 2) their sensing electrodes need not to be visible to the user, 3) they can be used to recognize both static and moving people, and 4) they can be implemented with large and cheap electrodes that can cover large areas. Indeed, as early as 1993, a simple electrode configuration for detecting the presence or movement of a person close to a robot was developed by Nils Karlsson et al. in [13], and later developed further in [11] and [12]. The purpose of this system was to stop a robot moving to ensure the safety of any person who came too close to it. Two years later after Karlsson's 'person detector', Thomas Zimmerman et al. from MIT Media Laboratory presented various ideas on capacitive applications in [56], including a two-dimensional *Finger-Pointing Mouse* and a *Person-Sensing Room*. The floor of the *Person-Sensing Room* was covered by a transmitting electrode and four receiving electrodes were placed on the walls. Using the floor electrode, a measurement signal was capacitively coupled to the subject and the strength of the emanated signal from the subject was measured by the wall electrodes. The room was able to use this information to locate a person two-dimensionally on the floor plan.

In 1999, Joshua Smith from MIT Media Laboratory published his dissertation thesis [41] covering electric field sensing with different kinds of sensing applications. Later, some of their most advanced applications, three dimensional *Field mice* [40] and *Gesture Wall* [42], were presented in depth. The *Field mice* was able to capture hand position and alignment above the measuring electrodes with shunt-mode measurements using a single transmitting electrode and three receiving electrodes arranged on a plane. The *Gesture Wall*, in contrast, used the transmit mode to transmit the measurement signal through the human body to four receiver electrodes placed at the corners of a screen. The users of the system were able to draw on the interactive screen by making hand gestures in front of the screen.

Recently, Henry Rimminen et al. have studied the use of electric fields for positioning people over a segmented floor electrode [35, 36]. Instead of using the transmit-mode measurement technique, he proceeded with the loading-mode measurement method, which measures the capacitance between a transmitter and the ground. Specifically, his systems scanned the floor area with the floor electrodes and toggled each electrode to transmit one measurement signal at a time [36]. When the electrode below the feet of the person was actuated, a larger than normal displacement current flowed from the body through the air to the surrounding grounded electrodes. The position of the person was determined using the physical locations of the electrodes and the level of current measured from each electrode. Although his systems achieved about the same level of positioning accuracy as our *TileTrack* system [48], the loading-mode measurement method used in his studies has not been demonstrated in user activity recognition applications, nor can it be modified to measure human height and multiple postures, as is possible with the transmit-mode systems presented in this chapter. Nevertheless, in [34] they demonstrated a fall detector that was able to reason when a person was lying on the floor.

5. Tile-based positioning

The test systems of publications [48], [50], and [51] use transmit-mode sensing to locate users at floor level. In contrast to the systems of Karlsson [13] and Zimmermann [56], the electrode arrangement in these studies consists of a segmented floor composed of multiple transmitters and one or more receivers placed in the vicinity of the transmitters. The segmented floor in all of these studies was built from raised floor tiles that had a conductive metal layer on the bottom surface of the tile (see Figure 2). Regardless of which receiver configuration was used, the same tracking algorithm was used in all of these studies.

Transmitter size The test systems of publications [48] and [50] used 60×60-cm commercial raised floor tiles that had steel reinforcement on the underside of the tiles. These tiles, shown on the left-hand side of Figure 3, were also used in [51] with the large test system built into the *TUT Smart Home*. In addition, however, multiple custom-built tiles with roughly 30×30-cm transmitting electrodes, shown on the right-hand side of Figure 3, were used to achieve more accurate positioning results. Indeed, the smaller transmitters can be placed in places where more accurate position information is necessary or preferable. However, dividing the area of

Figure 3. Two different types of tiles with different sizes of electrodes were used as transmitters in the implemented test systems of publications [48], [50], and [51]. The left-hand side of the figure shows the bottom surface of a standard, commercial 60×60-cm raised floor tile from the *TUT Smart Home*. The inset photo shows a snap fastener soldered to the middle of the electrode for cable connection and a screw drilled next to it to prevent the snap fastener from being flattened. The right-hand side of the figure depicts a custom-built floor tile with four roughly 30×30-cm copper foil transmitting electrodes. These electrodes are partly hidden by the two plywood boards on the sides that stiffen the tile and raise it to the same height as the commercial raised floor tiles. [51]

a large transmitter into four smaller segments also makes the measurements from these tiles four times slower with the algorithms we used, and increases the amount of wiring needed to connect the tiles to the measurement circuitry. Therefore, when deciding what size floor electrode should be used, the assumed usage frequency and application possibilities of the installation environment should be taken into consideration. It is also important to note that there is no upper limit to the size of the transmitter, so the compromise between accuracy, speed and the amount of wiring required can be made freely according to which room or area needs to be tracked.

Receiver configuration Both publications [48] and [50] used a vertical wire or a small plate electrode as a receiver but the large-scale test system of publication [51] relied only on receivers R1–R5 installed below the ceiling (R1–R2) or in common household items (R3–R5) as shown in Figure 4. Although the same algorithm was used in all of these studies and the slight variation in the receiver configuration does not affect the tile-based positioning method, it must be noted that all the used receivers must be installed close enough to the transmitters for the system to operate reliably. Based on the results of publication [48], the signal-to-noise ratio (SNR) decreases almost linearly according to the distance between the tracked person and the receiver. Therefore, in practice the receivers must be installed at a maximum of about three meters away from the transmitters to enable a good capacitive coupling between the transmitter and the person. This is also apparent from Figure 4, which shows the received signal strength in the *TUT Smart Home* as a percentage of the maximum signal. However, the measurement range of the receivers can be extended by ensuring that there is either no insulation, or only a thin layer of insulating material between the transmitter and the feet of the person, and by raising the emitted voltage.

Algorithm In tile-based positioning, a measurement signal is driven into the tracked person from the transmitting electrodes in the floor and the received signal strength, i.e. C_U^t, is

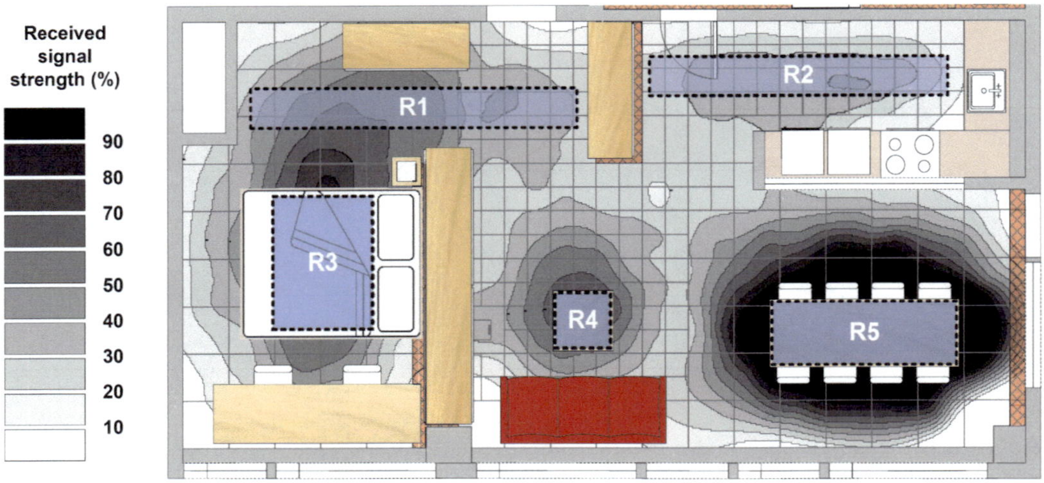

Figure 4. The variation in the the capacitive coupling between the tracked person and the receivers is demonstrated in the above figure with a percentual received-signal-strength map. The map was obtained by walking slowly around the *TUT Smart Home* and by recording the largest measured C_U^t for each position. The highest signals were recorded around the dining table and are represented with the darkest contour lines, but the positions of the other receivers are also easily distinguishable. Receivers R1–R3 are built of conductive textiles and are either suspended from the ceiling (R1 and R2) or placed under the bedlinen (R3). R4 and R5 are constructed of copper foil that has been attached to the undersides of the living and dining room tables. [51]

Posture	Transmitter size (cm)	μ (cm)	σ (cm)	90% (cm)	Max. (cm)
Standing	30×30	4.2	4.5	7.0	18.5
	60×60	5.8	4.2	11.3	14.3
Walking	30×30	9.4	6.5	17.3	51.1
	60×60	17.5	11.3	32.8	40.7

Table 1. Tile-based positioning results: the mean (μ) and standard deviation (σ) of the measured errors with the errors at different confidence levels. The maximum error in the table represents the 100% confidence level. [51]

measured by the receiver electrode. The measurements are taken in a time-multiplexed manner and a single tile measurement takes 11 ms. The 2D user position is calculated using the relative C_U^t values from each transmitter t according to the physical location of the electrodes. C_U^t increases when the feet of the person cover more of the electrode area, and, the greater the C_U^t, the closer to the centroid of the transmitter t the person is assumed to be.

Key results The practical test systems that were implemented demonstrate that this type of tracking system can be installed in the structures of the building and at least partly hidden from the users. As stated above, the tracking accuracy of these types of systems depends on the size of transmitter used. Furthermore, a standing person can be located with greater accuracy than a moving person, because of the fact that there is better electrical coupling between the person and the transmitters when they are in a standing posture with both feet

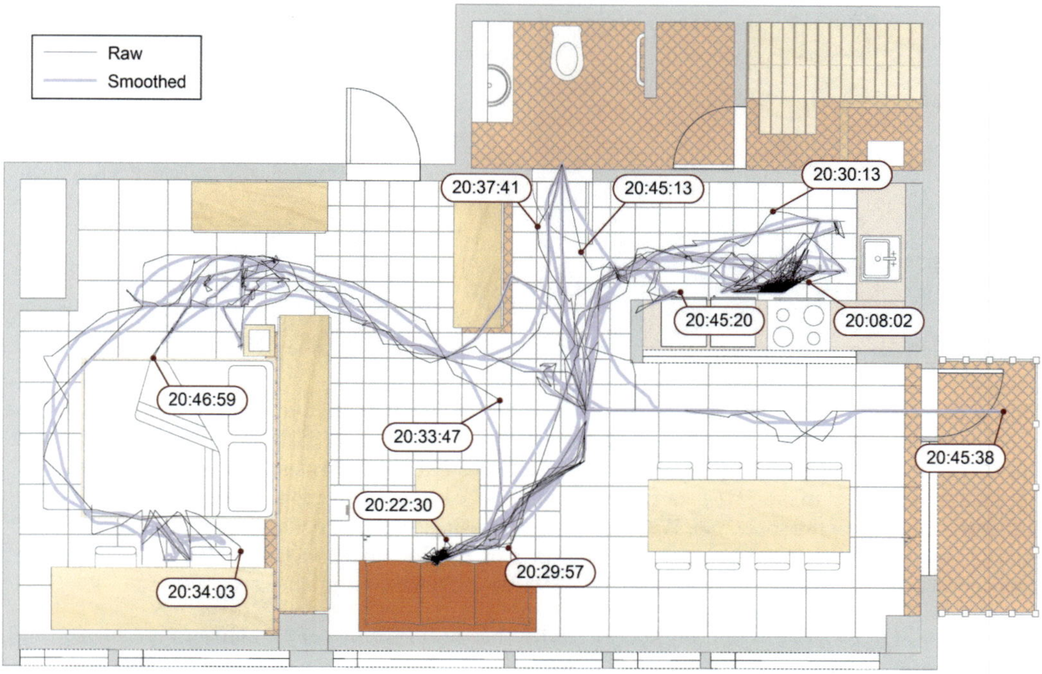

Figure 5. An example of the gathered walking tracks during the long-term living test in the *TUT Smart Home*. The data was gathered between 20:00 and 21:00 hours on the third day of living in the *TUT Smart Home*. The activities during this hour are analyzed in section 6 and shown in Figure 7. [51]

on the ground. Table 1 presents the measurement results with the two different sizes of transmitters measured without shoes. With shoes, the results can be expected to be somewhat poorer, although no exact measurements have been made [47].

When analyzing the results of Table 1, we see that the mean and standard deviation are almost the same for both transmitter sizes. Although the maximum errors differ significantly because of the weaker signal reception with the smaller transmitters, which have a thicker insulation layer between the transmitting electrode and the person, the smaller transmitters actually have about a 38% lower margin of error at the 90% confidence level.

As a part of the study presented in publication [51], a long-term test was performed in the *TUT Smart Home* to test the tile-based positioning system outside a laboratory environment. A test person was asked to live in the apartment for 14 days and his position was recorded while he was inside the home. Figure 5 shows a one-hour example of the data acquired during this test period on the 21st of October, 2009 between 20:00 and 21:00 hours, with the raw positioning data obtained from the hardware with a little filtering and with a smoothed track which depicts more precisely the true path of the body of the person. It is easy to see from this figure that the raw data obtained with the tile-based positioning system is rather erratic because it follows the individual steps of the person. All the gathered positioning data was released for public use and is freely downloadable at [17].

The hit rate of this system, analyzed from this one-hour data sample, was found to be 95.8%. It was calculated by dividing the number of times the system had not lost the track of the

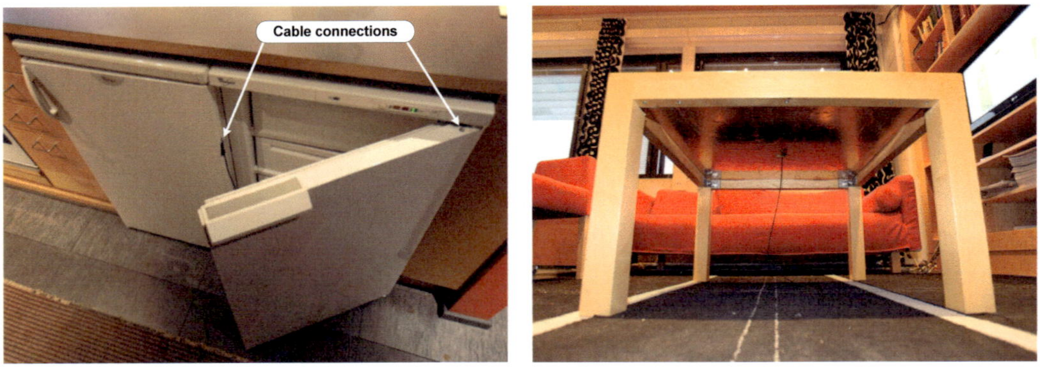

Figure 6. Both the transmitters and the receivers can be hidden in household items in order to recognize user touch and proximity to them. On the left-hand side, the metal front surfaces of the refrigerator (left appliance) and freezer (right appliance) were used as transmitting electrodes to detect when these appliances were used in the *TUT Smart Home*. On the right-hand side, a copper foil receiver is attached under the tabletop of the sofa table. With this receiver, the measured data was used both to detect contact with the table surface and to position the user on the floor in the vicinity of the table. The connecting cable for the receiver electrode, going under the floor through the carpet, is also visible. [51]

person, for at least a period of one second, with the total number of measurements performed. In other words, the system was not able to measure the person position for a total of about two and half minutes during the one hour sample. The places where the person was lost were distributed in the following way: a total of about one second in the dining room, 25 seconds in the hallway and two minutes and five seconds in the kitchen. In contrast, the false hit rate of the system, analyzed from this same one-hour sample, is 0.2%. This figure was obtained by calculating the number of times the system detected the person in the sensed floor area while the person was actually still in the bathroom or on the balcony, and then dividing that number by the total number of measurements performed during the periods of absence from the tracking area. In figures, the person was momentarily falsely recognized on the floor five times out of the 2404 measurements that were performed during the eight minutes and ten seconds of absence.

6. Contact sensing

The activities of people in a smart environment can be monitored through their interactions with the objects in the environment. For example, the system described in publication [51] can observe user contact with common household items such as a bed, sofa, table or refrigerator. When used in conjunction with the tile-based positioning method (see section 5) as in publication [51], the position of the person can be deduced simultaneously with the user interactions using a single measurement system.

Electrode configuration To measure interactions with the environment, transmit-mode sensing can be used to recognize user touch to both transmitting and receiving electrodes that are hidden in the objects of the environment. For example, sensing electrodes can be installed in beds, sofas, tables or electrical appliances as in publication [51]. The left-hand side of Figure 6 shows an example how the front doors of the refrigerator and freezer were connected to the measurement system in the *TUT Smart Home* and used for detecting when

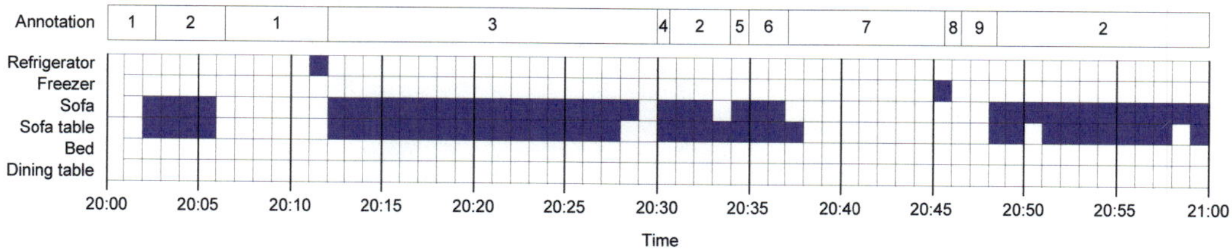

Figure 7. An example of one hour's worth of activity data acquired from the *TUT Smart Home* during a long-term living test. The one-minute-wide boxes are colored dark only if the test person touched the item during that one minute period. During this one hour, the test person 1) prepared food in the kitchen, 2) watched TV on the sofa or used a laptop on the sofa table, 3) ate food in the living room, 4) carried dishes to the kitchen, 5) checked the home's status from the bedroom computer, 6) undressed, 7) took a shower in the bathroom, 8) cooled off on the balcony, and 9) dressed in the bedroom. The walking paths during this one-hour period have already been shown in Figure 5. [51]

they were used. In the case of non-conductive objects, thin copper foil can be applied to the item, such as to the sofa table in the *TUT Smart Home* shown on the right-hand side of Figure 6.

Measurement method Human contact with household objects can be detected by measuring the C_U for the electrode attached to the object. The magnitude of C_U changes in direct proportion to the person's proximity to the electrodes. In other words, C_U is high when the absolute distance to the electrode is small and vice versa. However, the changes are non-linear with the distance because the measured capacitances increase quadratically as the distances between the person and the electrodes become shorter. In the test system described in publication [51], any contact with the object is given as a binary signal that is formed by comparing C_U to a predetermined constant threshold capacitance for the object. For example, the test system in publication [51] was defined to give a binary one when the person touched the object and a zero when the object was not touched by the person.

Key results As a part of the study in publication [51], a two-week long test was performed in the *TUT Smart Home* to test the contact-sensing system. Figure 7 presents an example of the activities detected in the *TUT Smart Home* during a one-hour period below the annotated events that were recorded by the test user on a portable voice recorder. This diagram shows that the annotation data matches well with the recorded contact data. Moreover, if the plotted tracks of the person shown in Figure 5 during this same one-hour period are analyzed together with the contact and annotation data, they fit seamlessly together. Therefore, we can conclude that by embedding the electrodes of a tile-based positioning system in the household objects of a smart environment, the interactions with the environment can be detected with a good accuracy. Furthermore, when the positioning data is analyzed together with the recorded interactions it is possible to infer the activities of the person under observation within a particular context.

7. Electric field ranging

Human position, leaning and hand or leg movements can be captured by measuring the capacitance between a transmitting floor electrode and a receiver electrode close to the person, when the person stands or moves over the floor electrode. This section presents such a passive, simple, and affordable electric field based ranging method that transforms the physical input

Figure 8. The electric field ranging method allows even slight body movements in the tracking space to be detected. By measuring the user distance from the vertical wires placed in each corner of the tracking space, the user position can be calculated and used in addition to indoor positioning, for example, to control a flight simulator game. [49]

into a two-dimensional position in an indoor environment. The method and a demonstration platform has been built and presented in detail in [49] and will be presented in this section with sufficient details about the system structure, capacitance model and results for the reader to understand the operation and possibilities of the measurement setup. Although this test system at the current state of development can only be used in small scale setups, it provides interesting possibilities to application developers, for example, in gaming, virtual exhibition, and virtual reality industries.

Instead of using multiple transmitters and a small number of receivers, it is possible to track a person with a single transmitter and multiple receivers. This is made possible by calculating the capacitance between the person and the receivers and converting it into distances. Moreover, this method allows even slight movements, such as leaning or moving an arm or a leg, to be discerned, assuming that an adequate signal can be received by the receivers. As was demonstrated in publication [49], this method allows the test system developed in publication [49] to be used not only for human position tracking, but also in gaming, virtual exhibitions, and virtual reality applications.

Construction The test system described in publication [49] consists of a single large transmitter, 180×180 cm in size with four vertical wire receivers placed at the corners of the transmitter. The transmitter consist of nine commercial raised floor tiles that have been electrically connected, and the receivers are made of standard power-line cables. The whole test installation has been placed next to the structures of a virtual reality cave, which has two background projection screens beside the transmitter. Figure 8 shows the physical construction of the test system with an example application of flight simulator control.

Measurement method This system measures the capacitances between the user, standing on a transmitting floor electrode, and four vertically aligned receiver wires placed in the corners of the tracking area. The system uses the transmit mode and measures one receiver at a time at 19.5 Hz. The measured capacitances are converted to absolute distances between the user and the receivers using a *capacitance-to-distance conversion function* in the horizontal dimension. If the distances between the user and the static receivers are known, a non-linear system of equations can be solved to calculate an unambiguous position for the user.

Specifically, the system measures the value of C_U^r for each receiver r at a time. A predetermined *capacitance-to-distance conversion function* is used to convert the measured C_U^r value to an absolute distance value d_r between the receiver r and the user. The non-linear system of equations is solved in real-time to find the best approximation for the 2D body position x by using all four of the computed distances d_r from the receivers using an iterative least squares (Gauss-Newton) algorithm. Also, because the C_U changes vary somewhat according to the person's height, feet size and footwear, the computed distances are corrected with a correction factor k whose value typically ranges between 0.6 and 1.5. Altogether, the measurement equation to be solved can be written as

$$d_r = k\|b_r - x\| + e_r, \tag{3}$$

where, b_r is the location of the receiver r and e_r is the measured error for the receiver.

Key results The test system which was constructed and described in publication [49] demonstrates that a person's body position and movements can be tracked passively with a simple electrode configuration and affordable electronics by converting the measured capacitances to absolute distances. The system works with different size people and in some applications the correction factor k could be used to adequately identify different-sized people or to identify the insulation properties of an individual's footwear. The test system can track a human body with relative accuracy of less than 10.4 cm and absolute accuracy of less than 15 cm.

Figure 9 shows two different ways that the system positions the user in a flight simulator scenario. First, the inner circular shape shows the limits to how much a test person was able to change the tracking result by swivelling in different directions with his arms at his sides. Second, the larger radial results show how the test user was able to position himself when he stepped in 8 different directions and bent his body in the direction of the step. The figure reveals that the test person was able to move his computed position away from the center position by an average of about 15 cm when swivelling and 70 cm when taking the single steps. In addition to these experiments, this system was tested for mouse cursor control, virtual-object viewing-angle manipulation, and drawing.

8. Height and posture sensing

It is difficult to unobtrusively recognise a user's height and posture passively without video cameras. In publication [47] we demonstrated a method for distinguishing a person's posture with a simple measurement system. The measurement method is based on transmit-mode measurements and instead of calculating the horizontal distance from the person to the receiver as in [49], the conversion is done vertically and the distance between the user and a receiver above the user is determined. Further, by first measuring the standing height of the person, the user's posture and vertical head position can be determined with these simple techniques.

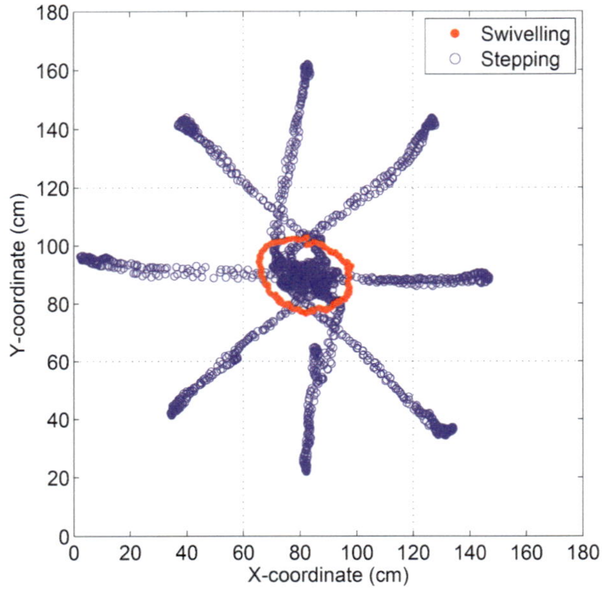

Figure 9. The response of the electric field ranging method was tested in publication [49] with two postures. First, a test user stood in the middle of the tracking area and swiveled through 360° while keeping both feet on the ground. Second, he stepped forward in eight different directions while keeping the other foot in the same place. Modified from [49].

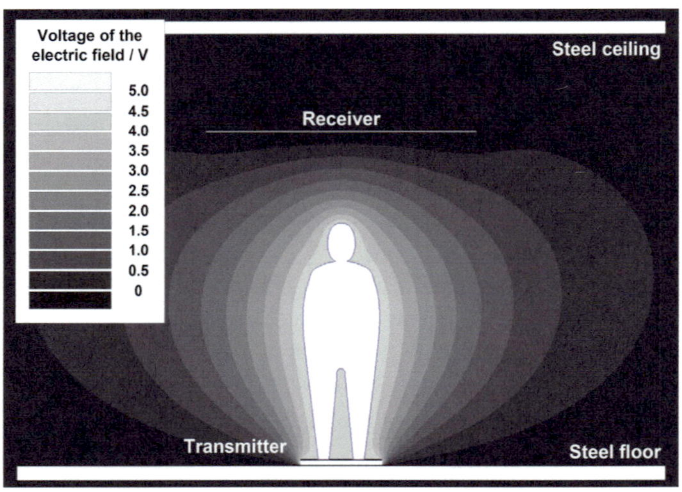

Figure 10. The electric field distribution around a 170-cm-tall human standing on a transmitting floor tile in a conductive building simulated with steel floor and ceiling. By measuring the capacitance between the body and the receiver above the person, both the human's height and posture can be determined. [47]

Measurement setup A system for measuring a person's height can be constructed from a single transmitter and a receiver in transmit mode. The transmitter under the person's feet is used to emanate an electric field around the body and the C_U is measured with a receiver

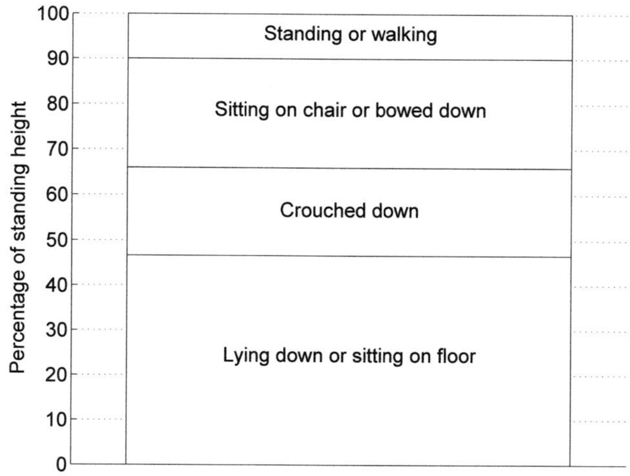

Figure 11. An example classification of different postures. [47]

set in a horizontal plane above the user. Figure 10 demonstrates the potential distribution in this type of a measurement setting in a well conducting building when the receiver has been installed at a height of 240 cm and the person standing on the transmitter is 170 cm tall.

Measurement method The height of a person is calculated using the measured C_U value by converting it to an absolute distance between the person and the receiver and by subtracting this from the height at which the receiver is installed. The conversion is done with a pre-calibrated *capacitance-to-height conversion function* that must be calculated for the operating environment before the system goes into operation. This conversion function can be derived either from simulation results by calibrating it for the given environment or by experimentally defining it in the operating environment. The head height of the person at any moment can be determined by calculating a percentual height $h_{\%}$ of a person from the standing height h_s with the equation

$$h_{\%} = \frac{measured\ height}{h_s} * 100\%. \tag{4}$$

Furthermore, this percentual height can be used to determine the posture of the observed person, for example, by using the posture classification presented in Figure 11. However, this classifier has not been verified, because the small number of test persons ($n = 14$) available in [47] were all used to create this classifier.

Key results Both the simulation and experimental results from publication [47] shown in Figure 12 demonstrate that C_U changes with the height of a user in a standing position. Moreover, since the simulation results and the actual results have a similar shape in Figure 12 but a somewhat different slope, the scaled simulation graph in Figure 12 shows how well the simulation and experimental results correlate with each other. This scaled graph was calculated from the simulation results using the equation

$$C_U^{Scaled} = 1.5C_U - 1.4 \tag{5}$$

where the constants have been experimentally defined.

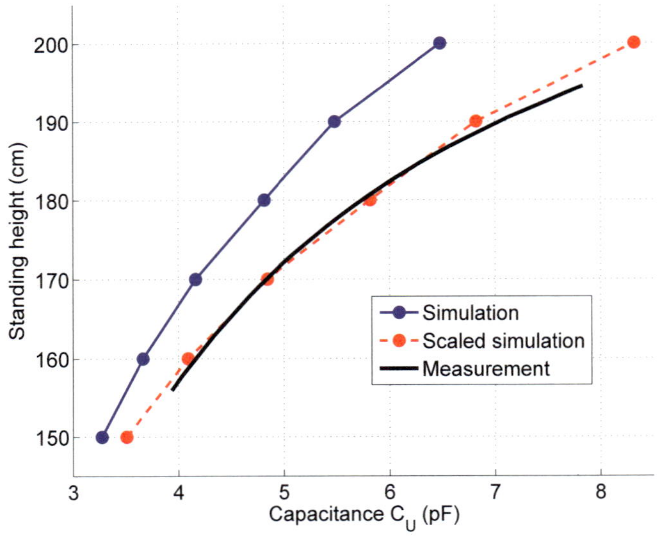

Figure 12. Simulated and measured *capacitance-to-height conversion functions* with scaled simulation results from publication [47]. [47]

Statistically, the experimental system described in publication [47] was able to calculate the height of a standing person with at least 7.1 cm accuracy. In all other postures ranging from standing erect to crouching as low as possible, the absolute accuracy of the system was always determined to be 21.7 cm or less. At 90% confidence level these values were 5.2 cm and 14.3 cm, respectively. If the relative change in the user's height from a standing posture is enough for a given application, the height change can always be measured to within at least 14.6 cm accuracy and at 90% confidence level to within 9.1 cm accuracy. Furthermore, if the system is calibrated for a single individual, the absolute and relative accuracies for a crouching person are 11.6 cm and 4.5 cm respectively at 100% and 90% confidence levels.

These unobtrusive height and posture measurement methods can be used in many context and activity recognition applications. Indeed, in smart environments the activities of the tracked persons can be reasoned with higher accuracy if the three dimensional position and posture of the person can be inferred. For example, if we know that a person is sitting next to a table, rather than standing, we can infer that the person might be eating or reading something. On the other hand, if the person is standing, we can infer that they might be laying the table. Similarly, if people are sitting on the floor and playing a board game, we may distinguish that from people standing in a circle and chatting with each other. Or, if a person is exercising on the floor and doing some sit-ups, we can differentiate that action from someone just lying on the floor. Therefore, the elderly could be monitored to support independent living at home and raise the alarm if the elderly person should fall and not get up again. Finally, smart homes could offer some unique services to their inhabitants like proactive home appliance control and real-time context delivery to family members and friends outside the home. Furthermore, if there are sufficient differences in the heights of the family members, they could be distinguished from each other, and so individualised services could be provided for them through the home user interfaces and actuators.

9. Discussion

This unobtrusive way of measuring a person's position, posture and contact with the environment in 3D will be helpful in creating new smart environments that are calm and non-distracting to their users. Moreover, these methods can be used outside smart environments in many different types of applications. In this section, we will examine the pros and cons of the methods, compare the methods to the state of the art technology, and present ideas for future work.

9.1 Accuracy and reliability

The results presented here indicate that the methods which have been developed cannot achieve the same level of accuracy as can be obtained with the methods used in the state-of-the-art active systems which can be applied in smart environments. For example, the two well-known ultrasound-based location systems, Cricket [32] and Bat [53], can position a person with about 10 cm and 5 cm accuracy, respectively. Likewise, a commercial Ubisense location system based on UWB (Ultra-wideband) technology can track a person with about 15 cm accuracy [16]. However, the test systems described in publications [48], [49] and [51] do demonstrate somewhat better accuracy than the the passive infrared system of Hauschildt et al. [4], which can only position a single person with an accuracy of 26 cm. On the other hand, our systems are not as accurate as the passive ultrasound system of Nishida et al. [24], which can track a human head with less than 5 cm error. However, it is important to note that in its current state of development the system of Hauschildt et al. suffers from reflection and dynamic background radiation issues, and the ultrasonic system of Nishida et al. would require the installation of hundreds of sensors in the ceiling of the *TUT Smart Home* in order to cover most of the rooms.

If the accuracies of the test systems presented in publications [48] and [51] are compared to the capacitive positioning system of Rimminen et al. [35, 36], we have a slightly better mean accuracy than Rimminen's system. Indeed, Rimminen et al. use 25×50-cm electrodes on the floor and get a mean accuracy of 21.2 cm for people walking, while our systems obtained a mean accuracy of 9.4 cm and 17.5 cm with the 30×30-cm-sized and 60×60-cm-sized floor electrodes respectively. Furthermore, the standard deviations in our systems are also somewhat lower: 13.2 cm for Rimminen's system versus 6.5 and 11.3 cm respectively for the small and large transmitters used in our systems. The main reason that our systems are more accurate than the Rimminen one, even with the larger electrodes, is probably due to the fact that Rimminen et al.'s system had wiring between the electrodes on the floor surface as shown in [33, p. 11]. This meant that the available electrode area on the floor surface was reduced, which led to an increase in the positioning errors.

The reliability of the tracking system built in the *TUT Smart Home* in [51] can be analyzed with the hit rate and false hit rate figures presented in section 5. To understand why the tracking system lost the track of the person in the *TUT Smart Home,* we must look at where the system lost track of the person. In section 5 we showed that, most of the time, the person was lost in either the kitchen (82.7% of the cases) or the hallway areas (16.6% of the cases). If we compare this distribution with the Figure 4, we can see that the places reducing the hit rate have a relatively low received signal strength. Thus, a low signal strength greatly affects the reliability of the tracking system. In addition, the contact of the person with grounded sources in the kitchen, i.e. the dishwasher, the sink and the stove, cause the transmitted signal to couple with the ground so it is not received by the system. Therefore, for much of the time when the person was considered lost he was preparing food and touching one or other of these grounded objects in the kitchen. This is also very clear from the publicly released test data [17].

The acquired hit rate of 95.8% is slightly better than the 91.6% achieved by Rimminen at al. [35]. Again, the reason for this difference is caused by the gaps between the floor electrodes in the UPM Kymmene Oyj sensor laminate used in their work [33]. Furthermore, our hit rate is better than the maximum hit rate of 87% obtained with the pressure sensitive floor in the Gator Tech Smart House [7]. Nevertheless, some ideas to further enhance the reliability of our system will be presented in section 10.

9.2 Construction, cost and complexity

The physical construction of the electrodes in all of the presented systems is fairly simple, because they consist of single layers of conductive material, which is easily available. For example, the floor electrodes could be built from aluminum foil placed underneath the floor surface, which could be either wood, plastic or ceramic. Similarly, the receiver electrodes could be built out of almost any conductive solids, textiles, or even from a sparse wire net. In fact, a 0.22 mm-thick wire net receiver was shown to be a feasible solution in publications [47] and [51] to create a barely noticeable receiver below the ceiling.

Unfortunately, the current design requires a single coaxial cable for each electrode, which means that a large amount of cabling is required to cover a house. In the case of the *TUT Smart Home*, about 400 meters of coaxial cable were installed to feed the measurement signals to and from the electrodes. One solution to this problem would be to replace the cabling and the electrodes with, for example, the thin sensor laminate made by UPM-Kymmene Oyj that was used in the work of Rimminen et al. [35, 36] and is shown in [33, p. 11]. This 150-μm-thick laminate was made from a 13-μm-thick aluminum film that was laminated between two layers of PET (polyethylene terephthalate) film. All the wiring to the 36$\times$30-cm-sized electrodes were incorporated into this laminate using etching techniques and were made accessible at the edge of the laminate using a standard connector.

Based on the large-scale tile-based positioning system implementation of the *TUT Smart Home* described in publication [51], we can calculate the total cost per square meter for a real-world capacitive positioning system. If we include all the materials and devices used to construct the positioning and contact-sensing system of [51] but exclude the cost of labour and the PC used for position calculations, we get a total price of about 5800 € for the roughly 53 m^2 area covered by system. This equates to a cost of 110 €/m^2, of which only about 16 €/m^2 goes on the electronics and cabling, and 4 €/m^2 on both the copper foil and silverized-textile receivers placed in the environment. Clearly, most of the costs arise from the raised floor tiles, which are relatively expensive, being about 108 €/m^2 and 68 €/m^2 for the commercial and the custom built floor tiles respectively. Thus, the type of flooring material greatly affects the total cost of these types of capacitive systems. By covering the floor with cheaper materials and using alternative construction for the floor electrodes, the total cost can be brought down to less than half of the implementation costs of the *TUT Smart Home* system.

The above compares well with the costs involved with other passive positioning system implementations. The capacitive positioning system of Rimminen et al., implemented with the UPM-Kymmene Oyj sensor laminate and a thin carpet covering the floor, costs only about 45 €/m^2 [33]. In contrast, the pressure-sensitive floor of Gator Tech Smart House costs about 140 €/m^2 [7] and can provide about the same level of accuracy than our tile-based system. The ultrasound sensors of Nishida et al. [24] can measure human position with much better accuracy, but their system requires a large array of sensors installed close to each other in the ceiling and the total cost of such a system could be estimated to be over 200 €/m^2. Conversely, the infrared sensors of Hauschildt et al. [4] could be significantly cheaper, as only one thermopile-array module would be required in each corner of a room. Indeed, if one module would cost about 100 € in large quantities [43], the covering of a 10 m^2 room would cost only 40 €/m^2. Nevertheless, the reflection and dynamic background radiation effects of thermopile detectors still would need to be solved so that they could be used reliably in real living environments.

9.3 Safety

The methods presented here can be regarded as completely safe for human beings because the electric and magnetic fields in the human body are well below the levels set by the International Commission on Non-Ionizing Radiation Protection (ICNIRP) [5]. The commission has set the limit for maximum internal electric field strength in the human body to 4.32 V/m at a 32 kHz frequency and the first harmonics of the square-wave signal which was used. Because the human body conducts these frequencies well, the potential difference between the feet and other parts of the body is only about one volt, as shown in publication [47]. Hence, the internal electric field within the human body remains under 1 V/m, which is well below the recommended limit.

Because the wavelength of the 32 kHz square-wave signal used in the practical implementations and its harmonics are in the order of magnitude of kilometers, while the dimensions of the floor electrodes which were used are measured in meters, the emanated magnetic field from the transmitter is virtually nil. Thus, the human body's exposure to magnetic fields from the systems is negligible. Finally, the current levels in the human body need to stay below the ICNIRP set reference levels for general public exposure [5]. In figures, the current in the neck was determined to be only 25 μA while the maximum current set by the ICNIRP for the 2.5–100 kHz frequency range is 6.4 mA. Thus, the current levels remain well under the set limit and do not pose any health hazards.

10. Future work

The test system of publication [51] installed in the *TUT Smart Home* can sometimes lose track of a person as already discussed in sections 5 and 9.1. In addition to the problem of the grounded objects electrically grounding the person in the kitchen area (see section 9.1), three other typical reasons for losing the track of the person were identified in publication [51]. These were 1) the insufficient signal strength over a certain area of the apartment, 2) a person moving too fast and 3) a person moving in an unusual way, such as leaping on the floor, jumping from one non-conducting chair to another or crawling on the floor in areas of low signal strength. Although fixing all these problems is challenging, the first two problems could be fixed, or at least alleviated in many ways. First, a better receiver network could be created simply by placing more receivers in the environment. If the receivers could be placed, for example, in the furniture, the contact with these items could also be recognized. Second, the structure of the floor tiles could be changed to yield a greater C_F^t. This could be done, for example, by moving the transmitting electrode closer to the surface of the tiles. In an installation in an old house this could be achieved by spreading the transmitting electrodes directly over the existing and non-conducting floor surface and by installing a new, thin floor surface directly on top of it. Also, the use of thinner carpets on the floor would help with this issue. Third, by decreasing the total number of transmitters, for example, by using only the larger 60×60-cm-sized electrodes, the scanning speed of the current implementation could be increased significantly. Fourth, new electronic circuitry that would enable faster scanning of the tiles could make a significant difference. Fifth, the tracking algorithm of the system could be enhanced significantly to help with the speed issue. Indeed, currently the transmitters under a person are not always scanned before that person moves to a new position.

The effects of shoes were not studied in detail in publications [47]–[51]. However, they can have a significant effect on the ability of the system to track people and on the accuracy of the results [47, 51]. This happens, because C_F^t gets smaller in inverse proportion to the distance between the feet and the transmitters, and the insulation characteristics of shoes cause negative effects. Indeed, the measured C_U can be only some tens of percent of its magnitude without shoes. Based on the short tests described in publication [47], the measured height of a person with shoes can vary between 50–80% from the measurements achieved when the person is not wearing shoes. Even though the correction factor k of Equation (3) partially quantifies the insulation characteristics of the shoes, other practical methods for quantifying the shoe insulation characteristics with tile-based positioning systems could be researched. However, if the transmitting electrodes were to be installed closer to the feet, as proposed above, this problem would also be alleviated.

11. Conclusion

This chapter studied capacitive user tracking methods for smart environments. The developed methods can be realized in an unobtrusive way to keep the living environment as a relaxing space. Thus, the people's lives in these environments can be supported and the inhabitants can concentrate on their primary goals rather than be burdened by the sensory actions of their living environment. Specifically, the chapter focused on the research work of Valtonen et al. and reported the used methods and key results of the developed and implemented tracking systems. Altogether, the methods presented in this chapter provide a reasonable level of accuracy for tracking people, and by using some of the developed sensing methods together, three-dimensional user tracking can be achieved in an affordable way.

The presented sensing methods could be used in many context and activity recognition applications for assisting the elderly at home. Indeed, by knowing the three dimensional human position and posture in a smart home, the home could reason the activity of the person with a much higher accuracy than without the position information. For example, the daily routines of the elderly could be monitored to support the independent living at home or to alert help if the elder should fall down and not get up. If family members have enough height difference, they can be recognized from each other and personal services could be provided. In addition, many applications in the areas of gaming and virtual reality could benefit from an unobtrusive and affordable human tracking system.

Author details

Miika Valtonen and Timo Vuorela
Tampere University of Technology, Finland

12. References

[1] *AD7746 datasheet* [2005]. Rev. 0.

[2] Augusto, J. C., Nakashima, H. & Aghajan, H. [2010]. Ambient intelligence and smart environments: A state of the art, *in* H. Nakashima, H. Aghajan & J. C. Augusto (eds), *Handbook of Ambient Intelligence and Smart Environments*, Springer US, pp. 3–31.

[3] Curran, K., Furey, E., Lunney, T., Santos, J., Woods, D. & McCaughey, A. [2011]. An evaluation of indoor location determination technologies, *Journal of Location Based Services* 5(2): 61–78.

[4] Hauschildt, D. & Kirchhof, N. [2010]. Advances in thermal infrared localization: Challenges and solutions, *Proc. of the International Conference on Indoor Positioning and Indoor Navigation*, pp. 1–8.

[5] ICNIRP [2010]. Guidelines for limiting exposure to time-varying electric and magnetic fields (1 Hz to 100 kHz), *Health Physics Society* 99(6): 818–836.
URL: *http://www.icnirp.de/documents/LFgdl.pdf*

[6] Junnila, S., Kailanto, H., Merilahti, J., Vainio, A.-M., Vehkaoja, A., Zakrzewski, M. & Hyttinen, J. [2010]. Wireless, multipurpose in-home health monitoring platform: Two case trials, *Information Technology in Biomedicine, IEEE Transactions on* 14(2): 447–455.

[7] Kaddoura, Y., King, J. & Helal, A. [2005]. Cost-precision tradeoffs in unencumbered floor-based indoor location tracking, *in* S. Giroux & H. Pigot (eds), *Smart Homes to Smart Care*, Assistive technology research series, IOS Press, pp. 75–82.

[8] Kaila, L. [2009]. *Technologies Enabling Smart Homes*, PhD thesis, Tampere University of Technology.
URL: *http://urn.fi/URN:NBN:fi:tty-200911107084*

[9] Kaila, L., Hyvönen, J., Ritala, M., Mäkinen, V. & Vanhala, J. [2009]. Development of a location-aware speech control and audio feedback system, *Proc. of the Seventh Annual IEEE International Conference on Pervasive Computing and Communications*, IEEE, pp. 1–4.

[10] Kaila, L., Vainio, A.-M. & Vanhala, J. [2005]. Connecting the smart home, *Proc. of the Networks and Communication Systems*, ACTA Press, pp. 445–450.

[11] Karlsson, B., Karlsson, N. & Wide, P. [2000]. A dynamic safety system based on sensor fusion, *Journal of Intelligent Manufacturing* 11(5): 475–483.

[12] Karlsson, N. [1994]. Theory and application of a capacitive sensor for safeguarding in industry, *Proc. of the IEEE Instrumentation and Measurement Technology Conference*, Vol. 2, pp. 479–482.

[13] Karlsson, N. & Järrhed, J.-O. [1993]. A capacitive sensor for the detection of humans in a robot cell, *Proc. of the IEEE Instrumentation and Measurement Technology Conference*, pp. 164–166.

[14] Koolwaaij, J., Tarlano, A., Luther, M., Nurmi, P., Mrohs, B., Battestini, A. & Vaidya, R. [2006]. Context watcher: Sharing context information in everyday life, *Proc. of the IASTED conference on Web Technologies, Applications and Services*, pp. 12–21.

[15] Lee, S., Kim, B., Kim, H., Ha, R. & Cha, H. [2011]. Inertial sensor-based indoor pedestrian localization with minimum 802.15.4a configuration, *Industrial Informatics, IEEE Transactions on* 7(3): 455–466.

[16] Liu, H., Darabi, H., Banerjee, P. & Liu, J. [2007]. Survey of wireless indoor positioning techniques and systems, *Systems, Man, and Cybernetics, Part C: Applications and Reviews, IEEE Transactions on* 37(6): 1067–1080.

[17] *Long-term living-test data archive for positioning data* [2009].
URL: *http://wiki.tut.fi/SmartHome*

[18] Malik, A. [2009]. *RTLS For Dummies*, Wiley Publishing.

[19] Mautz, R. & Tilch, S. [2011]. Survey of optical indoor positioning systems, *Proc. of the International Conference on Indoor Positioning and Indoor Navigation*, pp. 1–7.

[20] Mileo, A., Merico, D. & Bisiani, R. [2010]. Support for context-aware monitoring in home healthcare, *Journal of Ambient Intelligence and Smart Environments* 2: 49–66.

[21] Mozer, M. C. [2005]. Lessons from an adaptive home, *Smart Environments* pp. 271–294.

[22] Mulloni, A., Wagner, D., Barakonyi, I. & Schmalstieg, D. [2009]. Indoor positioning and navigation with camera phones, *IEEE Pervasive Computing* 8(2): 22–31.

[23] Ní Scanaill, C., Carew, S., Barralon, P., Noury, N., Lyons, D. & Lyons, G. M. [2006]. A review of approaches to mobility telemonitoring of the elderly in their living environment., *Annals of Biomedical Engineering* 34(4): 547–563.

[24] Nishida, Y., Murakami, S., Hori, T. & Mizoguchi, H. [2004]. Minimally privacy-violative human location sensor by ultrasonic radar embedded on ceiling, *Proc. of IEEE Sensors*, Vol. 1, pp. 433–436.

[25] O'Dowd, J., Callanan, A., Banarie, G. & Company-Bosch, E. [2005]. Capacitive sensor interfacing using sigma-delta techniques, *Proc. of the IEEE Sensors*, pp. 951–954.

[26] Paajanen, M. [2000]. Electromechanical film (emfi) Ů a new multipurpose electret material, *Sensors and Actuators A: Physical* 84(1–2): 95–102.

[27] Patel, S. N., Reynolds, M. S. & Abowd, G. D. [2008]. Detecting human movement by differential air pressure sensing in hvac system ductwork: An exploration in infrastructure mediated sensing, *Proc. of the 6th International Conference on Pervasive Computing*, Springer, pp. 1–18.

[28] Pavel, D., Callaghan, V. & Dey, A. [2010]. Looking back in wonder: How self-monitoring technologies can help us better understand ourselves, *Proc. of the Sixth International Conference on Intelligent Environments*, pp. 289–294.

[29] Pensas, H., Valtonen, M. & Vanhala, J. [2011]. Wireless sensor networks energy optimization using user location information in smart homes, *Proc. of the International Conference on Broadband and Wireless Computing, Communication and Applications*, pp. 351–356.

[30] Pirttikangas, S., Suutala, J., Riekki, J. & Röning, J. [2003]. Footstep identification from pressure signals using hidden markov models, *Proc. of the Finnish Signal Processing Symposium*, pp. 124–128.

[31] Plomp, J., Heinilä, J., Ikonen, V., Kaasinen, E. & Välkkynen, P. [2010]. Sharing content and experiences in smart environments, *in* H. Nakashima, H. Aghajan & J. C. Augusto (eds), *Handbook of Ambient Intelligence and Smart Environments*, Springer US, pp. 511–533.

[32] Priyantha, N. B. [2005]. *The Cricket Indoor Location System*, PhD thesis, Massachusetts Institute of Technology.
URL: *http://nms.csail.mit.edu/papers/bodhi-thesis.pdf*

[33] Rimminen, H. [2011]. *Detection of Human Movement by Near Field Imaging: Development of a Novel Method and Applications*, PhD thesis, Aalto University.

[34] Rimminen, H., Lindström, J., Linnavuo, M. & Sepponen, R. [2010]. Detection of falls among the elderly by a floor sensor using the electric near field, *IEEE Transactions on Information Technology in Biomedicine* 14(6): 1475–1476.

[35] Rimminen, H., Lindström, J. & Sepponen, R. [2009]. Positioning accuracy and multi-target separation with a human tracking system using near field imaging, *International Journal on Smart Sensing and Intelligent Systems* 2(1): 156–175.

[36] Rimminen, H., Linnavuo, M. & Sepponen, R. [2008]. Human tracking using near field imaging, *Proc. of the Second International Conference on Pervasive Computing Technologies for Healthcare*, pp. 148–151.

[37] Schantz, H. G., Weil, C. & Unden, A. H. [2011]. Characterization of error in a near-field electromagnetic ranging (nfer) real-time location system (rtls), *Proc. of the IEEE Radio and Wireless Symposium*, pp. 379–382.

[38] Seco, F., Plagemann, C., Jiménez, A. R. & Burgard, W. [2010]. Improving rfid-based indoor positioning accuracy using gaussian processes, *Proc. of the International Conference on Indoor Positioning and Indoor Navigation*, pp. 1–8.

[39] Skubic, M., Alexander, G., Popescu, M., Rantz, M. & Keller, J. [2009]. A smart home application to eldercare: Current status and lessons learned, *Technology and Health Care* 17(3): 183–201.

[40] Smith, J. R. [1996]. Field mice: Extracting hand geometry from electric field measurements, *IBM Systems Journal* 35(3.4): 587 –608.

[41] Smith, J. R. [1999]. *Electric Field Imaging*, PhD thesis, Massachusetts Institute of Technology.
URL: *http://sensor.cs.washington.edu/pubs/phd.pdf*

[42] Smith, J., White, T., Dodge, C., Paradiso, J., Gershenfeld, N. & Allport, D. [1998]. Electric field sensing for graphical interfaces, *Computer Graphics and Applications, IEEE* 18(3): 54 –60.

[43] Sparkfun electronics [n.d.]. Infrared thermometer – MLX90614. Accessed 16.4.2012.
URL: *http://www.sparkfun.com/products/9570*

[44] Stuntebeck, E. P., Patel, S. N., Robertson, T., Reynolds, M. S. & Abowd, G. D. [2008]. Wideband powerline positioning for indoor localization, *Proc. of the 10th international conference on Ubiquitous computing*, ACM, pp. 94–103.

[45] Suutala, J. & Röning, J. [2008]. Methods for person identification on a pressure-sensitive floor: Experiments with multiple classifiers and reject option, *Information Fusion* 9(1): 21–40.

[46] Tapia, E. M., Intille, S. S. & Larson, K. [2004]. Activity recognition in the home using simple and ubiquitous sensors, *Pervasive Computing* 3001: 158–175.

[47] Valtonen, M., Kaila, L., Mäentausta, J. & Vanhala, J. [2011]. Unobtrusive human height and posture recognition with a capacitive sensor, *Journal of Ambient Intelligence and Smart Environments* 3(4): 305–332.

[48] Valtonen, M., Mäentausta, J. & Vanhala, J. [2009]. Tiletrack: Capacitive human tracking using floor tiles, *Proc. of the Seventh Annual IEEE International Conference on Pervasive Computing and Communications*, pp. 28–47.

[49] Valtonen, M., Raula, H. & Vanhala, J. [2010]. Human body tracking with electric field ranging, *Proc. of the 14th International Academic MindTrek Conference: Envisioning Future Media Environments*, MindTrek '10, ACM, New York, NY, USA, pp. 183–186.

[50] Valtonen, M. & Vanhala, J. [2009]. Human tracking using electric fields, *Proc. of the Seventh Annual IEEE International Conference on Pervasive Computing and Communications*, pp. 1–3.

[51] Valtonen, M., Vuorela, T., Kaila, L. & Vanhala, J. [2012]. Capacitive indoor positioning and contact sensing for activity recognition in smart homes. Accepted to Journal of Ambient Intelligence and Smart Environments, IOS Press.
[52] Wang, J. J. & Singh, S. [2003]. Video analysis of human dynamics–a survey, *Real-Time Imaging* 9: 321–346.
[53] Ward, A., Jones, A. & Hopper, A. [1997]. A new location technique for the active office, *Personal Communications, IEEE* 4(5): 42–47.
[54] Weiser, M. & Brown, J. S. [1996]. Designing calm technology, *Powergrid Journal* 1.
[55] Yang, A. Y., Jafari, R., Sastry, S. S. & Bajcsy, R. [2009]. Distributed recognition of human actions using wearable motion sensor networks, *Journal of Ambient Intelligence and Smart Environments* 1: 103–115.
[56] Zimmerman, T. G., Smith, J. R., Paradiso, J. A., Allport, D. & Gershenfeld, N. [1995]. Applying electric field sensing to human-computer interfaces, *Proc. of the SIGCHI conference on Human factors in computing systems*, pp. 280–287.

Chapter 7

RSSI/DoA Based Positioning Systems for Wireless Sensor Network

Stefano Maddio, Alessandro Cidronali and Gianfranco Manes

Additional information is available at the end of the chapter

http://dx.doi.org/10.5772/50380

1. Introduction

The problem of localization of a mobile device has interested researchers since the beginning of XX century, as testified by the experiments of Bellini and Tosi [20]. This challenging research topic has gained even more momentum in recent years, particularly with the introduction of modern ICT technologies such as Wireless Sensor Network (WSN).

A Wireless Sensor Network is an infrastructure comprised of a set independent nodes able to sense (measure), process and communicate among themselves and toward a remote sink node which operates as data aggregator and forwards the information to the final user. WSN are already actively employed in unattended and non-invasive activities like prevention of art deterioration [1, 23], agricultural monitoring [18], environment monitoring [17, 22], surveillance application [2, 10]. In most cases, if not in all cases, it is necessary to report the measured data to the position of the observed phenomenon, otherwise the measurement would be meaningless. This means that the sensor nodes have to be aware of their position, and if this information is not known, a *localization service* must be implemented. The *position awareness* that comes from this service can boost specific routing operations (adaptability, latency, throughput) with the nodes able to independently determine the best modality to cooperate and communicate the data to the end user by means of a continuous exchange of messages.

The objective of a localization system is to assign a positional information to each node of the network, either in the form of a relative position to a known anchor reference of within a coordinate system [7]. A localization procedure can be described as the series of three steps:

- Signal observation
- Extract of position-related signal parameters
- Estimation of location coordinates

INTECH
open science | open minds

© 2012 The Author(s). Licensee InTech. This chapter is distributed under the terms of the Creative Commons Attribution License http://creativecommons.org/licenses/by/3.0), which permits unrestricted use, distribution, and reproduction in any medium, provided the original work is properly cited.

A localization strategy is *effective* if accurate for the specific sensing application, and it is *efficient* if operates with minimal (hardware, software, bandwidth) resources. The most popular example of localization system is given by the widespread Global Positioning System (GPS) [7, 12], which allows a mobile node to accurately compute its position using the distances from three or more satellites. Unfortunately, GPS is not always a feasible solution, for cost or power constraints. There are also applications, like communication in indoor area, where GPS fails for physical reasons.

The fundamental principles of localization in a GPS-denied condition have been thoroughly investigated in the literature, but a successful implementation of localization procedure is still challenged by many practical issues. Various strategies has been proposed, relying on several class of data sensors (sound and ultra-sonic, light/laser, inductive/proximity, etc). Among these, the systems capable of performing the positioning service without additional hardware – excluding the communication infrastructure – are particular efficient. COTS devices are already equipped with a *Received Signal Strength Indicator* (RSSI). This kind of device gives access to a power sensor at zero cost, typically in the form of a digital data available on a register. The RSSI collection represent the signal observation phase of the localization procedure.

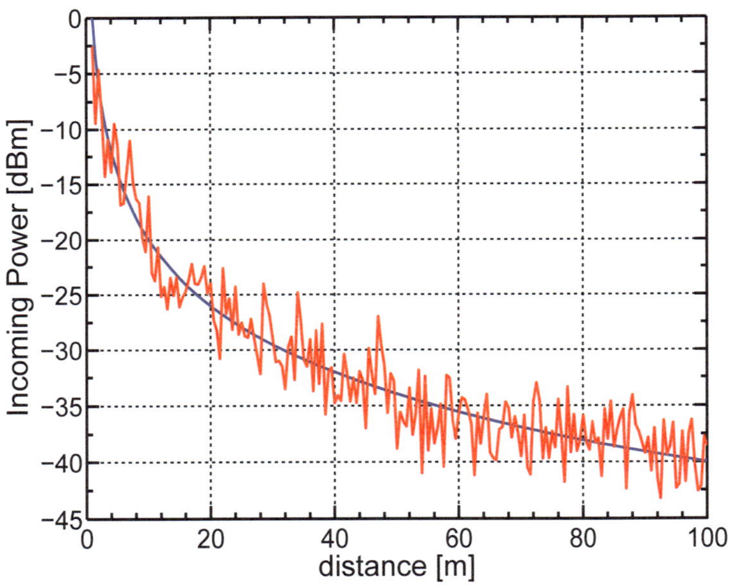

Figure 1. Distance Estimation with Power Measurement. The blue curve has a (theoretical) hyperbolic behavior. The red curve represents the actual behavior.

In the context of the Wireless Sensor Networks, RSSI measurements are already widely used to estimate the *range* – the relative distance – of the sensor nodes. The range estimation, which corresponds to the position-related signal extraction, is based on the fact that the power carried by an RF signal is inversely proportional to the traveled distance. This relationship is graphically described by the blue curve in Figure 1.

When at least three range measurements respect to three reference – non collinear – nodes are available, the estimation of location coordinates is obtained on the basis of trilateration algorithms, graphically described in Figure 2(a). The intersection of three circles, eventually in the mean square error sense, corresponds to the solution of a non-linear system, a trivial numeric elaboration.

Unfortunately, traveling radio signals are influenced by the indoor environment, being strongly affected by the reflected, refracted and scattered waves (multi-path) and disturbed by other devices communications (interferences). The actual power/distance relationship is more similar to the red noisy curve of Figure 1. Because of this noisy behavior, the power estimation is affected by a low accuracy, ultimately leading to a coarse position estimation, as depicted in Figure 2(b). This bad condition can be mitigated if more than three ranges are available, which leads to an overdetermined system (multilateration).

(a) Ideal case (b) Real case

Figure 2. Target localization by trilateration algorithm. The position of a node is determined with three range estimations respect to three reference nodes.

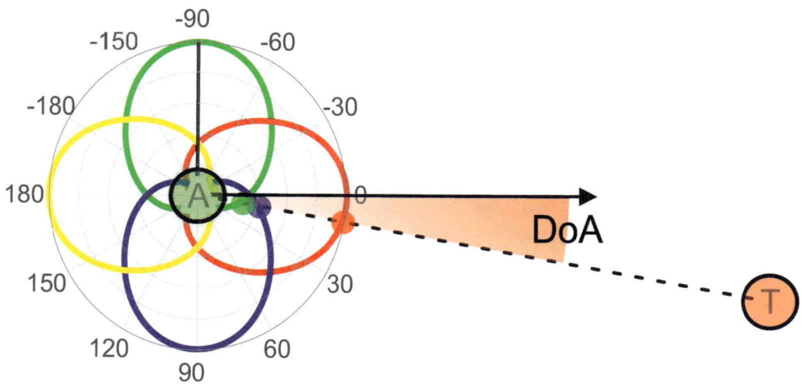

Figure 3. Direction of Arrival estimation with a multi-beam antenna system.

RSSI measurements can also be employed for *Direction of Arrival* (DoA) estimation. A positioning system based on DoA information does not rely on actual range measurements, i.e. it is *range-free* algorithm. The extraction of the signal DoA parameter is based on the antenna reception with a *multi beam system*, an antenna capable to radiate N directional beam patterns arranged in a sectorialised manner. The logic is to sense a power vector from different direction, as explained by Figure 3. Intuitively, if each beam is narrow enough, the antenna operates as a *spatial filter*, isolating a specific angular region. This estimation procedure is tolerant to noisy power measurements, because it has a range-free nature, thus it is particularly suitable for the a coarse power meter like RSSI. The critical condition for is the need of a *fast scanning* of the available beam, to ensure the same channel condition through the N antenna beams. For a DoA based algorithm, the estimation of position coordinates relies on the collaboration of at least two nodes operating as *beacons*, as explained in Figure 4. Each anchor estimates the DoA of the target respect to its relative reference, identifying a *line of bearing*. The intersection of two lines uniquely identify the target position. As for the multilateration, the collaboration of three or more anchors can enhance the estimation effectiveness.

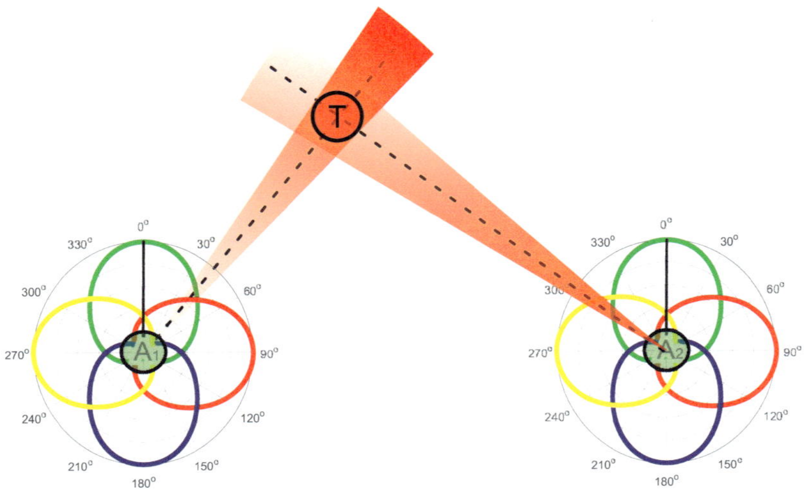

Figure 4. Target localization by DoA intersection. The estimated position does not rely on range measurement.

2. Node design

This section describes a brief summary of the typical hardware of a a sensor node suitable for the described RSSI/DoA-based localization system. The idea is to use two class of nodes. The master node, designed with a system consisting of a transceiver and a micro-controller as the core intelligence and equipped with the complex radiative system. The slave node, based on analogous but simplified design and equipped with a simple antenna. The master (anchor) node can be thought as the sink/coordinator, a specialized node, eventually placed by hand, serving as an access point. The slave (target) node, is the independent node, eventually free to move in the area within the communication range of at least one master node.

Recent advances in RFIC design opened the door to low-cost commercial transceiver technology. The CC2430 from Texas Instruments is a true System-on-Chip (SoC), an highly integrated RF chip built in 0.18 μm CMOS standard technology, with a compact 7×7 mm QLP48 package. CC2430 comes with an excellent transceiver, boosted with 2.4GHz DSSS

Figure 5. Node main board. The CC2430, the Ethernet module, the power plug and the I/O pins are visible.

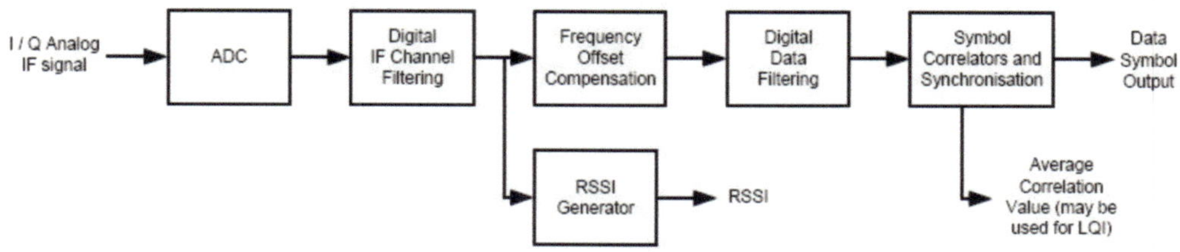

Figure 6. Internal structure of CC2430 demodulator, taken from CC2430 datasheet.

(Direct Sequence Spread Spectrum) operation which is IEEE 802.15.4 compliant, with an high sensitivity, suitable for operation with the ZigBee protocol. The embedded 8-bit core is compatible with the high performance and efficient 8051 industrial controller, with an embedded in-system programmable flash memory of 32, 64 or 128 Kbyte, 8 KB RAM and many other sensors and peripherals, such as analog-digital converter (ADC), several timers, AES128 coprocessor, watchdog timer, 32 kHz crystal oscillator with sleep mode timer, and 21 digital pins, which can be accessed by peripherals to manage I/O operations.

This SoC is also suitable for low-power operation, with a current consumption below 27 mA in either transmitting or receiving mode, while providing a wide supply voltage range (2.0 V - 3.6 V). A battery monitoring with temperature sensing is also integrated. Very fast transition times from low-power modes (0.3 μA in stand-by mode and 0.5 μA in powerdown mode) to active mode enables ultra low average power consumption in low duty-cycle systems, making CC2430 especially suitable for the applications which requires the battery's life is very long. Thanks to this impressive hardware architecture and the compatibility with the Zigbee protocol, this SoC is already a perfect candidate to build any kind of WSN node with only a few Components-Off-The-Shelf (COTS) external hardware. A detail of the CC2430's internal structure is shown as Figure6.

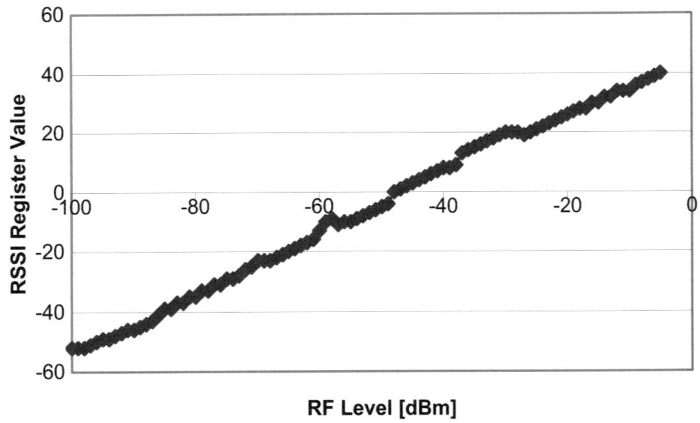

Figure 7. Typical RSSI behaviour, taken from CC2430 datasheet. $RSSI_{REG}$ vs actual received power on the RF pin.

2.1. Received Signal Strength Indicator

Among the other interesting features of this module there is the independent setting of the output power in accord with the external needs (through the action of the micro), and the built-in RSSI (Received Signal Strength Indicator) and LQI (Link Quality Indicator) indexes, always available to the micro, and so to the user via I/O ports. The built-in RSSI module operate averaging the received signal energy over an 8 symbol periods (128 μs) and return this value in the form of a formatted data register, in accord with the IEEE protocol. This 8-bit data is related to the effective incoming power through an equation provided by the manufacturer. The actual impinging power at the RF pin is expressed by (in dBm):

$$P_{in} = RSSI_{REG} + RSSI_{OFFSET} \tag{1}$$

that is, a linear combination of a constant bias value, and the sensed RSSI available on the register. Unfortunately this simple expression is a fitting where the $RSSI_{OFFSET}$ is found empirically around -45 dBm, but experiments reveals a variance of 2 or 3 dB. Even the linear behavior is a fitting, since experiment shows deviation even in cabled link. Nevertheless, the adoption of a built-in RSSI module permits hardware simplification and development time reduction, relaxing the hardware costs.

3. Switched beam antennas

The use of *smart antennas* improves the performance of wireless sensor network in several ways. Smart antennas technology has been introduced in the world of wireless communication system for two main reasons: alleviate the problems of limited performance of omni-directional radiators and gain the ability to perform operation otherwise impossible for canonical antenna.

The problem of limited resource in term of available power can be brilliantly solved with the aid of a spatial diversity system, an antenna system able to radiate power only where is needed, avoiding waste of power. Directional antenna allows a better efficiency for the power utilization, since the same received power is obtained with less transmission power, or, alternatively, greater transmission range with the same available power. The ability to reach longer range, focusing the available power only in the specific direction of the listener, is another great benefit paired with the reduced energy consumption. Another benefit of the spatial filtering nature of directive links, is the reduction of co-channel interference, since two transmitter in the same area can perform a communication task in two different directions on the same frequency band avoiding influence. This feature can solve the problem of clash and consequently idle time even in a dense transmission area. Another consequence of the latter is the spatial re-usability which can be exploited to increase network capacity and throughput [17].

In the general case, when the relative positions of transmitter and receiver are not known, a single directive antenna responsible for only a specific directional beam, is not enough. To cover the entire angular domain without losing the advantages of directive beams, a more complex structure is required, made of more elementary antennas appropriately arranged. The disadvantage is that this structure could become cumbersome if the size of the antenna is large. However, operating at a center frequency of 2.45 GHz, the need to compensate for the severe path loss multiple directional antenna makes the system a reasonable compromise. In addition, the radiative structure can be possibly be used shelter of the node itself. The intrinsically efficient power management have made directive antenna already suitable for cellular towers and base stations, where the benefits justify the costs (mobile phone tracking), but the use in WSNs is not equally widespread, mainly because of the need to design specific directional protocols. Nevertheless, directive antennas permits low-cost localization with no additional hardware, moving the balance of costs and benefits.

A *Switched beam antennas* consists of an antenna array fed by a beamforming network and it is capable of a predetermined set of beams which can be selected with an appropriate digital control [6]. This technology is complementary of the *adaptive beamformer*, which is an antenna array combined with a phase-shifting device, able to adaptively generate the required radiation pattern pointing in arbitrary direction. Nevertheless the SBA enables a low-cost, low-complexity solution for WSN based localization system. To ensure adequate reliability and accurate DoA estimation a SBA must fulfill two conditions:

- Each antenna element has to be in its maximum receiving condition when the other are in a null zone in order to have angularly uncorrelated signals at the input of the various elements,
- The opportune domain coverage has to be guaranteed in a cumulative sense.

Typically, the *cumulative* of the SBA radiation patterns has to be almost isotropic, for reason which will be clarified in the following sections.

3.1. Elementary antenna

The suitable elementary radiator of a SBA is *the printed antenna*, a planar radiator realized in the same technology of Printed Circuit Board, and based on the same inexpensive supports. Printed antennas have several attractive properties: they are lightweight, low-profile, compact, cheap, easy to fabricate and that they can be made conformal to the host surface. Patches can assume any arbitrary shapes, making them versatile in terms of resonant frequency, polarization and pattern shaping. Printed antennas operating in their fundamental resonant mode exhibit a directive pattern, behaving as a broadside radiator, with a gain ranging from 3 to 5 dB, and an half beam angle θ_{HP} ranging from 60 to 90 degrees. Other pattern configuration in term of directivity and θ_{HB} are possible by exploiting the higher resonant modes with the opportune feeding mechanism, or combining more than one in a (sub)array.

It is possible to demonstrate that the variance of the DoA estimation is proportional to RSSI variance [24], hence reducing σ_{RSSI} will directly reduce $\sigma_{\hat{\theta}}$, and therefore the DoA uncertainty. The radio channel dispersion, responsible of the measurement variance, is caused by time-varying multi-path propagation, which can be reduced by a proper choice of *antenna polarization*. It has been demonstrated that antennas operating in *Circular polarization* (CP) are effective in reducing this kind of variance and have been already exploited in wireless system operating indoors, and in radio-positioning applications.

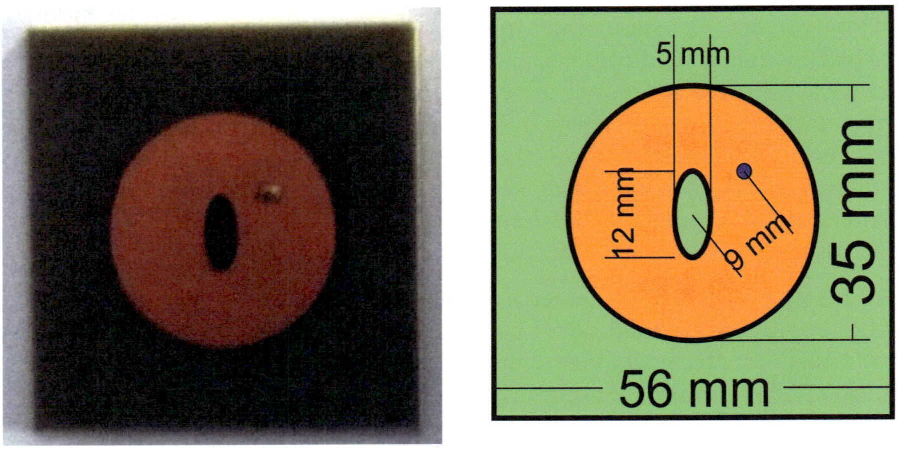

(a) element photograph (b) element design with quotes

Figure 8. An elementary patch radiator suitable for the Switched Beam Antenna.

A smart strategy to design compact CP antennas without an external splitting device is the modal degeneration. A *quasi-symmetrical* shaped patch antenna potentially support circular radiation with a single feed [4, 8]. The proposes radiative system is based on the *Elliptical Slitted Disk Antenna* (ESDA) a disc-based patch antenna working in the fundamental TM_{11} mode and exhibiting a boresight directional radiation pattern [16]. The elliptical slit at the center of the patch serves as the modal degeneration segment. Because of this perturbation, the antenna can sustain two orthogonal detuned modes which exhibit almost the same linearly polarized radiative behavior of a canonic disc working in the fundamental mode. CP radiation is achieved when these two modes, which are orthogonal, combine in phase quadrature with the same magnitude. This type of antenna has been already successfully employed in localization application [5, 6].

The antenna elements are printed over a common cheap FR4 substrate ($\epsilon_r = 4.4$, $h = 1.6\,mm$, 17 μm metalization thickness) shaped in a square geometry. Figure 8(a) shows a photograph of the antenna prototypes. Figure 9 shows the antenna reflection coefficient versus frequency. The 10 dB return loss bandwidth cover the entire ISM band. The antenna pattern is depicted in cartesian form in Figure 10. The Total gain, Left-Hand gain and Right-Hand gain component are shown. A peak gain of 3.75 dB is achieved in the boresight direction, while the principal lobe shows a θ_{HP} of almost 80 degrees. Co/Cross discrimination is around 20 dB at the gain peak condition.

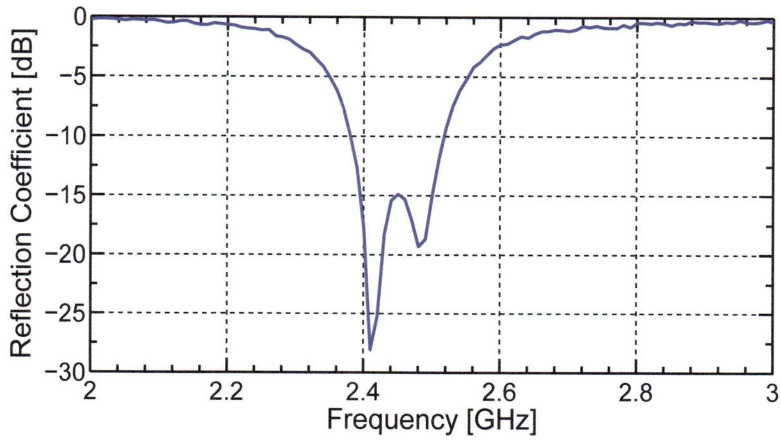

Figure 9. Antenna reflection coefficient versus frequency.

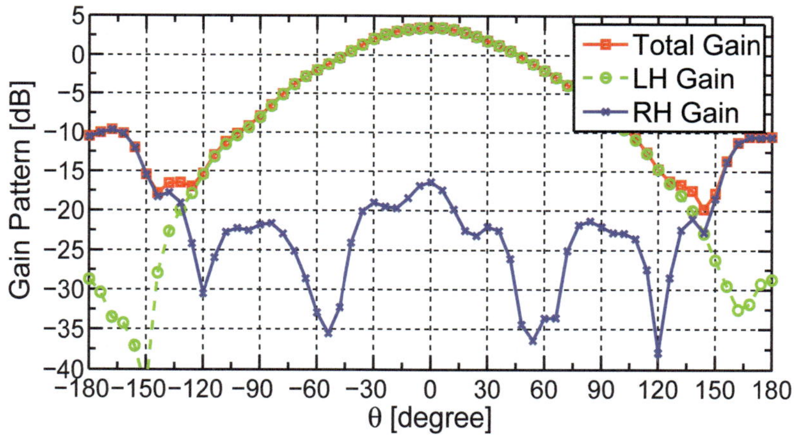

Figure 10. Total, LeftHand and RightHand Antenna Pattern versus Frequency.

3.2. Selection mechanism

The simpler form of RF selection logic is the Single Pole N through (SPNT) switch, directly feeding the N directional antenna (the through) with the the CC RF pin (the pole). The SPNT operation is controlled by the available digital lines taken from the programmable pins of the CC. The RSSI data collection is performed dynamically activating each antenna beam on a scheduling basis, at protocol level.

The adoption of a non-reflective switch is adequate for the SBA operation; the inactive faces are nominally terminated on matched dummy loads without significantly perturbing the

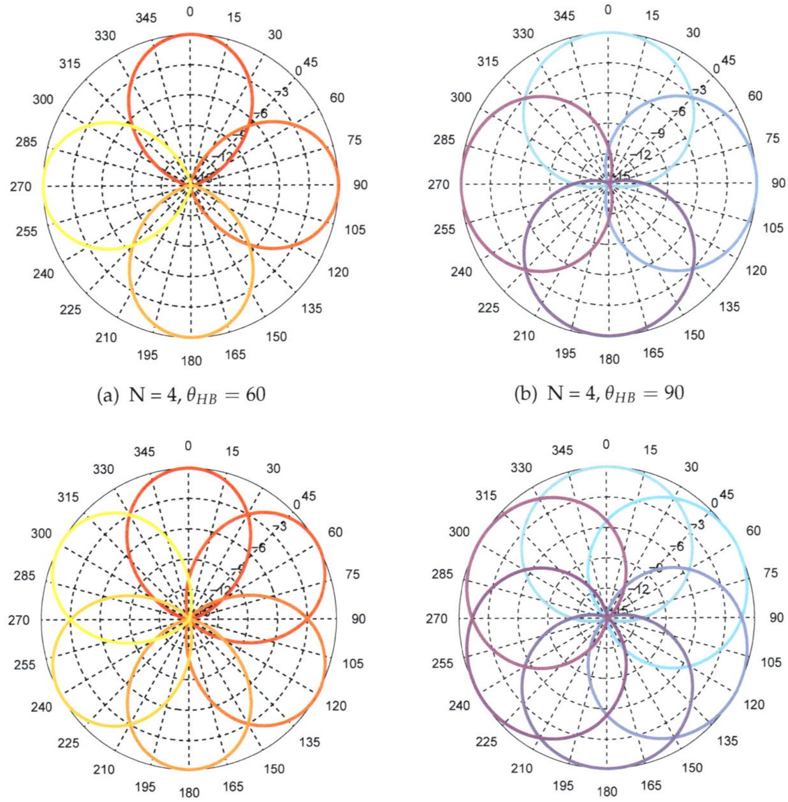

(a) N = 4, $\theta_{HB} = 60$ (b) N = 4, $\theta_{HB} = 90$

(c) N = 6, $\theta_{HB} = 60$ (d) N = 6, $\theta_{HB} = 90$

Figure 11. Regular arrangement of 4 and 6 antenna beams. Each pattern is modeled after Eq. 8, with $\theta_n = 2\pi n/N$.

radiation pattern of the active element. The drawback of this selection mechanism is the RF loss, which is proportional on N, i.e. on the antenna number. This fact poses a limit to the sensibility of the node. Another problem, is the low isolation of the channels, which could lead to coupling between the antennas, thus affecting the pattern generation, and corrupting the antenna polarization.

A suitable commercial SPNT family is provided by hittite [11]. The HMC family is composed by *non-reflective* SPNT which exhibits moderate insertion loss and adequate isolation for the application in exam. For example the HMC252QS24 SP4T, exhibits moderate insertion loss of 1 dB (over the channels) and an isolation in excess of than 35 dB, is adequate for the application of interest.

4. Localization algorithm

This section describes the algorithmic approach suitable for the proposed localization system. The performances of the localization are based on the SBA characteristics and on the radio channel conditions.

The simplest radio link model is the Friis transmission formula, which relate the received power to the transmitted power as a function of the distance between the source and the destination, as well as of many other link parameters [3, 19]:

$$P_{Rx} = P_{Tx} \frac{G_{Rx}(\bar{\theta}) G_{Tx}(\bar{\theta})}{d^2} \left(\frac{\lambda}{4\pi} \right)^2 \tag{2}$$

where P_{Tx} and P_{Rx} are the transmitted and receiver power, G_{Tx} and G_{Rx} are the transmitter and receiver antenna gains – evaluated in $\bar{\theta}$, the direction of the link – and d is the distance between the units. This expression is valid under the *matching condition* hypothesis, i.e. when each antenna is matched to the respective transceiver and they are also matched in polarization sense. The distance term is referred to the wavelength of the signal, which for the ISM central frequency is equal to:

$$\lambda_0 = \frac{c}{f_0} = \frac{c}{2.45\,GHz} = 0.12\,[m] \tag{3}$$

The free space model is approximatively valid only for communication in the far-field condition, when the distance $d/\lambda_0 >> 1$. The Friis formula is most commonly re-casted in dB form, and in this form it is straightforward to add the mismatches as subtraction terms:

$$P_{Rx} = P_{Tx} + G_{Rx}(\theta) + G_{Tx}(\theta) - R_{LOSS} - T_{LOSS} - X_{LOSS} - PATH_{LOSS} \tag{4}$$

Where T_{LOSS} and R_{LOSS} are the reflection loss of transmitting and receiving antennas, and X_{LOSS} is the antenna polarization mismatch, and each quantity is intended in dB sense. The use of CP antenna is and aid to avoid polarization mismatch since CP antenna communicate independently form their relative orientation. The various loss terms are constant for a fixed pair of transceivers, while the $PATH_{LOSS}$ term is:

$$PATH_{LOSS} = n_p \log(4\pi d/\lambda_0) \tag{5}$$

where n_p, called *loss exponent*, is equal to two only for an unobstructed free space condition. In a more realistic scenario – for example in a complex indoor area – the RF link model cannot be taken as simple as the one expressed by eq. 2. It is still possible consider the model valid, but in a statistical sense, and to better account the higher loss rate, the n_p exponent is raised to three, four or even more. DoA estimation is a range-free algorithm, so it can be implemented with minimal assumptions on the propagation model that relate the RSSI to the distance.

The angular diversity mechanism is exploited by a progressive sensing of the incoming message over the set of available antenna beams. Consider the link of an anchor A and a target T. Upon a request message sent by the anchor, the node T responds with a opportunely repeated broadcast message. The anchor, which is now in monitoring phase, sense the incoming power through the set of its N prefixed beams. Operating on these measurements, node A is able to estimate the DoA of the T respect to its A-centric reference system.

Given the small antenna dimensions, all the faces are at about the same distance from the target: this is the common simplification made in the array theory. According to eq. 4, the received power P_n correspondent to the nth beam is:

$$P_n = P_0 + G_n(\theta) \tag{6}$$

where P_0 is the power of the target's message impinging on the SBA – comprehensive of all the loss terms – while $G_n(\theta)$ is the gain of the nth beam evaluated in the transmission angle, specified to a reference fixed to the anchor node.

Considering Figure 10, an analytical model suitable for the description of the antenna beam is the regular cardioid. A cardioid pattern is identified by its maximum gain, the half-power radiation angle, and the nominal pointing direction specified by θ_0:

$$G(\theta) = G_{max}\left(\frac{1}{2}\left(1 + \cos(\theta - \theta_0)\right)\right)^m \tag{7}$$

Where the exponent m is inversely related to the aperture of the beam. This model, while simple, is not unrealistic and it is suitable for numerical implementation [14]. Supposing a regular known arrangement, the nominal pointing direction θ_{n0} of each SBA element is known, and so that the gain pattern of the set of N elements, expressed in dB form:

$$G_n(\theta) = G_{max} + 10m_n \log\left(0.5 + 0.5\cos\left(\theta - \theta_{n0}\right)\right) \tag{8}$$

The radiation pattern of the T node is supposed to be omni-directional, thus it adds the same contribute for each reaching antenna.

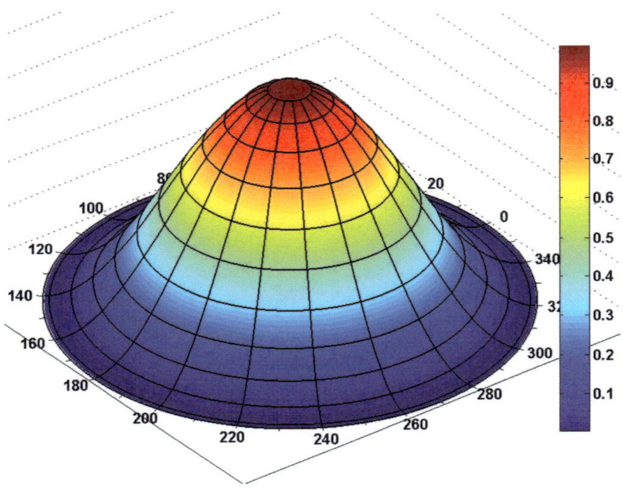

Figure 12. 3D plot of the analytical cardioid model expressed by Eq. 7.

The available data for the algorithm is ultimately a vector of RSSI samples, which can be considered a *function* of the incoming data, as expressed in 2.1:

$$RSSI_n = f\left(G_n(\theta) + P_0 + W_n\right) \qquad n = \{1, \ldots, N\} \tag{9}$$

where W_n is the unavoidable noise term. In a realistic scenario, the noise term accounts also for the RSSI BIAS, with a constant but unknown term, characteristic of the specific device – i.e. $< W_n >= \text{RSSI}_{BIAS}$, $< W_n^2 >= \sigma_W^2$. Typically, the RSSI register is given as an integer. A simple but not unrealistic model for the RSSI is:

$$RSSI_n = \lfloor G_n(\theta) + P_0 + W_n \rfloor \qquad n = \{1, \ldots, N\} \tag{10}$$

where $\lfloor\cdot\rfloor$ is the floor function. Finaly, the RSSI vector are collected in sampled form, with a discretization step ΔT:

$$RSSI_n[k] = RSSI_n(T_0 + k\Delta T) \tag{11}$$

The following section deals with some possible algorithms for the Direction of Arrival estimation, formally identified with $\hat{\theta}$, on the basis of M repetitions of the RSSI vectors. The SBA patterns are supposed to be known and the eventually difference between the nominal and the actual pattern can be considered part of the measurement noise – the *Degree of Imperfection* (DOI).

4.1. Strongest beam

The simplest localization algorithm is based on the classification of the RSSI vector elements. [6, 13]. The Direction of Arrival – which in this case can be more properly defined as the Sector of Arrival – is identified with the antenna domain of the strongest beam, where the n-th antenna domain is defined as the angular range where the n-th antenna gain is higher than the other antennas:

$$\mathcal{D}_i = \{G_j(\theta) > G_i(\theta), \quad j \neq i\} \tag{12}$$

The formal angular estimate is obtained using the reference direction of the antenna that receives the maximum RSSI level, averaged over k = 1...M measures. This estimator comes at zero computational cost, since it consists only in a search and sort of the RSSI vector, and it can be handy as preprocessing stage.

4.2. Least square error

Another estimation strategy is based on the difference between the expected and the actual value [9]. Considering Eq. 10, an *error function* can be defined:

$$\text{err}(\theta) = ||\text{RSSI} - G(\theta)||. \tag{13}$$

This cost function has the form of the N-dimensional error norm, which clearly has a minimum in $\theta = \bar{\theta}$, and this minimum would be nominally equal to the a vector equal to P_0, in the ideal noiseless case. The formal DoA $\hat{\theta}$ can be found by minimizing this cost function in the sense of least square error:

$$\hat{\theta} = \underset{\theta}{\text{argmin}} \left(||\text{RSSI} - G(\theta)||\right) \tag{14}$$

As a side not, the actual minimum of $||\text{RSSI} - G(\theta)||$ is a vector composed of N repetitions of P_0, therefore it can be used to obtain a range estimation, with a reduced uncertainty with respect to the case of the canonical range estimation approach.

4.3. MUSIC

One of the best algorithm for the DoA estimation with a SBA is based on the *Multiple Signal Classification* (MUSIC) algorithm [13, 21], an array signal processing based on the spectral decomposition of the covariance matrix of the power readings on each face, exploiting the signal space projection properties. Classical MUSIC algorithm for DoA assumed a complex signal – i.e. module and phase – but a variant where no phase information is required is possible, which means an algorithm based on RSSI only [14, 15, 24].

The MUSIC assumes readings expressed by the following linear model:

$$\begin{bmatrix} y_1[k] \\ \vdots \\ y_N[k] \end{bmatrix} = \begin{bmatrix} G_1(\theta) \\ \vdots \\ G_N(\theta) \end{bmatrix} x[k] + \begin{bmatrix} w_1[k] \\ \vdots \\ w_N[k] \end{bmatrix} \tag{15}$$

where $y_n(m)$ is the power in linear form estimated on face n upon reception of the mth message – $y_n = 10^{(RSSI_n/10)}$ –, while $x(m)$ is the impinging signal affected by a N-dimensional $W(k)$ noise vector.

The data correlation matrix – R_{yy}– estimated on based of K repetitions, is given by:

$$\hat{R}_{yy} = E\left[y[k]y[k]^{\top}\right] = \frac{1}{K}\sum_{k=1}^{K} y[k]y[k]^{\top} \tag{16}$$

It is possible to demonstrate that:

$$\hat{R}_{yy} = \sum_{m=i}^{M} \sigma_x^2 G(\theta) + \sigma_w^2 I \tag{17}$$

where $\sigma_x^2 = E|x_n^2[k]|$ is the power of the nth signal and $\sigma_w^2 = E|w_w^2[m]|$ is the noise power. Thus, applying the *single value decomposition*, the set of the following matrices is obtained:

$$R_{yy} = USU^* \tag{18}$$

Considering only an incoming signal, a partition of the space spanned by the columns of $U = [U_x, U_w]$ is obtained: We refer to U_x as the *signal subspace*. Similarly, U_n is the *noise subspace*. Since U is a unitary matrix the signal and noise subspaces are orthogonal, so that $U_x U_n = 0$. Thus we define a *pseudo-spectrum*:

$$\mathrm{P}_{\mathrm{MUSIC}}(\theta) = \frac{1}{G(\theta)U_n} \tag{19}$$

that exhibits a sharp maximum for angle supposedly close to the true DoA, formally identified with:

$$\hat{\theta} = \underset{\theta}{\operatorname{argmax}}\, \mathrm{P}_{\mathrm{MUSIC}}(\theta)\,. \tag{20}$$

5. Simulated experiment

In this section, a set of simulated localization experiments is presented. Each communication process involved in the simulation is modeled after eq. 10, considering also the case of the incoming power lower than receiver sensitivity. The geometrical quantities are analytically determined, and the communication noise is generated as a random number with a Gaussian statistic. Each simulation is parametrized by a set of geometrical, physical and statistical parameters. For the sake of simplicity, the only algorithm considered in the following sections is MUSIC, since it has the best performance/cost trade-off.

If the incoming power is lower than receiver sensitivity, which is around -94 dBm for the system in exam, the only possible conclusion is that the transmitter node is out of the receiver range.

5.1. Single node

To validate our choice, the position of single node located in a unobstructed square area of 10×10 m is considered. At the the corners of the area, four anchors are placed at the distance of 1 m to the boundary wall. Each anchor is equipped with a N-SBA ($N > 2$). Each elementary anchor node is supposed to be a patch nominally identical to the ones describes in section 3.1, characterized by a nominal gain G_0, a nominal half beam angle θ_{HB} (degree) and a nominal Front to Back Ratio $F2B$ (dB). The elements are equally distributed around the node ($\theta_n = 2\pi n/N$), and thus the same goes for the radiated beams.

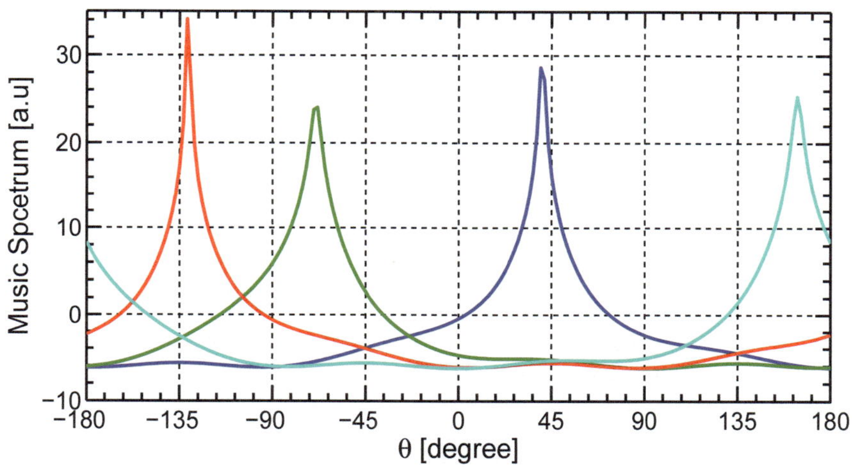

Figure 13. The four Music spectrum for the four anchors.

The target node, equipped with a 2D-omnidirectional antenna, can assume any arbitrary position in the area, identified by the 2D coordinates $(\bar{x}, \bar{y})$. The goal of the localization experiment is the given a position estimatation, indicated with $(\hat{x}, \hat{y})$, under various conditions of the anchors and the channel. The performance of the localization are evaluated on the basis of a localization error:

$$\epsilon = (\bar{x}, \bar{y}) - (\hat{x}, \hat{y}) \tag{21}$$

The first step of the localization procedure is the estimation of the four DoA's, labeled $\hat{\theta}_{i=1,4}$, individuated by the target respect to the four anchors. To fix the idea, the target coordinates are $P = (3, 4)$, a central area of the room. Each Anchor, composed by $N = 4$ antenna elements, performs the communication task on the basis of a scheduled activity. The RSSI collection and the following elaboration leads to a MUSIC spectrum, defined in the 1D angular range of the target. The four music spectra are illustrated in Figure13, where is evident the typical sharp behavior of MUSIC, consequence of its definition (see Eq. 19).

The formal estimation is identified with the maximum of the spectrum, which define a line of bearing. Instead of simply consider the intersection of four line of bearing, another approach is proposed. A modified spectrum is derived after the music estimation. The analytical expression of this auxiliary spectrum is:

$$MAP_i(x,y) = \exp\left(-\left(\frac{\theta_i(x,y)-\hat{\theta}}{\sigma_\theta}\right)^2\right) \quad (22)$$

Where $\theta_i(x,y)$ is a function, defined respect to the i-th anchor, which map each point of the area to the azimuth angle θ seen by the anchor. This map is a function of the area and the anchor position, which are defined before the experiment, thus the map is build in a off-line preprocessing phase. The term σ_θ is taken proportional to the standard deviation of the RSSI message, in order to take account of the intrinsic uncertainty of the link model. This spectrum redefinition leads to the four *music maps*, plotted in Figure14. Each map has a radial behavior, with a linear region of high probability around the line of bearing, and a decreasing behavior departing from this straight line. The product of the four spectra generate a global music spectrum, which synthesize the information of the four anchors. The peak of the spectrum identifies the estimated position of the target, as depicted in Figure15. The advantage of this approach is that a single unreliable DoA estimation is filtered out by the product, while the intersection of two reliable estimation is exalted.

5.2. Sensor network

In typical wireless sensor network application, the localization involves computing the position of the set of numerous nodes in a 2D space. The nodes are supposed to be non interagent, i.e. they not collaborate for the localization service, but they neither interfere. According to the previous analysis we will use the same approach based on the music map. The procedure was evaluated by generating a 13×13 nodes network, regularly spaced in a grid of $8m \times 8\,m$ centered in the same $10 \times 10\,m$ area.

A first simulation is depicted in Figure16. The pathloss exponent is $n_p = 2$ and the σ_{RSSI} is 3 dB. The SBA anchors are made of $N = 4$ antenna elements are characterized by a gain $G_n = 3\,dB$ and a $\theta_{HB} = 50$, while the target is equipped with an regular 2D-omnidirectional antenna ($G_T = 0\,dB$). For each node of the network, a local error is computed as the euclidean norm of the distance between actual and estimated position. In Figure17 an histogram of the global computed errors is depicted. To summarize, with the described condition, a mean absolute error of $0.75\,m$ ($mean_x = 0.51\,m$, $mean_y = 0.50\,m$) and a max absolute error of $1.80\,m$ ($max_x = 1.5\,m$, $max_y = 1.3\,m$) is evaluated. As expected, the minimal error condition is around the center of the area, where the four anchors show the best collaboration effect. The nodes located too close to one of four anchors suffers for a biased estimation, as testified by the error arrows. This was expected, since the combination of the four map is unbalanced. Nevertheless, the error seems systematic, thus eventually resolvable.

To pursue a deeper investigation, a set of experiment was taken with a statistical variation of σ_{RSSI}. The localization performance are expected to aggravate as the standard deviation – and so the noise level – increases. In Figure18(a), the mean and max errors are computed as σ_{RSSI} increases from 0 to 15 dB. From a simple inspection is clear that both of them grow with a regular monotonic behavior, maintaining almost the same ratio of the max error ranging from 2 to 3 times the level of the mean error.

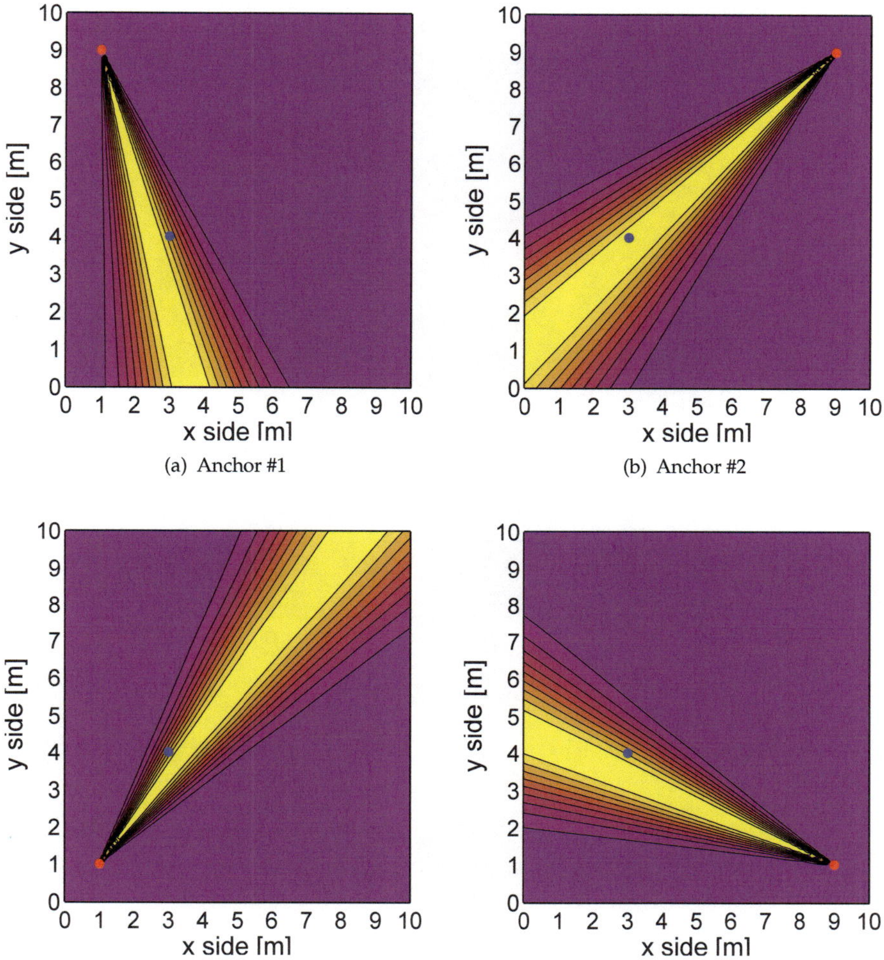

(a) Anchor #1 (b) Anchor #2 (c) Anchor #3 (d) Anchor #4

Figure 14. Four 2D Music maps estimate the target DoA referred to the the four anchor nodes.

Another set of simulation was taken to confirm the behavior of Figure18(a). The simulation is similar, but with a different θ_{HB} value. A narrower beam is expected to improve the localization results, as the angular filtering effect helps the localization mechanism. Figure18(b) and Figure18(c) presents the case of $\theta_{HP} = 50, 70$. The plots confirm the spatial filtering behavior, where a narrower and wider beam lead respectively to a better and worst localization performance, in terms of both mean and max error. Analyzing this data, one can think that a very narrow beam could lead to an even better performance.

Unfortunately, an excessive directivity, hence an extreme beam narrowness, is not always appropriate. As a demonstration, Figure19(a) shows the mean and max localization error for a set of parametric simulation where the free parameter is θ_{HB}, while the σ_{RSSI} is set at 6 dB – adequate to describe a noisy indoor case – and the other geometric and physic

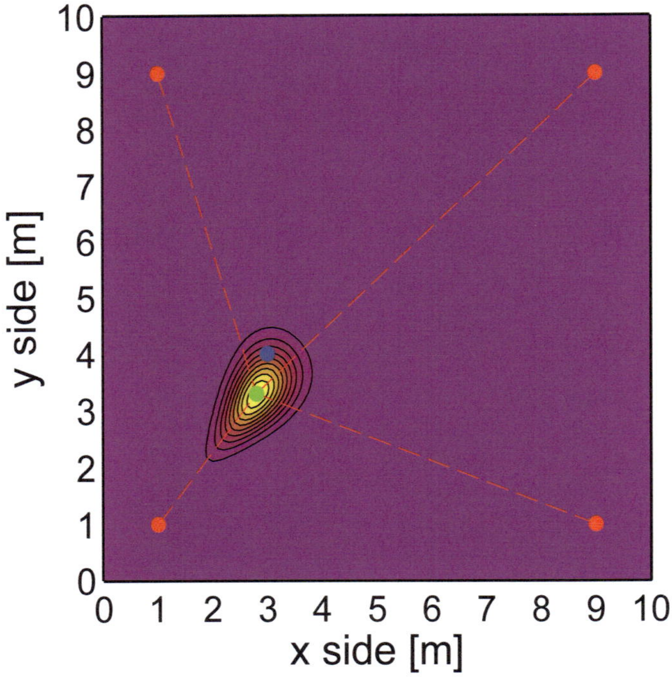

Figure 15. Localization based on the global music map with the four lines of bearing identifying the estimated position.

parameters are the same of the previous simulation. The localization error increases with θ_{HB}, confirming increasing error as the antenna beams enlarge, as expected for this class of DoA estimator. Paradoxically, when the beam is excessively narrow, the error cross a *critical point* where it experiments a absolute minimum, and then grows abruptly reaching an even higher peak. The physical reason lies in the fact that with extremely directive antenna, the angular range between two adjacent narrow beam peaks is low in absolute value, leaving an *uncover* area, where the signal became too weak, leading to an RSSI vector with a very low magnitude – ||RSSI|| – and thus heavily corrupted by the random noise. For the positioning algorithms this is an unbearable condition, where the estimation is affected by a very low accuracy. Figure11(b) and 11(c) depict an optimal situation, where the cumulative pattern covers the entire 2π domain with a maximum small ripple of 3 dB. The case of Figure11(a), while still satisfying, begins to show the problem of excessive directivity. As a further proof of this behavior, the same simulated experiment described by Figure19(a) is repeated in a bigger area of $20 \times 20\,m$. The results of the simulation are depicted in Figure19(b), where the mean and max error are almost doubled.

To avoid this issue, the directivity by itself is not a solution. A better condition is achieved if each antenna element is directive, but at the same time the cumulative pattern of the entire antenna set should show an isotropic coverture, as expressed in section 3. This ensure an RSSI vector whose norm ||RSSI|| is always meaningful, stronger than the noise. To demonstrate this

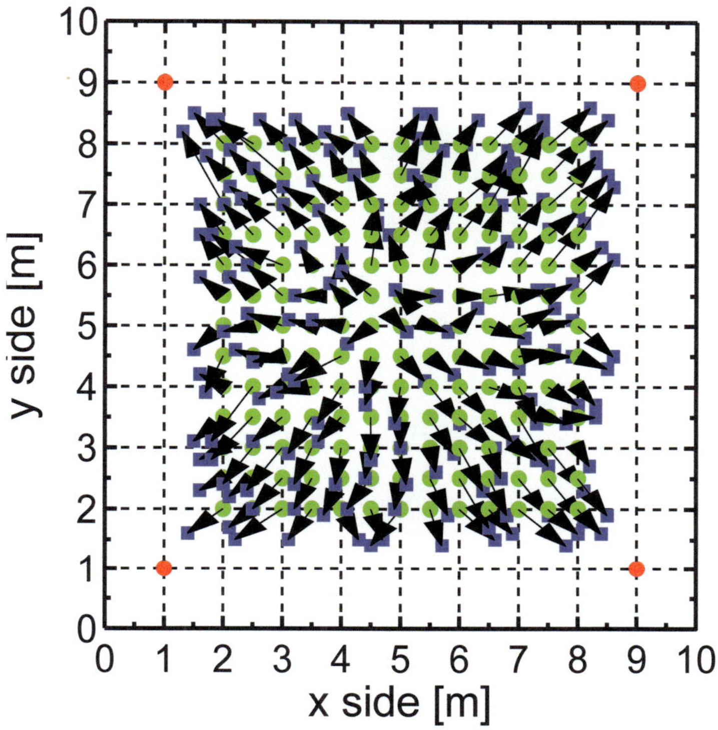

Figure 16. Localization of wireless senor network in a indoor area. The dots represent the actual position, while the square are the estimated positions.

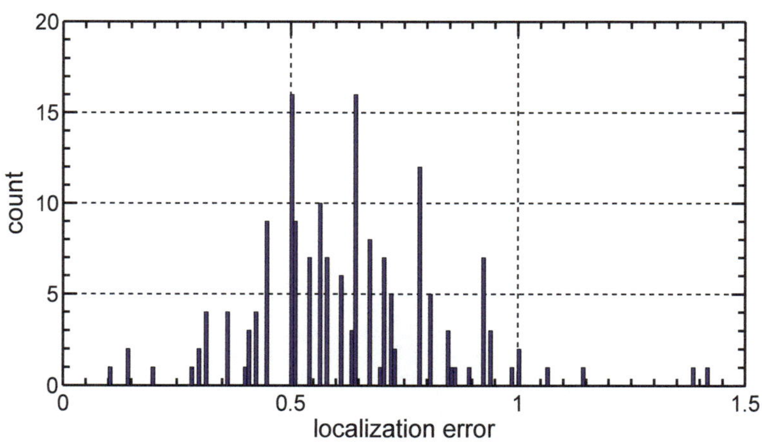

Figure 17. Localization error histogram for the simulation results of Figure16.

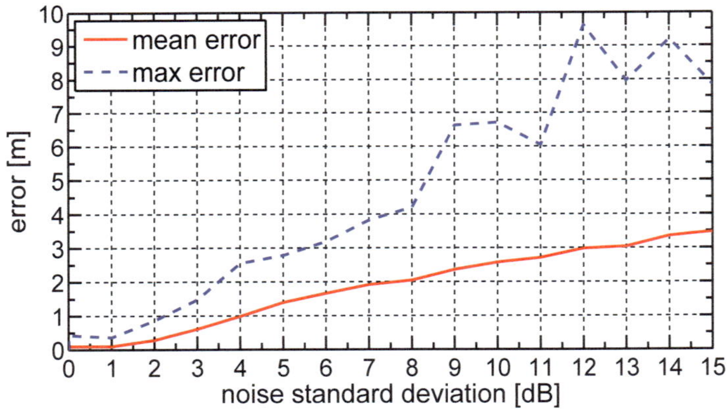

(a) $\theta_{HB} = 30$

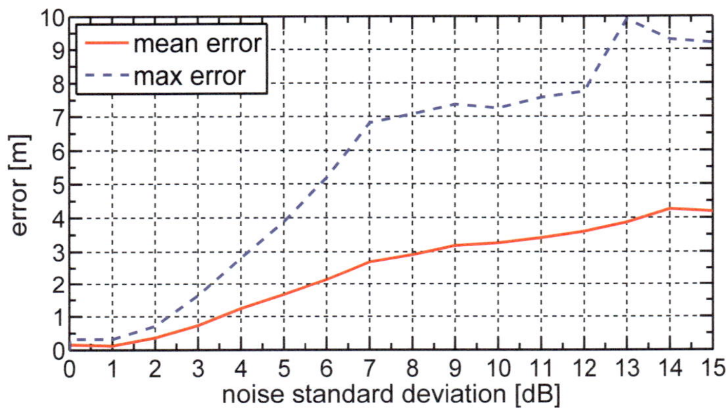

(b) $\theta_{HB} = 50$

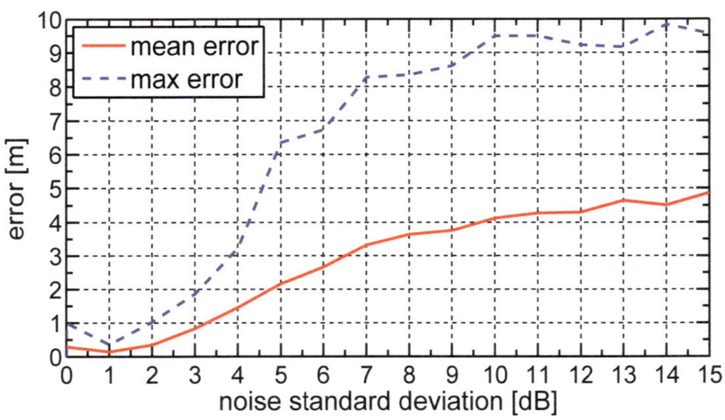

(c) $\theta_{HB} = 70$

Figure 18. Various simulated experiments showing the influence of σ_{RSSI} and θ_{HB} on the mean and max error.

condition, another set of simulation is presented in Figure20. In this case all the conditions are taken constant, with a θ_{HB} equal to 20 degrees and $\sigma_{\text{RSSI}} = 6\,dB$. The free parameter is the antenna number N. As expected, the mean decreases as N grows, and so does the max error, confirming the positive contribute of an increased set of antenna. When N passes a threshold, the contribute of another antenna is ininfluent, leading to a saturated error level. Not only: too much antenna element make the SBA cumbersome, making the SBA unfeasible for various aspects. As a final proof, the same simulation described in Figure20 was repeated with N = 6.

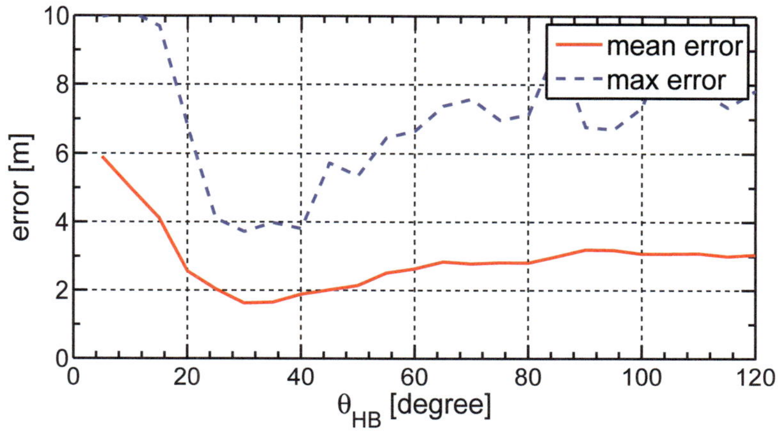

(a) $10\,m \times 10\,m$

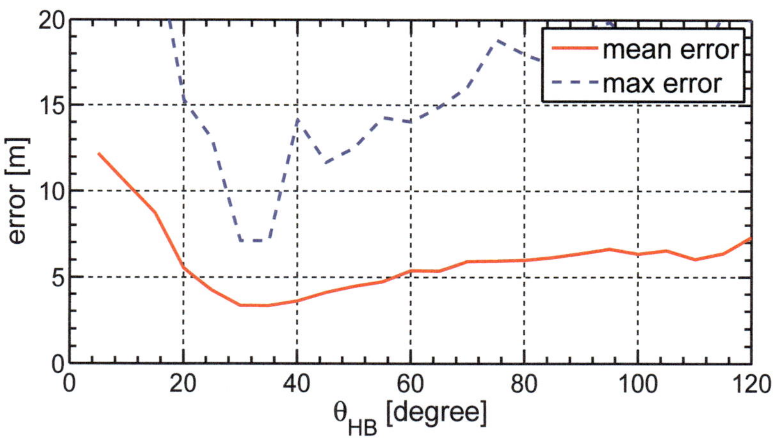

(b) $20\,m \times 20\,m$

Figure 19. Mean and max localization error vs θ_{HB}. N = 4, $\sigma_{\text{RSSI}} = 6\,dB$.

A similar error plot is obtained, but with a lower critical point and a more extreme divergence. With a choice of $\theta_{HB} = 20$, the optimal performance of $mean_{err} = 35\,cm$ are achieved. With the aid of these kind of plot is possible to choose the better parametric condition to ensure a satisfing error level for the application in exam.

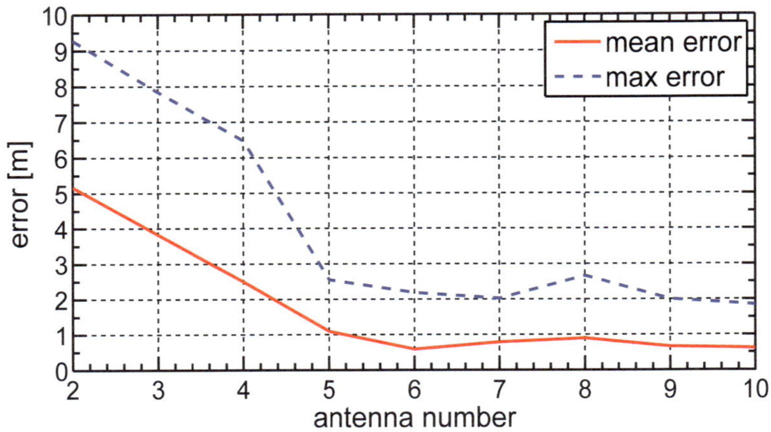

Figure 20. Mean and max localization error vs N. $\theta_{HB} = 20$ degrees, $\sigma_{\text{RSSI}} = 6\,dB$.

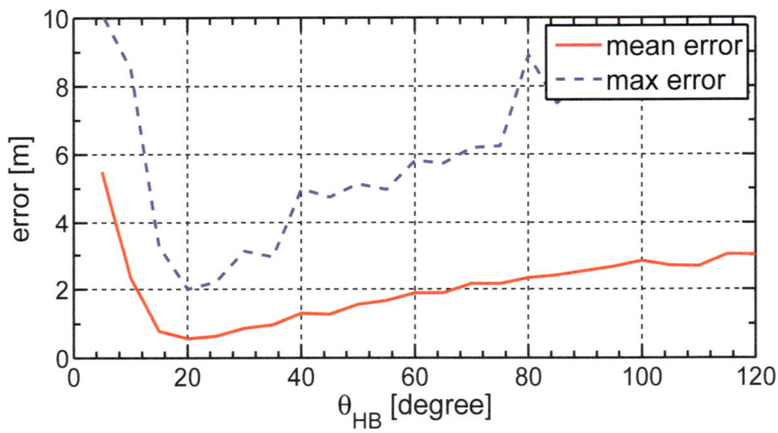

Figure 21. Mean and max localization error vs θ_{HB}. N = 6, $\sigma_{\text{RSSI}} = 6\,dB$.

6. Conclusions

In this paper, the concepts for an indoor localization system suitable for a Wireless Sensor Network in a GPS-denied scenario was presented. The localization of sensor nodes in a necessary tasks to give position-awareness to the nodes, a physical information leading to a vast range of impacting applications. The problem was addressed in terms of systemic approach, suitable hardware, consequent algorithm and estimated performance.

The proposed localization system is based on the Direction of Arrival approach, implemented manipulating Radio Signals. The position of a generic sensor node in the network is estimated on the basis of its RF communication with a master node, without the need of other spatial sensors. Equipped with a complex switched beam antenna, The master node serves as anchor. The localization algorithm relies on the collaboration of a set of anchor nodes, responsible of independent DoA estimations. The proposed algorithm operates on the RSSI-meter, the coarse power meter embedded in almost all the modern commercial wireless devices. In particular, the CC2430 System-on-a-Chip was individuated as the best hardware for the master and the sensor nodes.

Range measurement are extremely inaccurate in indoor area, making traditional range-based algorithm like trilateration fail in hostile area like complex indoor environments. The RSSI data, while affected by numerous noise sources, are suitable for the Direction of Arrival algorithm, which is a *range-free* algorithm, an approach which does not rely on range estimation. The results showed in this paper have demonstrated that a WSN system made of COTS-based nodes is suitable for in-door positioning service. A wide dynamic of cases, with a punctual correlation to the physic and statistical parameters, gives a exemplary design principle for the structure of the nodes, with particular emphasis on sub-antenna system.

Author details

Stefano Maddio, Alessandro Cidronali, Gianfranco Manes
Department of Electronics and Telecommunication - University of Florence, Via S. Marta 3, 50139 Florence, Italy

7. References

[1] Akyildiz, I., Su, W., Sankarasubramaniam, Y. & Cayirci, E. [2002]. A survey on sensor networks, *Communications magazine, IEEE* 40(8): 102–114.

[2] Arora, A., Dutta, P., Bapat, S., Kulathumani, V., Zhang, H., Naik, V., Mittal, V., Cao, H., Demirbas, M., Gouda, M. et al. [2004]. A line in the sand: a wireless sensor network for target detection, classification, and tracking, *Computer Networks* 46(5): 605–634.

[3] Ash, J. & Potter, L. [2004]. Sensor network localization via received signal strength measurements with directional antennas, *Proceedings of the 2004 Allerton Conference on Communication, Control, and Computing*, pp. 1861–1870.

[4] Carver, K. & Mink, J. [1981]. Microstrip antenna technology, *Antennas and Propagation, IEEE Transactions on [legacy, pre-1988]* 29(1): 2–24.

[5] Cidronali, A., Maddio, S., Giorgetti, G., Magrini, I., Gupta, S. & Manes, G. [2009]. A 2.45 ghz smart antenna for location-aware single-anchor indoor applications, *Microwave Symposium Digest, 2009. MTT'09. IEEE MTT-S International*, IEEE, pp. 1553–1556.

[6] Cidronali, A., Maddio, S., Giorgetti, G. & Manes, G. [2010]. Analysis and performance of a smart antenna for 2.45-ghz single-anchor indoor positioning, *Microwave Theory and Techniques, IEEE Transactions on* 58(1): 21–31.

[7] Djuknic, G. & Richton, R. [2001]. Geolocation and assisted gps, *Computer* 34(2): 123–125.

[8] Garg, Bahl, I. [2001]. *Microstrip Antenna Design Handbook*, Artech House.

[9] Giorgetti, G. [2007]. *Resource-Constrained Localization in Sensor Networks*, PhD thesis, Universita' Degli Studi di Firenze, Italy.

[10] He, T., Krishnamurthy, S., Stankovic, J., Abdelzaher, T., Luo, L., Stoleru, R., Yan, T., Gu, L., Hui, J. & Krogh, B. [2004]. Energy-efficient surveillance system using wireless sensor networks, *Proceedings of the 2nd international conference on Mobile systems, applications, and services*, ACM, pp. 270–283.

[11] Hittite [2012]. *http://www.hittite.com/*.

[12] Hofmann-Wellenhof, B., Lichtenegger, H. & Collins, J. [1993]. Global positioning system. theory and practice., *Global Positioning System. Theory and practice., by Hofmann-Wellenhof,*

B.; Lichtenegger, H.; Collins, J.. Springer, Wien (Austria), 1993, 347 p., ISBN 3-211-82477-4, Price DM 79.00. ISBN 0-387-82477-4 (USA). 1.

[13] Krim, H. & Viberg, M. [1996]. Two decades of array signal processing research: the parametric approach, *Signal Processing Magazine, IEEE* 13(4): 67–94.

[14] Maddio, S., Cidronali, A., Giorgetti, G. & Manes, G. [2010]. Calibration of a 2.45 ghz indoor direction of arrival system based on unknown antenna gain, *Radar Conference (EuRAD), 2010 European*, IEEE, pp. 77–80.

[15] Maddio, S., Cidronali, A. & Manes, G. [2010]. An azimuth of arrival detector based on a compact complementary antenna system, *Wireless Technology Conference (EuWIT), 2010 European*, IEEE, pp. 249–252.

[16] Maddio, S., Cidronali, A. & Manes, G. [2011]. A new design method for single-feed circular polarization microstrip antenna with an arbitrary impedance matching condition, *IEEE Transactions on Antennas and Propagation* 59(2): 379–389.

[17] Malhotra, N., Krasniewski, M., Yang, C., Bagchi, S. & Chappell, W. [2005]. Location estimation in ad hoc networks with directional antennas, *Distributed Computing Systems, 2005. ICDCS 2005. Proceedings. 25th IEEE International Conference on*, IEEE, pp. 633–642.

[18] Mukhopadhyay, S. [2012]. *Smart Sensing Technology for Agriculture and Environmental Monitoring*, Vol. 146, Springer Verlag.

[19] Patwari, N., Ash, J., Kyperountas, S., Hero III, A., Moses, R. & Correal, N. [2005]. Locating the nodes: cooperative localization in wireless sensor networks, *Signal Processing Magazine, IEEE* 22(4): 54–69.

[20] Schantz, H. [2011]. On the origins of rf-based location, *Wireless Sensors and Sensor Networks (WiSNet), 2011 IEEE Topical Conference on*, IEEE, pp. 21–24.

[21] Schmidt, R. [1986]. Multiple emitter location and signal parameter estimation, *Antennas and Propagation, IEEE Transactions on* 34(3): 276–280.

[22] Werner-Allen, G., Lorincz, K., Ruiz, M., Marcillo, O., Johnson, J., Lees, J. & Welsh, M. [2006]. Deploying a wireless sensor network on an active volcano, *Internet Computing, IEEE* 10(2): 18–25.

[23] Yick, J., Mukherjee, B. & Ghosal, D. [2008]. Wireless sensor network survey, *Computer networks* 52(12): 2292–2330.

[24] Zekavat, R. & Buehrer, R. [2011]. *Handbook of Position Location: Theory, Practice and Advances*, Vol. 27, Wiley-IEEE Press.

Chapter 8

One Stage Indoor Location Determination Systems

Abdullah Al-Ahmadi and Tharek Abd. Rahman

Additional information is available at the end of the chapter

http://dx.doi.org/10.5772/50522

1. Introduction

Recent advances in communication technologies have great impact on location determination systems. Location determination systems are deployed in almost every building, from hospitals where the location of patients and doctors or any medical equipment can be determined, or sending information to customers based on their location and capability, to organize the traffic and reducing congestion in the highways.

Global Positioning System (GPS) is a standard location determination system that enables the users to locate themselves in outdoor environments. Unfortunately, GPS does not work well in indoor locations because it requires a line-of-sight between the mobile station and the satellites. Hence, an alternative approach is to use a specialized positioning infrastructure that was built exclusively for positioning purposes.

Although these systems generally provide high accuracy rate, but in many situations, they suffer from high cost, and they require an extensive work to build and for maintenance. On the other hand, recent researchers have taken advantages of the available wireless LANs infrastructure, which was built solely for the communication purposes in the first place, and try to develop their positioning systems on the top of the wireless LANs [4]. The main advantages of this approach are low cost, easy to maintain and the availability of WLAN in almost every building.

These systems are commonly called *Off-the-shelf* positioning systems. The basic idea of this approach is to collect radio *fingerprints* at random or predetermined *reference points* to build a *radio map*. These fingerprints are usually a collection of received signal strengths (RSS) combined with their (x, y) coordinates. In the next phase, the system reads the current RSS with unknown location and search for the nearest value in the radio map. The search for the nearest value may include a single or a set of values.

Various positioning systems have been developed using different positioning techniques such as Proximity Sensing, Lateration and Angulation techniques. But due to the complex nature of the indoor environments, it represents an obstacle for such techniques to be applied for indoor

INTECH
open science | open minds

© 2012 The Author(s). Licensee InTech. This chapter is distributed under the terms of the Creative Commons Attribution License http://creativecommons.org/licenses/by/3.0), which permits unrestricted use, distribution, and reproduction in any medium, provided the original work is properly cited.

positioning systems. This obstacle is known as the multipath phenomenon that includes reflection, diffraction and scattering.

The improvements of the fingerprinting approach are the way of representing the reference points in the radio map. These reference points could be represented either by a single value or by a collection of RSSs values. Another way to improve this approach is to collect a large number of fingerprints or to use other properties of the radio signals such as the Bit Error Rate (BER), Signal-to-Noise Ratio (SNR), Time of Arrival (ToA), Time Difference of Arrival (TDoA), Angle of Arrival (AoA) or Phase of Arrival (PoA). Some research have suggested that by combining two or more radio signal properties, this resulted in a better performance but an increased in system complexity.

Figure 1 shows the structure of the conventional location determination systems and the single-phase systems. In the conventional systems, an off-line training phase is required in order to build a radio map, the radio map is constructed either by empirically collecting a single or multiple RSS fingerprints at predetermined or random anchor points, or by using a model based indoor propagation model [15]. In the on- line phase, the location is estimated by matching the RSS fingerprint stored in the radio map with the RSS measured in the real time. Examples of such systems can be found in [2, 4].

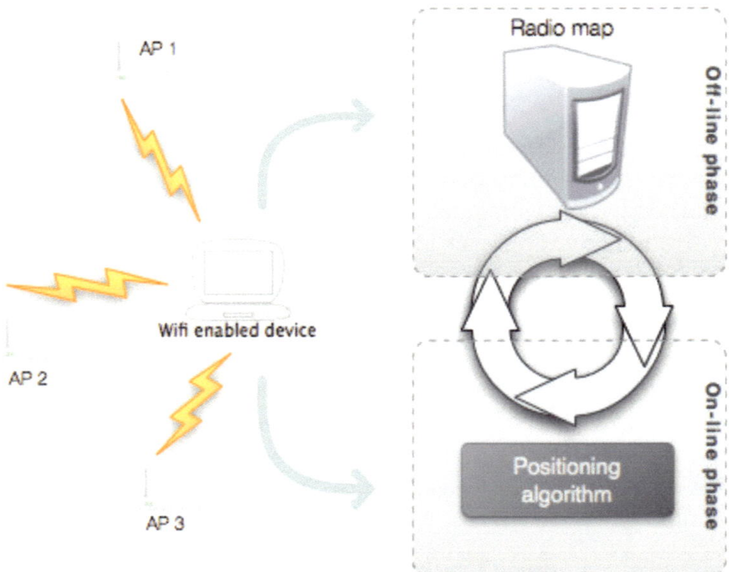

Figure 1. Structure of location determination systems showing conventional system

Single-phase systems do not require an offline phase; instead, they use the online RSS fingerprints to estimate the target's location. Usually such systems compromise the offline stage for the system's accuracy. A zero configuration system proposed in [21] uses the online RSS readings between WLAN Access Points (APs) and between mobile terminals and their AP neighbors. The system suggests a singular value decommission (SVD) technique to implement a mapping between RSS fingerprints and the true location. Table 1 shows a comparison between conventional and single-phase location determination systems [35].

System	Type	Observable	Accuracy
RADAR[4]	Conventional	RSS	2 - 3 m
Ekahau	Conventional	RSS	3.1 m
Nibble[6]	Conventional	SNR	10 m
Lim H. et al.[21]	Single-phase	RSS	2.57 m
Mazeulas et al. [25]	Single-phase	RSS	~ 4 m.

Table 1. Comparison between conventional and single-phase location determination systems

Mazeulas et al. [25] presented a single phase indoor location determination system that first searches for the best propagation model that is proper for the current environment, and then estimates the target's location from the online RSSs obtained by the mobile target from the available APs using lateration techniques. The system estimates the distance d from each AP by calculating the maximum likelihood as follows:

$$\hat{d} = 10^{\left(\alpha - P_{R_i}\right)/10n_i} \tag{1}$$

where α is a constant depends on various indoor phenomenon's such as multipath and shadowing, P_{R_i} is the RSS from the ith AP and defined as follows:

$$P_R = \alpha - 10 \cdot n \cdot \log_{10}(d) + X \tag{2}$$

where n is the multipath exponent. The system achieved an average location error lower than 4 m. Off-the shelf location determination systems use the existing wireless LAN infrastructure deployed in almost every building. The main advantage of these kinds of indoor location determination systems is the ease of installation and deployment. A popular location determination system of this category is the Horus system proposed in [37], it is a probabilistic positioning system that uses a location clustering technique to reduce the computational cost by grouping the locations in the radio map based on the APs covering them. During the on-line stage, the system uses discrete space estimator which estimates the target's location x by finding value that maximizes the probability of obtaining location x given a signal strength vector s, $P(x|s)$. Correlation between RSS fingerprints from each AP was introduced to improve the system performance.

Specialized location determination systems do not use the building WLAN infrastructure, instead they developed a sophisticated devices such as Active Badges [36], Bats [10] and Crickets [30]. The Cricket location system is an example of specialized indoor positioning systems, it consists of a number of crickets, which serve as listeners that scan for data coming from anchor points called beacons deployed through the building. The system uses a combination of ultrasound and radio signals emitted from these beacons to locate the target. Since the RF signals are faster the ultrasound impulse, it is used as indication for the arrival of this ultrasound impulse, which can be used to calculate the distance between listeners and beacons [20]. A sensor fusion technique was proposed in [26] in order to improve the accuracy of the Cricket system by using multi-sensor of four listeners covering a horizontal plane angled at 90° from each other, a 0.3 m accuracy was achieved using the sensor fusion technique compared to 10.8 m location error in the original Cricket system. Although the small location errors achieved by specialized systems, the high cost of such systems represents an

System	Type	Accuracy	Cost
Cricket [30]	Specialized	Low	High
Mitilineos et al. [26]	Specialized	High	High
Bats [10]	Specialized	Medium	High
Horus [37]	off-the-shelf	Medium	Low

Table 2. A comparison between specialized and off-the-shelf location determination systems

obstacle for such systems to be used. Table 2 shows a comparison between specialized and off-the- shelf location determination systems.

Our main objective in this research is to design and evaluate an indoor location determination system that is capable to locate a mobile terminal in a multi-floored building using probabilistic Bayesian graphical models, the proposed system will be compared with similar positioning systems.

2. Positioning systems overview

Pahlavan et al [28] presented a general block diagram for the common components of a positioning system as shown in Figure 2. Firstly, the different positioning systems use different types of received signals ranging from radio signals, ultrasound to infrared. Location sensors such as mobile terminals collect these signals in order to produce an informative data. These informative data can be in form of Received Signal Strength (RSS), Angle of Arrival (AOA) or Time of Arrival (TOA). The produced data will be used to compute the location of the mobile terminal using positioning algorithms, which can be either a Deterministic or Probabilistic method. The estimated location can be symbolized by an (x, y) coordinates or a descriptive location. Finally, the display system displays the estimated mobile terminal's location in a textual of graphical form.

3. Received signal technologies

There are different sensing technologies used in indoor positioning systems. These technologies are affected by the indoor multipath phenomenon, which includes diffraction, reflection and scattering. The most common used sensing technologies are [34]:

3.1. Radio Frequency (RF)

The ability of RF signals to penetrate the walls and floors attract researchers to develop their positioning systems based on RF technology. RF Signals have also a good coverage area of 10 to 30 m compared to other technologies, which means fewer numbers number of sensors are required to cover a certain area. RF signals also have a high speed of 3×10^8 m/s. Since most buildings are equipped with wireless LAN technology, such systems can then be developed on top of these networks without extra equipment. This indicates the ability to develop a low cost positioning systems. Another advantage to using WLAN based positioning systems is that most of these networks operate in 2.4 GHz unlicensed frequency, which can reduce the interference with other devices [34].

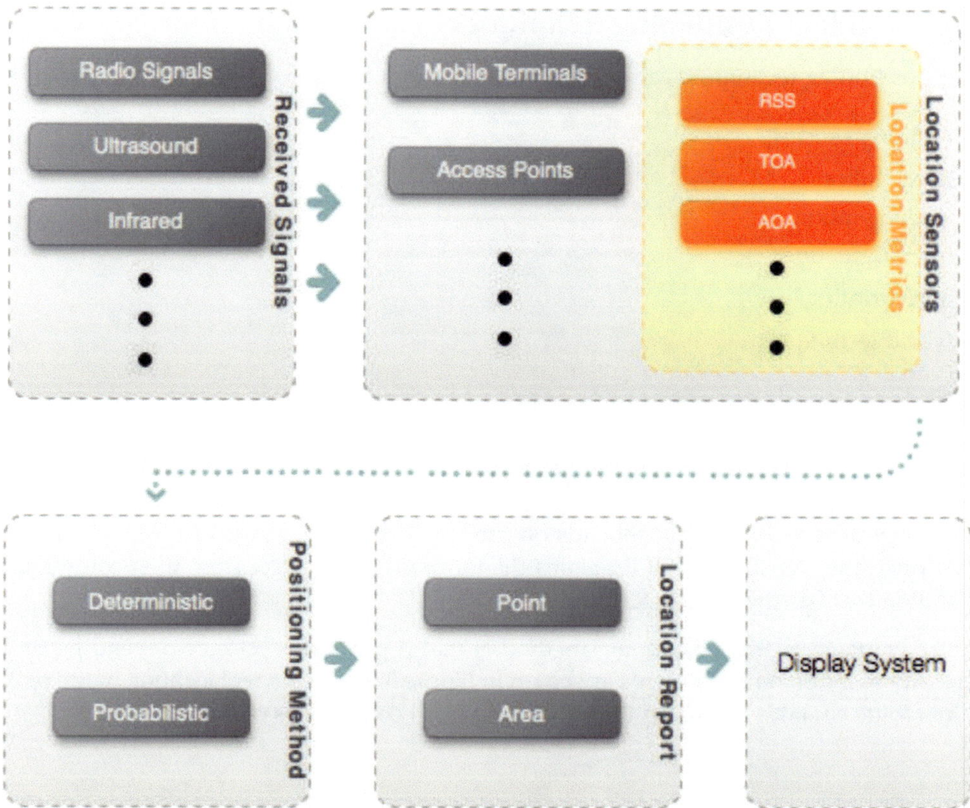

Figure 2. The basic structure for positioning systems [28].

3.2. Infrared

Although infrared signals have the same high speed as the RF signals of 3×10^8 m/s, Infrared signals are interfered with the ambient light. The properties of Infrared signals which are inability to penetrate walls and a limited range of 5 m may be considered as an advantage in some systems where it can provide coarse grained area accuracy by implementing special devices in each room. These devices are called beacons, they transmit signals every 10 seconds, which makes these devices consume low power.

The disadvantages of developing an Infrared based positioning system lies in maintenance time required to keep these beacons work properly and the high installation cost.

3.3. Ultrasound

Since ultrasound wave travels at a low speed of about 345 m/s, it is used in positioning systems by measuring the travel time between the transmitter and the receiver. These signals usually operate between 40 and 180 kHz. The same as Infrared signals, Ultrasound waves have a short coverage range of about 3 to 10 m and could be reflected by the walls. In addition, Ultrasound waves also affected by the environment temperature.

4. WLAN indoor positioning techniques

Indoor positioning techniques can be summarized into four main positioning methods [3]:

(a) Proximity sensing.

(b) Triangulation

  (i) Lateration

  (ii) Angulation

(c) Fingerprinting

(d) Hybrid techniques

4.1. Proximity sensing

Proximity sensing is considered to be the simplest positioning technique because it does not require any modification to the existing network infrastructure, it can be either used in the cellular networks or WLAN. It depends on the small coverage range of the radio, the idea behind proximity sensing is that it obtains the location of the target from the position of the base station that has the highest RSS.

The disadvantage of this positioning method is that it provides accuracy depending on the AP density in indoor environments as shown in Figure 3 where the real location of the mobile terminal (blue circle) is estimated to be the same (x, y) coordinates of AP4.

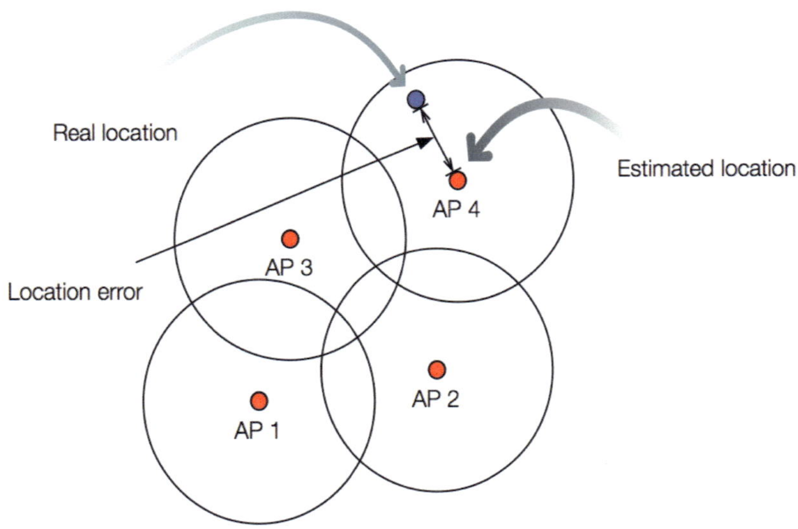

Figure 3. Proximity sensing technique showing the real location in blue and the estimated Bayesiand location in red [3].

4.2. Triangulation

The triangulation methods calculate the location of the mobile terminal either by a set of radial distances (Lateration) or a set of angles (Angulation).

4.2.1. *Lateration*

Lateration positioning method measures the distance between the mobile terminal and a set of at least three reference points (RP) as shown in Figure 4. Different signal metrics are used to estimate the location such as Time of Arrival (TOA), Time Difference of Arrival (TDOA) and others.

Lateration technique assumes that the distance d_i between the mobile terminal and a number $i = 1, 2, \ldots, n$ of RPs is known. In Figure 4(a) where there is only one RP, then the estimated location is considered to be any location point on the circle's perimeter. In Figure 4(b), the intersect between the two circles representing RP_1 and RP_2 reduces the mobile terminal's location uncertainty to only two possible locations. By adding another RP as shown in Figure 4(c) it can produce a single location estimation in which can be calculated by using the Euclidian distance equation:

$$d_i = \sqrt{(X_i - x)^2 + (Y_i - y)^2} \quad (3)$$

where (X_i, Y_i) is the coordinates of the ith RP and (x, y) is the coordinates of the mobile terminal.

4.2.2. *Angulation*

Angulation technique calculates the angle θ_i between the target the the ith RP. θ is called Angle of Arrival (AOA) or Direction of Arrival (DOA). Unlike the lateration technique, this method requires at least two RPs to locate a mobile terminal as shown in Figure 5.

In Figure 5(a), θ_1 is the angle of the transmitted signal calculated at the RP. Although it is known the direction from which this signal has been sent, but the distance is unknown and considered to be at any point along the line between the mobile terminal and the RP. By adding another RP as in Figure 5(b), then the estimated location is the intersection between the two lines.

Triangulation techniques usually require a line -of-sight between the transmitter and the receiver, which is unavailable in indoor environments most of the times. Therefore, they cause the multipath phenomenon where the signals are received from multiple sources. This disadvantage of triangulation techniques prevent the researchers from developing indoor positioning systems based on this techniques.

4.3. Fingerprinting

Fingerprinting techniques is also called pattern recognition techniques. In general, every fingerprinting techniques woks in two stages, *Offline* and *Online* stages.

In the offline stage, the test bed is covered by a set of predetermined or random points called *reference points*. At each reference point, the user must collect a set of readings, each set contains the coordinates of that point and *signal to noise ratio* (SNR) or in the most popular systems the *received signal strength* (RSS) values from multiple APs and then store these readings in a server - in case of network based systems - or in the target device. In the online

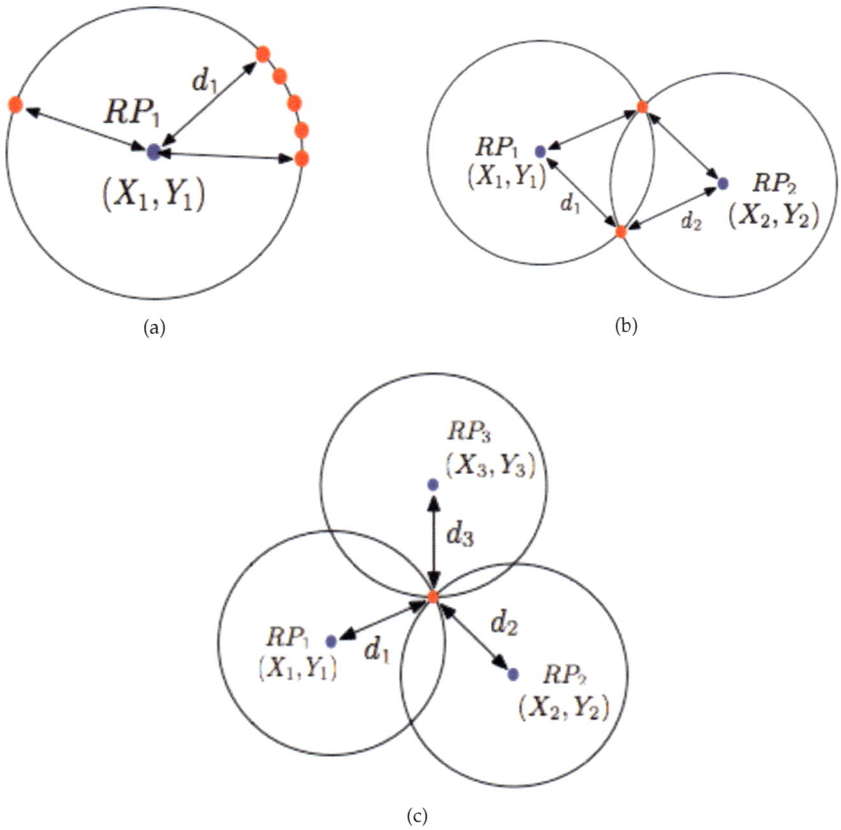

Figure 4. Lateration positioning method where the blue circles represent the reference points and the red circles symbolize the mobile terminal in (a) single reference point (b) two reference points and (c) three reference points[3].

stage and when the target's location is needed, the target collects a set of RSS readings and try to match them with the stored fingerprints from the offline stage.

5. Category of indoor positioning systems

Indoor positioning systems can be divided into two main categories. Either by the infrastructure they implemented in or by the positioning algorithms. Each of these categories are subdivided into many sections.

5.1. Based on infrastructure

Figure 6 shows the categories of indoor positioning systems based on the their infrastructure. They are divided into infrastructure based and infrastructure-less or decentralized. In the Infrastructure based positioning systems, the target's location is determined using the installed network infrastructure in the testbed whereas the decentralized indoor positioning systems locate the target's location in an ad-hoc network setup.

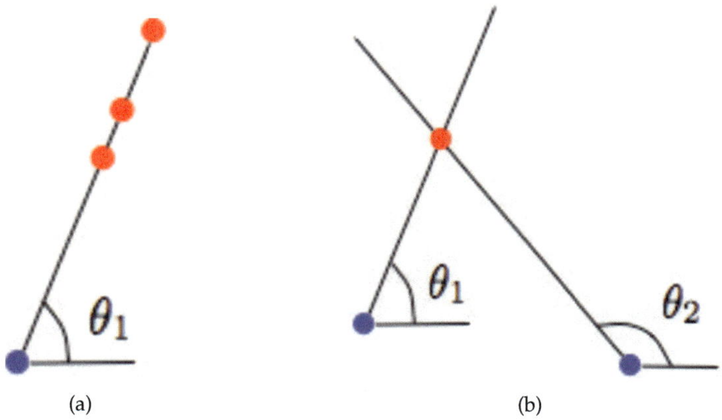

Figure 5. Angulation technique where the blue circles represent the reference points and the red circles symbolize the mobile terminal in (a) one reference point and (b) two reference points [3].

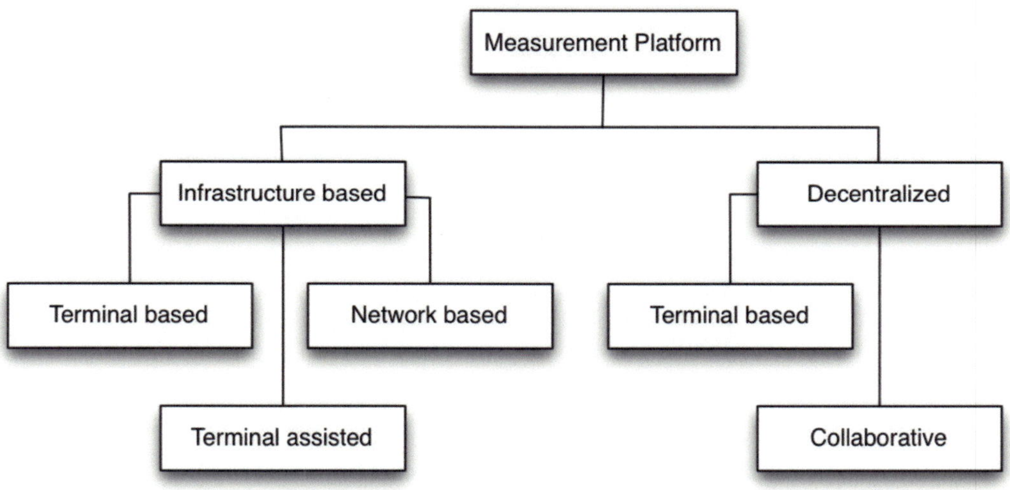

Figure 6. Category of Positioning systems based on their measurement platform [16].

5.1.1. Infrastructure based positioning systems

This category can be divided into three subcategories as shown in Figure 6. The main differences between these three categories are based on the device used to transmits the signal, the devices used for measurements and estimation [16].

In terminal based systems [29]. The signals are sent by base stations, the mobile terminals are then collect the signals, store them and estimate their location. The advantage of these systems is the privacy they provide where the location of the mobile terminal is exclusive to the users only. In terminal assisted systems [6], the signals are also sent by the base stations but the mobile terminals only collect the signals and send them to a network server where the estimation process occur. Finally, in network based systems [19], the signals are sent and collected by both base stations and mobile terminals. The data are then sent to a network server where the data will be stored and location will be estimated. This property can be beneficial in reducing the computation cost and power consumption on the mobile terminals.

5.1.2. *Decentralized positioning systems*

In decentralized systems, a special devices act as base stations spread all over the targeted areas in a grid like [30] or randomly distributed [22] in an ad-hoc setups. The purpose of developing such systems is to enable localizing without a prior knowledge about the building layout. This is important in situations where the WLAN infrastructure of a building get damaged because of fire. This category is divided into two categories, *terminal based* where the beacons send signals to a server terminal to calculate the target's location and *collaborative* systems in which the beacons send the signals in order to perform the estimation process.

5.2. Based on positioning algorithm

This category is divided into two subcategories [17], *deterministic* and *probabilistic* algorithms. The main difference between these two subcategories is the way they model the signal properties.

5.2.1. *Deterministic systems*

In deterministic methods, the estimated locations are represented by a single value such as the average RSS.

Nearest Neighbor in Signal Space (NNSS) is one example of deterministic methods, the target's location is estimated by applying the Euclidian distance algorithm between the nearest value of the signal property stored in the radio map and the current one. The drawback of this method lies in some conditions where the replication of the same stored values for different locations due to multipath phenomenon.

k-Nearest Neighbor (k-NN) was introduced to overcome the limitation of NNSS algorithm where k is set of number of signal properties. k-NN works by first searching for the k-values in the radio map having the smallest error mean with the current signal property [17].

5.2.2. *Probabilistic systems*

T. Roos *et al* [31] have introduced a probabilistic approach for the location estimation problem. The approach is based on calculating the conditional probability distribution of getting a location l given a signal value SV using the Bayes' theorem:

$$P(l \mid SV) = \frac{P(SV \mid l)\,P(l)}{P(SV)} \tag{4}$$

The Bayes' theorem consist of three probability distributions:

(a) Posterior distribution $P(l \mid SV)$: is the knowledge about unknown parameters. It is the product of the prior distribution and the likelihood function [8].

(b) Prior distribution $P(l)$: represents the previous knowledge about the random variable l before obtaining any new information.

(c) Likelihood Function $P(SV \mid l)$: the probability value for random variable SV after obtaining additional information about the location variable l.

The Bayes' theorem will search for the location l which will maximize the posterior distribution $P(l \mid SV)$ and consider this value as the estimated location:

$$argmax_l \left[P(l \mid SV)\right] = argmax_l \left[\frac{P(SV \mid l)\, P(l)}{P(Sv)}\right] \quad (5)$$

6. Characteristics of RSS in indoor environments

Signal strength in indoor environments is difficult to predict due to some multipath phenomenon such as reflection, diffraction and scattering. The indoor multipath phenomenon occurs when the signals are sent from the transmitter arrive at the receiver from multiple directions. Generally, there are three main phenomenon as shown in Figure 7

1. Reflection: it occurs when the signal waves collide on a smooth surface object that has dimensions larger than the signal's wavelength.
2. Diffraction: when the signal waves hit an objects with sharp edges, it causes them to diffract off these objects to various directions.
3. Scattering: when the signals impinge on an object that has a rough surface causing them to scatter.

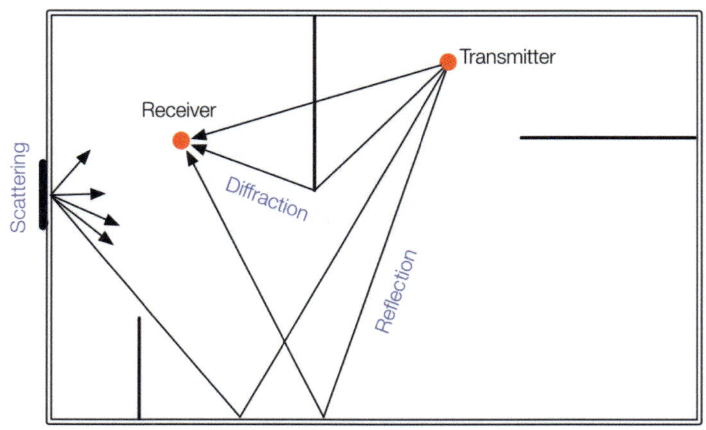

Figure 7. Indoor Multipath Radio Propagation

The RSS distribution in indoor environments are believed to follow a lognormal distribution [32] due to the similar values for the mean, median and mode. Figures [8, 9] and Table 3 , show the RSS histograms at the same location and the statistics from two APs in the first floor of WCC building, respectively.

There are three factors that have an impact on the RSS propagation in multi-floored buildings [33]:

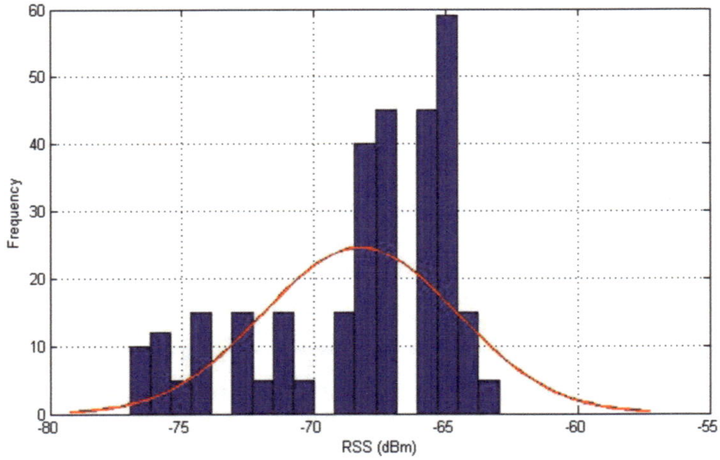

Figure 8. RSS histogram at fixed location for five minutes from AP1

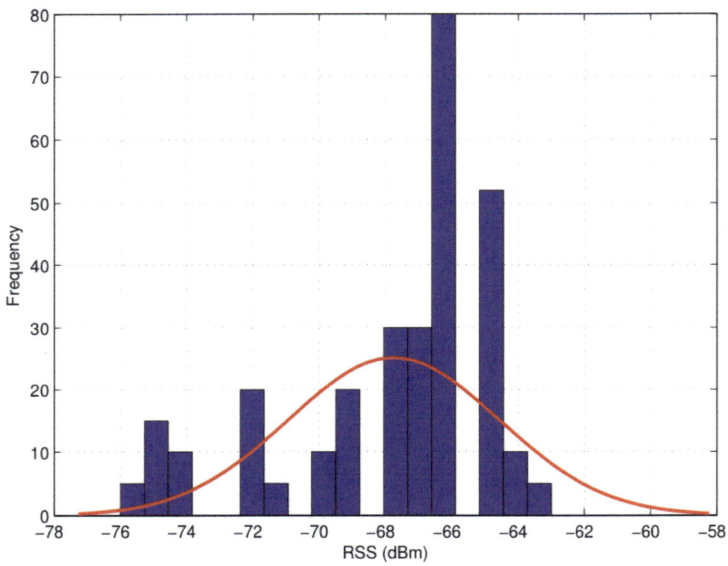

Figure 9. RSS histogram at fixed location for five minutes from AP2

- **Floor Attenuation Factor**: the radio signal that arrives at the receiver after passing through floors.
- **Multiple diffraction at window frames**: the diffracted signals at window frames from different floors.
- **Reflected signals from nearby buildings**: the reflected signals from adjacent buildings.

A concrete floor can reduce the RSS approximately 15 dBm to 35 dBm [18], to investigate this, we conduct a measurement at two vertically location in two different floors from the same AP, the AP5 is placed in the center of the building as shown in Figure **??**, Table 5 shows the

	AP1	AP2
Minimum (dBm)	-77	-76
Maximum (dBm)	-63	-63
Mean (dBm)	-68.25	-67.75
Median (dBm)	-67	-66
Mode (dBm)	-65	-66

Table 3. Statistical values for two APs in the first floor showing the similarity between the mean, median and mode.

	Centre 1st floor	Centre 2nd floor
Mean (dBm)	-72.18	-47.21
Median (dBm)	-72	-47
Std. Deviation	3.592	1.467

Table 4. The floor attenuation factor effect in the centre of the building from AP5

statistics of FAF effect at the centre of the building from AP5, the attenuation on the RSS in the first floor was about 25 dB as shown in Figure 13. The APs installed are DWL-2000 APs operate at frequency between 2.4 GHz and 2.4835 GHz.

Figure 10. First floor layout with 3 APs

Figure 13 shows the effect of FAF in addition to the diffracted signals arriving from AP4 which is placed near a window in second floor, the receiver was in a vertical place from AP4 in first floor. From Table 5, the attenuation on the RSS near windows was similar to the attenuation achieved in the centre.

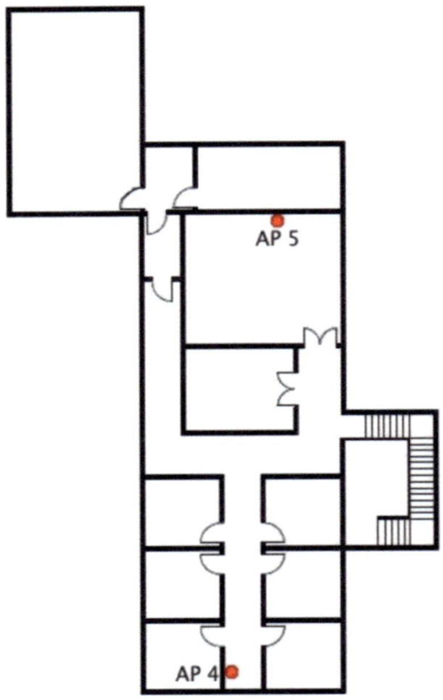

Figure 11. Second floor layout with 2 APs

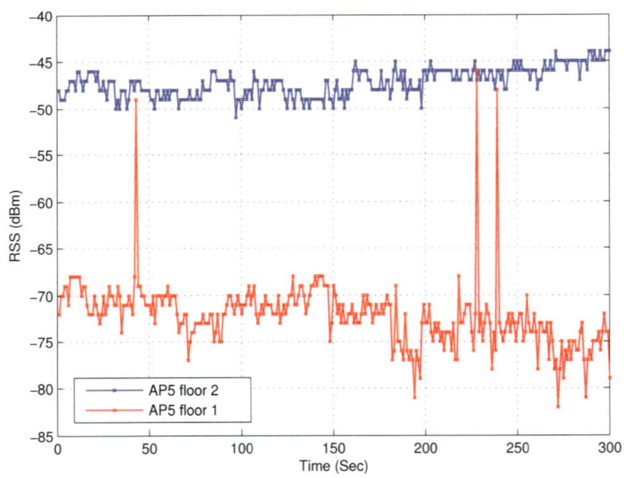

Figure 12. Floor attenuation factor at two vertically location in two different floors from AP5

7. Probabilistic Bayesian graphical models

7.1. Bayes' theorem

Bayes' theorem describes the relationship between the conditional probability and the joint probability of random variables [27]. Let α and β be two random variables in which $P(\beta) > 0$,

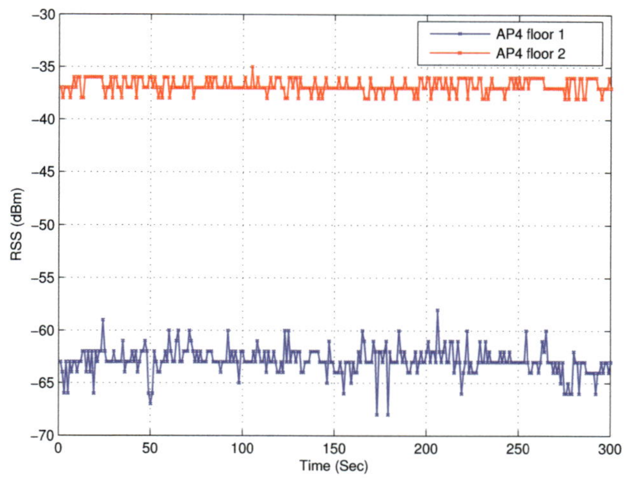

Figure 13. Floor attenuation factor at two vertically location in two different floors from AP4.

	Window 1st floor	Window 2nd floor
Mean (dBm)	-62.89	-36.79
Median (dBm)	-63	-37
Std. Deviation	1.315	0.6753

Table 5. The floor attenuation factor effect in the centre of the building from AP5

then the conditional probability of an event α given event β is :

$$P(\alpha \mid \beta) = \frac{P(\beta \mid \alpha) P(\alpha)}{P(\beta)} \tag{6}$$

$P(\alpha \mid \beta)$ is called a posterior distribution which is a result of the prior distribution $P(\alpha)$ multiplied by the likelihood $P(\beta \mid \alpha)$, the prior distribution represents the previous knowledge about a random variable before obtaining any new information, where the likelihood is the probability value for a certain random variable after obtaining additional information.

7.2. Bayesian networks

A Bayesian Network [13] represents a set of probability distributions, Figure 14 shows a simple graphical model, the nodes symbolize random variables α, β and γ, where the arrows represent the relationships between these random variables. The joint density for random variables in Figure 14 is:

$$P(\alpha, \beta, \gamma) = P(\beta \mid \alpha, \gamma) P(\gamma \mid \alpha) P(\alpha) \tag{7}$$

From Equation 7, the random variable α is considered to the a parent for node γ, and γ is a child for node α. A parent variable is the direct influence on its children, the joint density between parent nodes and their children could be expressed as follows:

$$P(\Delta) = \prod_{\delta \in \Delta} P(\delta \mid parent(\delta)) \tag{8}$$

In some graphical models, we use a plate to handle the replication of random variables. In

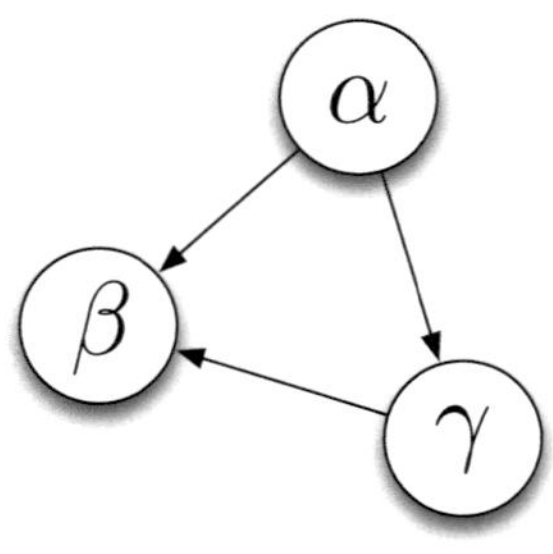

Figure 14. A simple graphical model showing three random variables represented by circles and relationships symbolized by arrows

Figure 15, we show the same graphical model where the replication of random variable μ in 15(a) was handled in 15(b) by a plate notation to some index l.

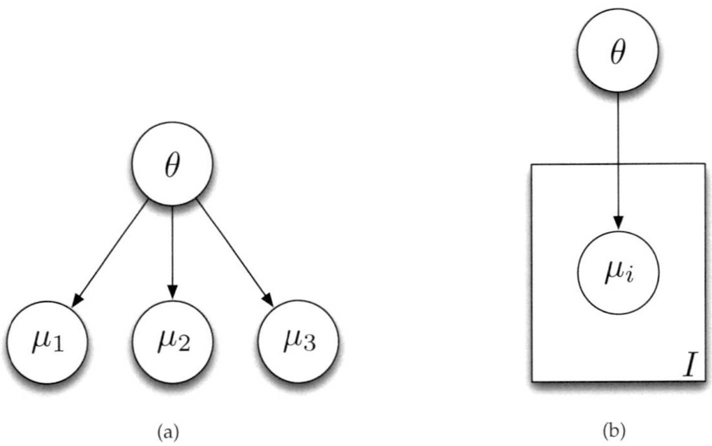

Figure 15. Bayesian graphical models with (a) replication on the children nodes $(\mu 1, \mu 2, \mu 3)$ and in (b) with the plate notation to some index I.

7.3. Markov Chain Monte Carlo sampling techniques

Figure 16 shows an example of a Markov Chain (MC) sampling technique, the rejection sampler, here we want to draw samples from a target distribution $P(z)$ which is not a standard distribution, therefore, we draw samples from a proposal distribution $Q(z)$, which we are able to evaluate from, up to some normalizing constant c where:

$$cQ(z) \geq P(z) \tag{9}$$

then, a candidate sample (z, b) is randomly generated in the area below $cQ(z)$, if this sample lied under $P(z)$ then it will be accepted [23].

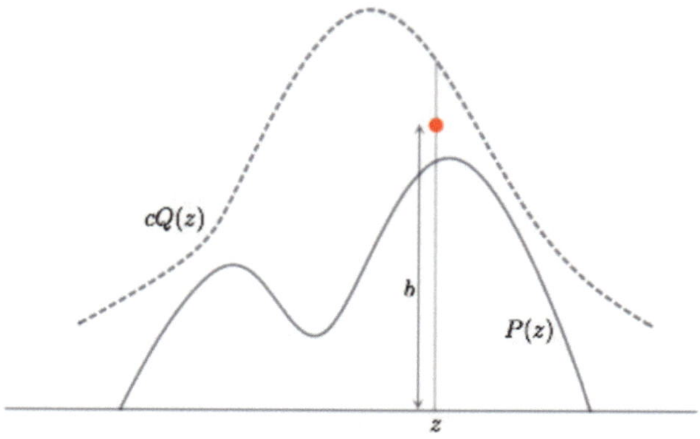

Figure 16. An example of the rejection sampler were point z will be accepted if it lies under $P(z)$ and rejected otherwise [23]

Although the rejection sampler technique is considered to be simple to implement, it suffers from some limitation in cases where a bad choice of Q, or a proper Q with a poor constant c. This will lead to a high rejection rate. To solve this problem, we will use Markov Chain Monte Carlo (MCMC) techniques where the target distribution will eventually converge to the proposal distribution.

7.3.1. The Gibbs sampling technique

Geman and Geman [9] introduced the Gibbs sampler where the samples are drawn sequentially from the full conditional distribution. Suppose we want to draw samples from:

$$P(\Gamma) = P(\gamma_1, \gamma_2, \ldots, \gamma_z) \tag{10}$$

then, the Gibbs sampler replaces the value of γ_i from a sample value drawn from the conditional distribution $P(\gamma_i \mid \Gamma)$ as follows [5, 12]:

$$\begin{aligned}
\gamma^{(\pi+1)} &\sim P\left(\gamma_1 \mid \gamma_2^{(\pi)}, \gamma_3^{(\pi)}, \ldots, \gamma_Z^{(\pi)}\right) \\
&\vdots \\
\gamma_\tau^{(\pi+1)} &\sim P\left(\gamma_\tau \mid \gamma_1^{(\pi+1)}, \ldots, \gamma_{\tau-1}^{(\pi+1)}, \ldots, \gamma_Z^{(\pi)}\right) \\
&\vdots \\
\gamma_Z^{(\pi+1)} &\sim P\left(\gamma_Z \mid \gamma_1^{(\pi+1)}, \gamma_2^{(\pi+1)}, \ldots, \gamma_{Z-1}^{(\pi+1)}\right)
\end{aligned} \tag{11}$$

7.3.2. The Metropolis-Hasting sampling technique

The Metropolis-Hasting sampling techniques was introduced in [11], it solves the limitation of the rejection sampler where the rejected samples will not be discarded but they are weighted according to an acceptance rate α [23].

Parameter	Specification
Environment	Single/Multi-floor
Estimation technique	Fingerprinting techniques
Fingerprint type	RSS
Sensing device	Mobile terminal
Calculation device	Mobile terminal
Packet scanning	Passive
Estimation algorithm	Bayesian graphical model
Location report	Physical

Table 6. Proposed system specifications

Figure 17 shows an illustration of the Metropolis-Hasting sampler. Suppose we want to draw samples from target distribution $P(z)$, then a candidate sample z^* is drawn from the proposal distribution $Q(z, z^t)$. Later α will determine whether this candidate sample z^* is accepted or weighted as follows:

$$\alpha = \min\left(1, \frac{P(z^*)}{P(z^t)} \frac{Q(z^t; z^*)}{Q(z^*; z^t)}\right) \tag{12}$$

if the candidate sample was accepted then z^{t+1} is set to z^* otherwise it will be set to the same state z^t.

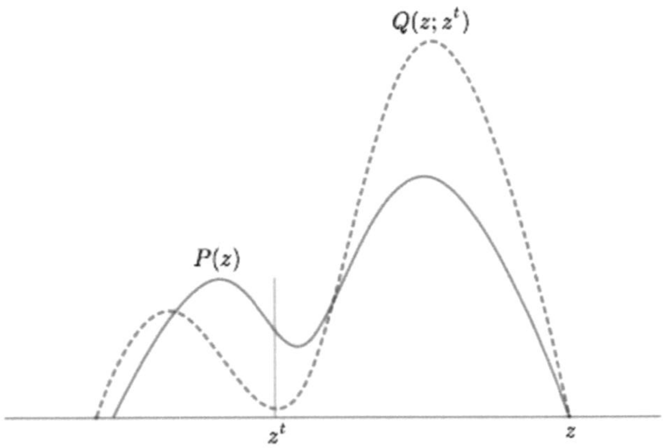

Figure 17. The Metropolis-Hasting sampling technique

8. Model design and measurement setup

8.1. System specification

The proposed system is capable to locate a target in a single or multi- floor buildings using RSS fingerprinting technique, both sensing and calculating processes are done by a mobile terminal equipped with a WLAN card. The system collects RSS fingerprints form the APs and feed them to a BGM which in turn tries to infer the target's location. Table 6 shows the specifications of the proposed system.

Hardware	Specification	
Notebook	Brand	Apple MacBook
	Model	Late 2008
	WLAN card	AirPort Extreme card
	IEEE standards	IEEE 802.11 a/b/g/n
	Operating system	Windows XP SP3
Access Point	Brand	D-Link
	Model	DWL-2000AP
	Operating frequency	2.4 GHz - 2.4835 GHz
	Transmit power	15 dBm

Table 7. Experimental hardware specifications.

8.2. Experimental hardware

A notebook is used as our mobile terminal, the notebook is a MacBook with an AirPort Extreme card that supports IEEE 802 11 a/b/g/n standards running Windows XP SP3. The mobile terminal collects RSS fingerprints from five D-link DWL-2000 APs operate from 2.4 GHz to 2.4835 GHz, each AP has 15 dBm transmission power. Table 7 shows the experimental hardware specifications.

8.3. Experimental test bed

Figure 10 shows the first floor layouts for WCC building at UTM and the second floor layout in Figure 11, the building has 2 floors, the first floor is about 36 m by 30 m and the second floor is 21 m by 28 m. The building is equipped with five APs $(AP1, AP2, ..., AP5)$, three in the first floor and two in the second floor, the building's walls are made of concrete and some plaster board walls, the walls thickness is 15 cm and the floor thickness is 80 cm.

8.4. Experimental software

For the Feed and Infer algorithm, we have to use two software applications for each part of the algorithm. For feeding part, the system requires RSS fingerprints at random locations to be collected in order to feed the BGM, therefore we developed *UTM WiFi Scanner*, a network sniffer written in c sharp based on *InSSIDer* by *MetaGeek*.

In the inferring part, *WinBUGS* [7] (Bayesian inference Using Gibbs Sampling) is used to estimate the target's location using Bayesian graphical models, WinBUGS uses Markov Chain Monte Carlo (MCMC) sampling techniques to estimate the posterior distribution, the current RSS fingerprint will be used in WinBUGS as a likelihood for the graphical model.

8.5. Model design

A single unshaded circle represents a continuous random variable and a shaded circle symbolizes a discrete random variable, double circle refers to a logical variable while a square represents a constant, we will provide the prior distributions for each random variable in our model. Figure 18 shows the proposed model which we introduced in [2], Nodes X_i and Y_i

represent the user's location at the ith point and they are assigned to a continuous uniform distribution as follows:

$$X_i \sim dunif\,(0, L) \tag{13}$$

$$Y_i \sim dunif\,(0, W) \tag{14}$$

where L and W represent the length and the width of the test bed respectively. Node Z_i is the floor number and it is assigned to a discrete uniform distribution since we are not interested in the height of the APs instead the floor number of which the ith AP is located in:

$$Z_i \sim DiscreteUinf\,(1, N) \tag{15}$$

but since WinBugs does not support discrete uniform distributions, we had to construct Z_i as a categorical distribution as follows:

$$p\,[i] \leftarrow 1/N \tag{16}$$

$$Z_i \sim dcat\,(p\,[])$$

Categorical distribution is a generalization of the Bernoulli distribution with sample space $\{1, 2, \ldots, n\}$.

$$D_{ij} = \log\left(1 + \sqrt{\left(X_i - \bar{x}_j\right)^2 + \left(Y_i - \bar{y}_j\right)^2}\right) \tag{17}$$

D_{ij} is the Euclidean distance between the jth AP and the ith RSS fingerprint, we exploited the fact that the RSS distribution in indoor environments follows a log-normal distribution [14], we also added 1 to the equation because we do not want to have zero as an argument of the log function.

$$S_{ij} \sim dnorm\left(m_{ij}, \tau_j\right) \tag{18}$$

where

$$m_{ij} = b_{0j} + b_{1j}D_{ij} + b_{2j}Z_i + b_{3j}w\,FAF \tag{19}$$

and

$$\tau_j \sim dgamma\,(0.1, 0.1) \tag{20}$$

The random variable m_{ij} is the mean for the normal distribution assigned to S_{ij} which symbolizes the RSS obtained at the ith location point from the jth AP. m_{ij} is a regression model with four parameters $(b0, b1, b2, b3)$ and four independent variables D_{ij},Z_i,w_i and, FAF (Floor Attenuation Factor). w is a binary variable that takes two values, 0 if the collected RSS is in the same floor with the AP which will cancel the effect of FAF and 1 otherwise.

8.6. Feed and infer algorithm

In Figure 19, we show the flow chart of the proposed feed and infer algorithm, it starts by defining a Bayesian model shown in Figure 18 using WinBUGS, defining a model requires the specification of the location variables X_i, Y_i and Z_i, the floor attenuation factor, the signal strength and the parameters of the regression model.

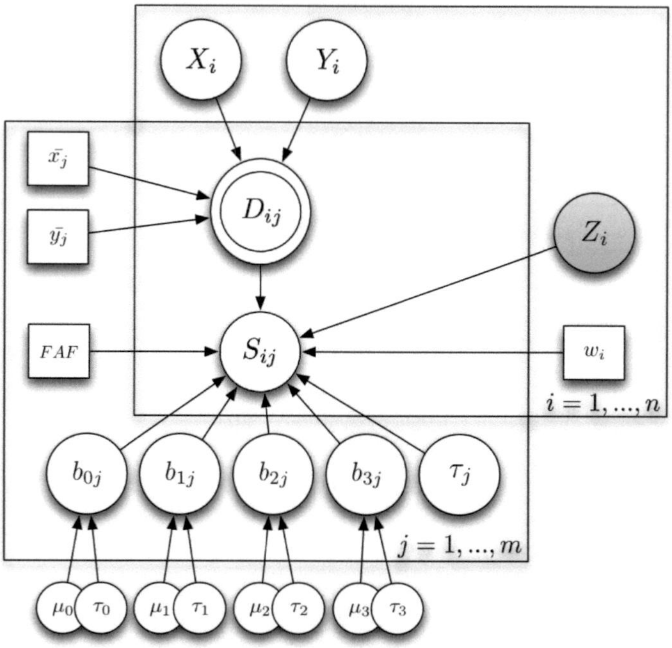

Figure 18. The proposed Bayesian model using WinBUGS plate notation.

In the next step, setting a value for the plate index i which represent the size of the RSS fingerprints collected, $i = 1$ will be set as initial value and will be increasing depending on the RSS sampling time. Next, checking the model error is done by the specification tool in WinBUGS, the specification tool also allows us to specify the initial values for random variables b,τ and μ. After compiling the model, we then specify the size of the burn-in samples which are the samples that will be initially generated and then ignored to allow the Markov chain to reach the stabilization state.

Next, using the inference samples tool in WinBUGS, we choose the random variables that will be later evaluated. Then, we draw samples for the random variables specified in the previous step using update model tool, also in this step we may choose the over-relax which means generating multiple random values and selecting the sample that is negatively correlated with the current sample [27]. Using save state tool, we feed the values again to the model and update the index $i = i + 1$. After collecting sufficient number of RSSs, we check the posterior summary using check state tool and produce the visual kernel estimate of the posterior distribution using the density tool.

8.7. Data analysis

8.7.1. Moving target

Figures 20 and 21 show the location error and the accuracy cumulative distribution function for the three windows $(c1, c2, c3)$ for a moving target inside WCC building, the target was moving throughout the corridors in two different floors. The proposed model started

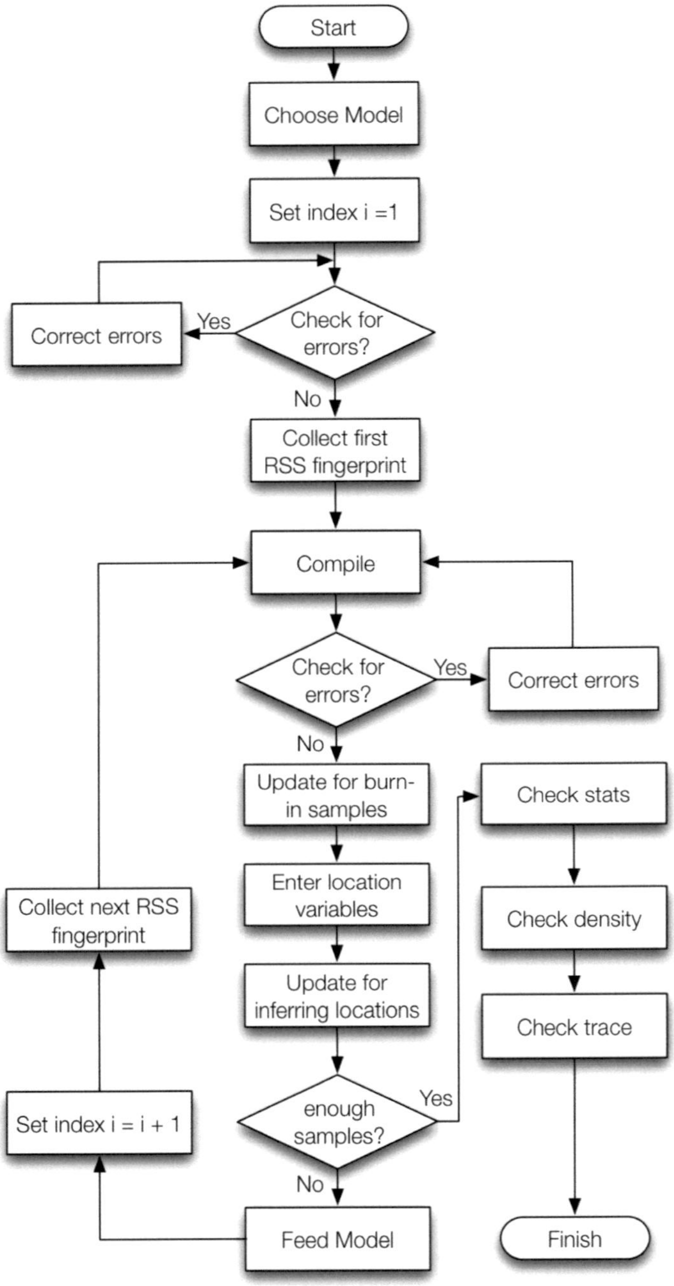

Figure 19. The feed and infer algorithm flowchart.

performing poorly with the first collected RSS, then the accuracy improved while the number of RSS increased, hence we divided the system performance into three windows $(c1, c2, c3)$, each window contains different portions of RSS samples. Window $c1$ contains all the 90 RSS fingerprints that were collected while testing the system, the mean accuracy obtained is about

Window	Number of values	75% percentile	Std. Deviation	Mean
$c1$	90	8.482	4.470	6.385
$c2$	50	5.946	2.689	4.214
$c3$	21	2.601	0.7292	2.272

Table 8. Location error statistics for the three windows

6.38 m. A better accuracy of 4.2 m was obtained from the second window $c2$ which starts from the 50th fingerprint and discarding the previously collected RSSs. In window $c3$, only last 21 RSS samples were included, the location error was much improved with mean error of 2.27 m. Table 9 shows the location error statistics for the three windows.

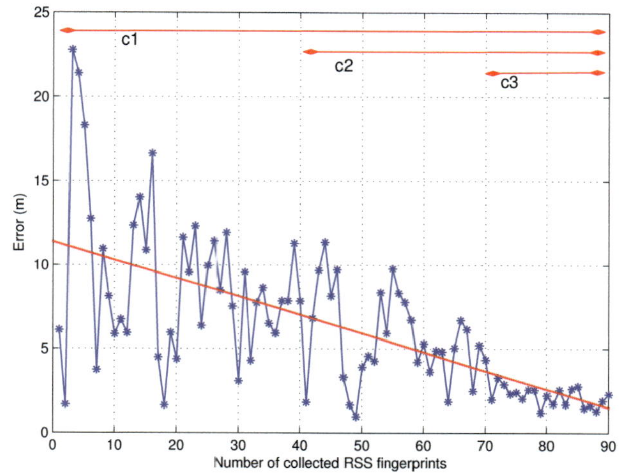

Figure 20. Location error results with all RSS fingerprints included.

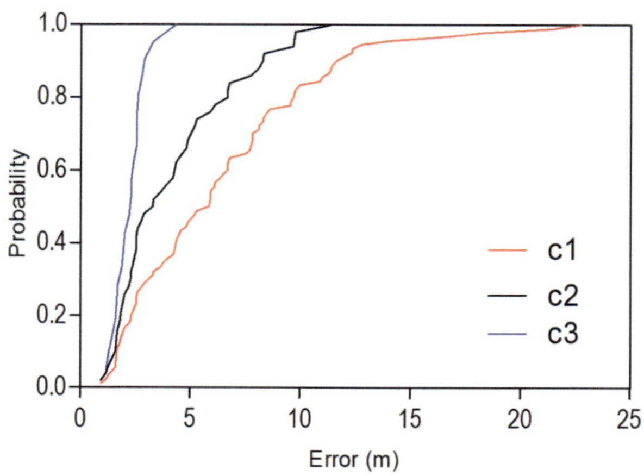

Figure 21. Cumulative distribution function of location error for the three windows.

	Location 1	Location 2	Location 3
Minimum	1.23	1.034	1.822
75% Percentile	2.613	2.809	2.195
Std. Deviation	0.4605	0.8748	0.3154
Mean	2.205	1.989	2.139

Table 9. Location error statistics at three fixed locations

8.7.2. Fixed target

Now we shall consider the system performance when the target is not moving, the system was tested at three random fixed locations, Figures [22-24] show the estimated location of 20 RSS fingerprints at the same location. The system performed slightly different at each location, the mean accuracy acquired was 2.2 m, 1.9 m, and 2.1 m at locations 1, 2 and 3 respectively. Table 9 shows the accuracy statistics at the three fixed locations.

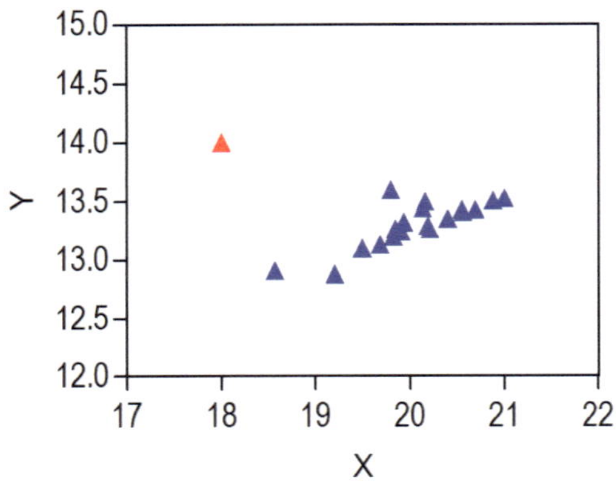

Figure 22. Location error at random location 1

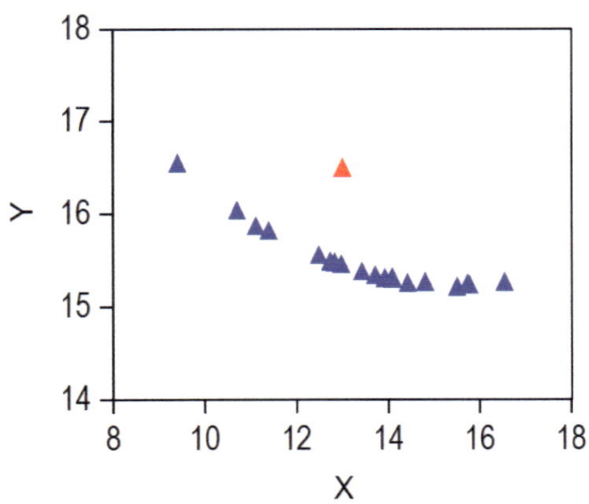

Figure 23. Location error at random location 2

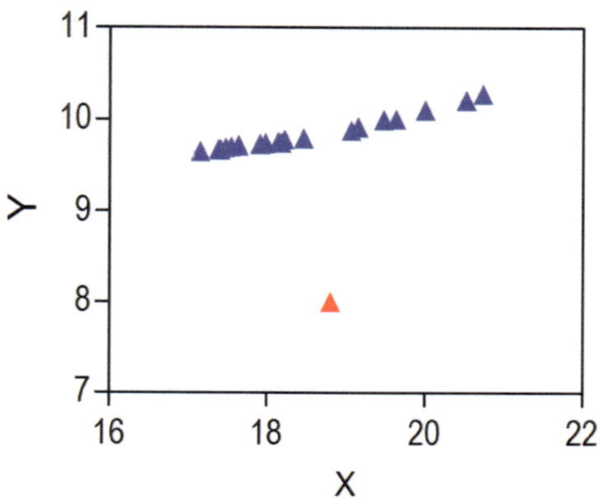

Figure 24. Location error at random location 3

	Madigan	Hyuk L.	Proposed system
Single-Phase	Partially	Yes	Yes
Accuracy	20 feet	2.57 m	2.27 m
Multi-Floor	No	No	Yes

Table 10. Comparison with other indoor location determination systems

In Table 10 we compare the proposed system with other well known single-phase systems, a Bayesian model proposed by [24] and a zero configuration model by [21].

9. Conclusion

This chapter presented a single-phase location determination system using Bayesian graphical models. The proposed system [1] does not require an offline phase to build the radio map. Instead, it uses the online RSS gathered in real time to estimate the user's location in multi-floor environments.

The results showed that the system was capable to locate a mobile target in a multi-floor environment without the need for a time consuming offline training stage to build the radio map. Instead of using Monte Carlo sampling techniques such as rejection sampling which suffer from low performance in complex Bayesian networks, MCMC sampling techniques were used to sample from the posterior distribution for the location random variables X, Y and Z.

Rather than using a single sampling technique, the system uses a collection of MCMC sampling techniques to draw samples from the posterior distribution. The Bayesian graphical model presented a visual approach to visualize the relationships between the random variables.

The Feed and Infer Algorithm presented a way to directly sample from the posterior distribution each time the Bayesian network was fed with a new inferred value from the previous step in order to facilitate the elimination of the training stage. Although the performance of this algorithm was not good enough in the first two windows $(c1, c2)$, the final systems performance was based only on the third window $c3$ that showed an excellent mean accuracy of about 2.27 m.

Author details

Abdullah Al-Ahmadi and Tharek Abd. Rahman
Wireless Communication Centre (WCC), Faculty of Electrical Engineering, Universiti Teknologi Malaysia (UTM), Johor 81310, Malaysia.

10. References

[1] Al-Ahmadi, A., Rahman, T., Kamarudin, M., Jamaluddin, M. & Omer, A. [2011]. Single-phase wireless lan based multi-floor indoor location determination system, *Parallel and Distributed Systems (ICPADS), 2011 IEEE 17th International Conference on*, pp. 1057 –1062.

[2] Al-Ahmadi, A., Rahman, T. et al. [2010]. Multi-Floor Indoor Positioning System Using Bayesian Graphical Models, *Progress In Electromagnetics Research B* 25: 241–259.

[3] Axel, K. [2005]. *Location-Based Services: Fundamentals and Operation*, John Wiley & Sons, Hoboken, NJ, USA.

[4] Bahl, P. & Padmanabhan, V. [2002]. RADAR: An in-building RF-based user location and tracking system, *INFOCOM 2000. Nineteenth Annual Joint Conference of the IEEE Computer and Communications Societies. Proceedings. IEEE*, Vol. 2, Ieee, pp. 775–784.

[5] Bishop, C. & SpringerLink [2006]. *Pattern recognition and machine learning*, Vol. 4, Springer New York.

[6] Castro, P., Chiu, P., Kremenek, T. & Muntz, R. [2001]. A probabilistic room location service for wireless networked environments, *Ubicomp 2001: Ubiquitous Computing*, Springer, pp. 18–34.

[7] Cowles, M. [2004]. Review of WinBUGS 1.4, *The American Statistician* 58(4): 330–336.

[8] Gelman, A. [2002]. Posterior distribution, *Encyclopedia of environmetrics* .

[9] Geman, S. & Geman, D. [1993]. Stochastic relaxation, Gibbs distributions and the Bayesian restoration of images*, *Journal of Applied Statistics* 20(5): 25–62.

[10] Harter, A., Hopper, A., Steggles, P., Ward, A. & Webster, P. [2002]. The anatomy of a context-aware application, *Wireless Networks* 8(2): 187–197.

[11] Hastings, W. [1970]. Monte carlo sampling methods using markov chains and their applications, *Biometrika* 57(1): 97.

[12] Hrycej, T. [1990]. Gibbs sampling in Bayesian networks, *Artificial Intelligence* 46(3): 351–363.

[13] Jordan, M. [2004]. Graphical models, *Statistical Science* 19(1): 140–155.

[14] Kaemarungsi, K. & Krishnamurthy, P. [2004]. Properties of indoor received signal strength for WLAN location fingerprinting, *Mobile and Ubiquitous Systems: Networking and Services, 2004. MOBIQUITOUS 2004. The First Annual International Conference on*, IEEE, pp. 14–23.

[15] Kjaergaard, M. [2007]. A taxonomy for radio location fingerprinting, *Proceedings of the 3rd international conference on Location-and context-awareness*, Springer-Verlag, pp. 139–156.

[16] Kj¿rgaard, M. [2007]. A taxonomy for radio location fingerprinting, *in* J. Hightower, B. Schiele & T. Strang (eds), *Location- and Context-Awareness*, Vol. 4718 of *Lecture Notes in Computer Science*, Springer Berlin / Heidelberg, pp. 139–156.

[17] Kolodziej, K. & Hjelm, J. [2006]. *Local positioning systems: LBS applications and services*, CRC/Taylor & Francis.

[18] Komar, C. & Ersoy, C. [2004]. Location tracking and location based service using IEEE 802.11 WLAN infrastructure, *European Wireless*, pp. 24–27.

[19] Krishnan, P., Krishnakumar, A., Ju, W., Mallows, C. & Gamt, S. [2004]. A system for LEASE: Location estimation assisted by stationary emitters for indoor RF wireless networks, *INFOCOM 2004. Twenty-third AnnualJoint Conference of the IEEE Computer and Communications Societies*, Vol. 2, IEEE, pp. 1001–1011.

[20] Kupper, A. [2005]. *Location-based services*, Wiley Online Library.

[21] Lim, H., Kung, L., Hou, J. & Luo, H. [2005]. Zero-configuration, robust indoor localization: Theory and experimentation, *work* 2005: 1818.

[22] Lorincz, K. & Welsh, M. [2007]. MoteTrack: a robust, decentralized approach to RF-based location tracking, *Personal and Ubiquitous Computing* 11(6): 489–503.

[23] MacKay, D. [2003]. *Information theory, inference, and learning algorithms*, Cambridge Univ Pr.

[24] Madigan, D., Einahrawy, E., Martin, R., Ju, W., Krishnan, P. & Krishnakumar, A. [2005]. Bayesian indoor positioning systems, *INFOCOM 2005. 24th Annual Joint Conference of the IEEE Computer and Communications Societies. Proceedings IEEE*, Vol. 2, IEEE, pp. 1217–1227.

[25] Mazuelas, S., Bahillo, A., Lorenzo, R., Fernandez, P., Lago, F., Garcia, E., Blas, J. & Abril, E. [2009]. Robust indoor positioning provided by real-time RSSI values in unmodified WLAN networks, *Selected Topics in Signal Processing, IEEE Journal of* 3(5): 821–831.

[26] Mitilineos, S., Kyriazanos, D., Segou, O., Goufas, J. & Thomopoulos, S. [2010]. Indoor Localization With Wireless Sensor Networks, *Progress In Electromagnetics Research* Vol. 109: 441–474.

[27] Ntzoufras, I. [2009]. *Bayesian modeling using WinBUGS*, John Wiley & Sons Inc.

[28] Pahlavan, K., Li, X. & Makela, J. [2002]. Indoor geolocation science and technology, *Communications Magazine, IEEE* 40(2): 112–118.

[29] Prasithsangaree, P., Krishnamurthy, P. & Chrysanthis, P. [2002]. On indoor position location with wireless LANs, *Personal, Indoor and Mobile Radio Communications, 2002. The 13th IEEE International Symposium on*, Vol. 2, IEEE, pp. 720–724.

[30] Priyantha, N. [2005]. *The cricket indoor location system*, PhD thesis, Massachusetts Institute of Technology.

[31] Roos, T., Myllymaki, P., Tirri, H., Misikangas, P. & Sievanen, J. [2002]. A probabilistic approach to WLAN user location estimation, *International Journal of Wireless Information Networks* 9(3): 155–164.

[32] Sklar, B. [2002]. Rayleigh fading channels in mobile digital communication systems. I. Characterization, *Communications Magazine, IEEE* 35(7): 90–100.

[33] Tan, S., Tan, M. & Tan, H. [2002]. Multipath delay measurements and modeling for interfloor wireless communications, *Vehicular Technology, IEEE Transactions on* 49(4): 1334–1341.

[34] Tauber, J. [2002]. Indoor location systems for pervasive computing, *Massachusetts Institute of Technology. Area exam report* .

[35] Wallbaum, M. & Diepolder, S. [2005]. Benchmarking wireless lan location systems, *Proceedings of the 2005 Second IEEE International Workshop on Mobile Commerce and Services (WMCS 2005), Munich, Germany*, pp. 42–51.

[36] Want, R., Hopper, A., Falcao, V. & Gibbons, J. [1992]. The active badge location system, *ACM Transactions on Information Systems (TOIS)* 10(1): 91–102.

[37] Youssef, M. & Agrawala, A. [2008]. The Horus location determination system, *Wireless Networks* 14(3): 357–374.

Section 3

Coupled GPS and Other Sensors

Chapter 9

Activity-Artifact Flow of GPS/INS Integration for Positioning Error De-Correlation

Xiaoying Kong, Li Liu and Heung-Gyoon Ryu

Additional information is available at the end of the chapter

http://dx.doi.org/10.5772/50516

1. Introduction

Vehicle positioning systems are composed of positioning sensors and positioning algorithms. Positioning sensors provide direct or indirect position, velocity, attitude and timing information. Positioning algorithms transfer sensor direct readings to desired positioning information. Positioning sensors are classified into absolute sensors and dead reckoning sensors (Sukkarieh, 2000). Absolute sensors directly provide the relationships between vehicle position and external positioning references. Dead reckoning sensors measure the vehicle's position and orientation increment to its last moment without external references. Global Navigation Satellite System (GNSS) is a widely used absolute positioning system using external satellite references. Inertial Navigation System (INS) is an example of dead reckoning positioning system. INS uses inertial measurement unit (IMU) to measure the self-contained position changing rate, then computes the desired positioning information. INS has been employed in navigation for rockets, missiles, aircrafts, land vehicles, ocean vessels and robots.

These two types of positioning sensors have their own characteristics. For example, Global Positioning System (GPS), one of the GNSSs, provides positioning and timing information to world-wide users. GPS has played an important role in national security, economic growth, transportation safety, and critical national infrastructure (GAO, 2009). However GPS sometimes suffers accuracy and availability problems. INS provides self-contained high frequency positioning information by integration of accelerations and rotation rates. During the integration process, any bias and errors in IMU sensors will be amplified over time. This results in the unbounded growth of positioning errors in the entire real time computing process.

Considering these sensor characteristics, many efforts have been made to aim for higher quality positioning achievement. There are two levels of research efforts. At low level or

INTECH
open science | open minds

© 2012 The Author(s). Licensee InTech. This chapter is distributed under the terms of the Creative Commons Attribution License http://creativecommons.org/licenses/by/3.0), which permits unrestricted use, distribution, and reproduction in any medium, provided the original work is properly cited.

component level, hardware and infrastructure improvement, sensor error modelling techniques have been studied (Nebot et al., 1996; Nebot et al., 1999; Kong et al., 1999). At high level or system level, integrations of two types of sensors are developed as positioning system architectures (Nebot et al., 1996; Scheding, 1997, Scheding et al., 1997, Sukkarieh, 2000).

Research and development efforts on GPS quality improvement include: launching new satellites (GAO, 2009; Committee on Oversight and Government Reform, 2009); deploying Differential GPS (GPS) and Wide Area Augmentation System (WAAS) infrastructures to improve GPS accuracy and availability (Wikipedia Differential GPS, 2012; Federal Aviation Administration, 2012); modelling GPS multi path effects (Gaylor et al., 2005; Wu & Hsieh, 2010); mitigating GPS errors in high-interference environment (Groves, 2005); modelling ionospheric effects (Rose et al., 2009); analyzing unintentional interference (Owen & Wells, 2001); modelling GPS signal blockage (Gaylor et al., 2005); and modelling GPS external behaviours (Nebot et al., 1996; Kong et al., 2010).

Efforts on INS improvements include: developing new types of INS using new mechanisms such as developing new optical IMU; developing low cost IMU such as MEMS (Microelectromechanical systems) IMU and associated algorithms to extend user services (Geiger et al., 2008; Sahawneh & Jarrah 2008); developing high quality IMU testing facilities; modelling IMU biases; modelling INS errors in initial alignment stages (Nebot et al., 1996; Kong et al., 1999); improving INS error propagation models during positioning missions (Kong et al., 1999).

Besides the positioning sensor research, on system level, integrating two types of sensors to form an improved positioning system has been a major effort in positioning research. In literature, GPS/INS integration approaches have been widely adopted and can be classified to uncoupled, loosely coupled, and tightly coupled integrations (Sukkarieh, 2000). In uncoupled systems, GPS and INS work independently. Loosely coupled systems integrate GPS and INS provided positioning information without feedback to GPS or INS. In tightly coupled system, direct measurements from GPS or INS, and computed information are integrated. Real time feedbacks are provided to GPS or INS. Recently, ultra-tight integration has emerged. On ultra-tight integration architecture, INS, GPS phase, frequency and code tracking loops are integrated deeply to improve GPS performance (Alban et al., 2003; Lashley et al., 2010).

The GPS/INS integration is usually implemented using a filter such as Kalman filter. From the view point of filter mechanism, the above GPS/INS integration approaches could also be classified to feedforward filter and feedback filter (Maybeck, 1979; Sukkarieh, 2000). In a feedforward filter approach, the filter estimates the desired positioning or error states. These filter outputs are not fed back to GPS or INS measurements for real time correction. The positive aspect of feedforward filter is that if the filter fails, the sensor measurement information is still available. But the INS errors will grow over time due to the integration of INS bias drift (Maybeck, 1979). A feedback filter allows filter estimations to feedback to sensors for measurement correction. The INS information is bounded over time by real time error feedback and correction. Under the feedback and feedforward filter types, there are direct filter and indirect filter depending on the filter outputs. Direct filters output positioning readings directly. Indirect filters output positioning error estimates.

We use an entity-relationship model to analyze and classify the GPS/INS integration concepts as in Fig. 1.

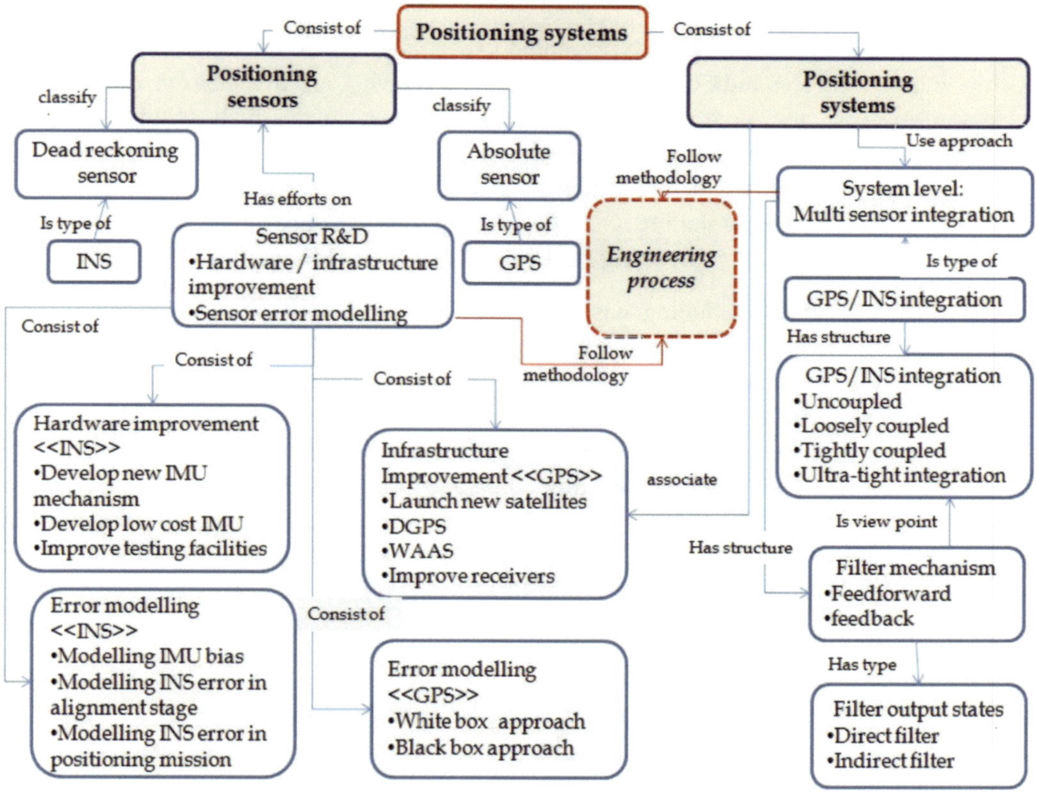

Figure 1. Classifications of positioning system research using GPS/INS integration

This chapter presents a GPS/INS integration approach for positioning error de-correlation by analyzing system filter structure and performance. It is from the view point of implementing feedforward and feedback structures. The aim of this approach is to improve the positioning quality of the GPS/INS integration system in accuracy and availability aspects. The contributions of this research fill in the gap in the above classification map in both positioning sensor level and system level. A new engineering process based on the practice in this research is proposed in this chapter. The new engineering process is added into the map as a new classification entity.

To present this research in this chapter, a bottom-top approach is used to describe this GPS/INS positioning system integration process. In the following sections, we start from component level presentation by analysing sensor models in Section 2. System level design and system performance verification are presented in Sections 3 and 4. Section 3 presents the feedforward filter approach in frequency domain and time domain. Feedback filter approach is discussed in Section 4. Based on the engineering practices in Sections 2, 3 and 4, an engineering process with research activities and associated artifacts is formed and summarized. Section 6 draws the conclusion.

2. Modelling GPS error using shaping filter

In the GPS/INS integration approach presented in this chapter, GPS is chosen to be the main positioning sensor. INS aids GPS. Nowadays GPS receivers are available in market with relatively low cost. INS accuracy level and cost level vary in relatively wide range. GPS receivers could be purchased first. INS accuracy level could be decided by analyzing GPS/INS filter performance before purchasing decision making. Once a GPS receiver is ready for system implementation, GPS positioning performance could be analyzed by modelling GPS errors. GPS positioning quality could be improved then by using INS aiding.

In literature, GPS error modelling has been focusing on understanding GPS signal error sources. Major efforts are on modelling the characteristics of GPS multi path effects (Gaylor et al., 2005; Wu & Hsieh, 2010), ionospheric effects (Rose et al., 2009), unintentional interference (Owen & Wells, 2001), high-interference environment (Groves, 2005), and GPS signal blockage (Gaylor et al., 2005). These approaches look at the internal aspects of GPS errors. We refer to these as "white box" modelling approach. Another type of research looks at the external behaviours of GPS errors. GPS errors are modelled using shaping filter in frequency domain (Maybeck, 1979; Nebot et al., 1996; Kong et al., 2010). This approach could be referred as "black box" approach. In this chapter, we use black-box approach to understand GPS errors.

GPS measured information includes time and pseudo range to satellite. Latitude and longitude of GPS receiver's position are calculated using pseudoranges to a number of satellites. It was found that the errors of GPS calculated positions show the characteristics of colour noise (Nebot et al., 1996). In the approach of this chapter, we wrap the entire calculated position errors as a black-box from external user's view point. The entire GPS calculated position errors are modelled as colour noise using shaping filter.

Shaping filter models colour noise as a linear system driven by white noise (Maybeck, 1979). Measurement errors are the output of the shaping filter. The form of a shaping filter could be examined using Power Spectral Density (PSD) in frequency domain. GPS error modelling using shaping filter with PSD is explained as follows.

From external user's view point, GPS measurement errors include satellite signal biases and receiver noises. On satellite bias side, satellite constellation could be changed if some satellites are de-orbited or new satellites are lunched (GAO, 2009). Selective Availability (SA) was intentionally introduced to degrade positioning accuracy (Wikipedia, Differential GPS, 2012). From GPS receiver's side, user receivers could be purchased from different vendors with different quality. The factors from the two sides will affect the GPS measurement errors' frequency characteristics. Therefore GPS errors using shaping filter should be modelled with a particular receiver under the current satellite constellation. We take two cases as examples for shaping filter analysis. Stand-alone GPS is considered for vehicles in areas without the coverage of DGPS and WAAS.

Case 1: Stand-alone GPS signal with Selective Availability error; receiver type 1.

Case 2: Stand-alone GPS signal without Selective Availability; receiver type 2.

By examine the PSD curves of GPS position errors in Case 1, we found the GPS error shows the characteristics of a second order system. Transfer positioning errors to east-north-down frame. For axis x (east) as an example, the shaping filter of GPS error can be presented as in Fig. 2 (a).

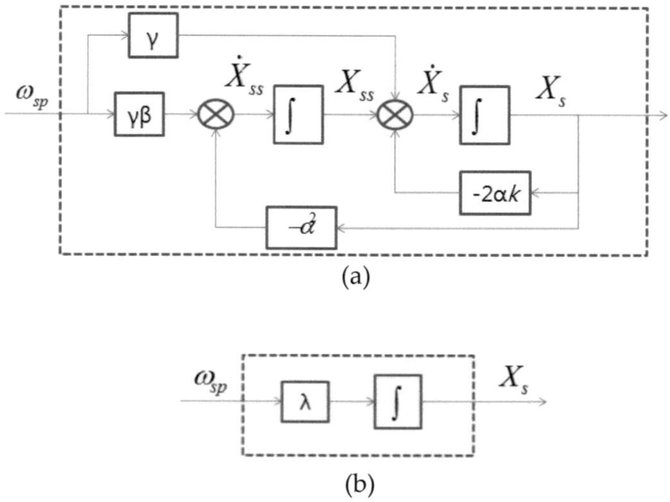

Figure 2. (a) Shaping filter with a second order system (b) Shaping filter with a first order system

In this shaping filter structure, ωsp is the input of the shaping filter. ωsp is a unity white noise. *Xs* is the output of the shaping filter. *Xs* represents the GPS colour noise. α, β, γ and k are filter parameters obtained by experimental data sets during PSD curve fitting (Kong et al., 2010). This shaping filter could be presented in state space as the following second order system:

$$\begin{aligned} \dot{X}_s(t) &= -2\alpha k X_s(t) + X_{ss}(t) + r\omega_{sp}(t) \\ \dot{X}_{ss}(t) &= -\alpha^2 X_s(t) + r\beta\omega_{sp}(t) \end{aligned} \tag{1}$$

For Case 2, the shaping filter is experimentally obtained as a first order system as in Fig. 2 (b). The state space form of Case 2 is presented as in Equation (2).

$$\dot{X}_s(t) = \lambda\omega_{sp}(t) \tag{2}$$

The aim of the shaping filter is to duplicate the GPS measurement errors. The GPS positioning could be constructed using an augmented system with the shaping filter. The augmented system is constructed with a mx1 state vector $x(t)$, a px1 measurement vector $z(t)$, a qx1 white noise vector $w(t)$ on process model. Using the second order shaping filter as an example, the augmented system is as Fig. 3. The augmented system is presented using the following equation:

$$\begin{aligned}\dot{x}(t) &= F(t)x(t) + G(t)w(t)\\ z(t) &= H(t)x(t) + n_s(t) + \tau_{gps}(t)\end{aligned} \tag{3}$$

where x is the system state. w is white noise. z is the system measurement. z is corrupted by colour noise ns and white noise τ_{gps} . ns is presented as:

$$n_s(t) = X_s(t) \tag{4}$$

where Xs is the output of the shaping filter described in Fig. 2 (a).

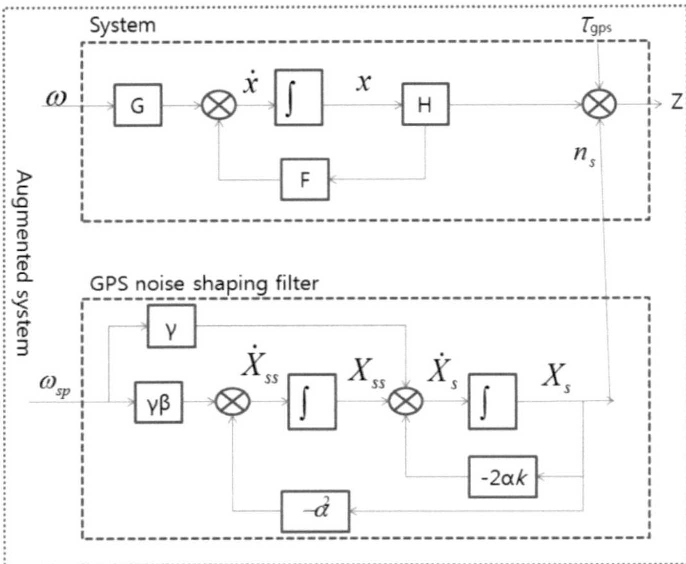

Figure 3. Augmented system with shaping filter duplicating GPS measurement noises

As in Fig. 3, the augmented state x_s can be defined as follows:

$$x_s(t) = \begin{bmatrix} x(t) \\ X_s(t) \\ X_{ss}(t) \end{bmatrix} \tag{5}$$

The process model of the augmented system can be described as:

$$\dot{x}_s(t) = \begin{bmatrix} F(t) & 0_{m\times2} & 0_{m\times2} \\ 0_{1\times m} & -2\alpha k & 1 \\ 0_{1\times m} & -\alpha^2 & 0 \end{bmatrix} \begin{bmatrix} x(t) \\ X_s(t) \\ X_{ss}(t) \end{bmatrix} + \begin{bmatrix} G & 0_{q\times1} \\ 0_{1\times q} & r \\ 0_{1\times q} & r\beta \end{bmatrix} \begin{bmatrix} \omega(t) \\ \omega_{sp}(t) \end{bmatrix} \tag{6}$$

The observation model of the augmented system is:

$$z(t) = \begin{bmatrix} H(t) & I_p 0_p \end{bmatrix} \begin{bmatrix} x(t) \\ X_s(t) \\ X_{ss}(t) \end{bmatrix} + \tau_{gps}(t) \tag{7}$$

Once the augmented system is constructed, INS could be used to aid de-correlating the GPS errors. In the following sections, de-correlating approach is described using two types of filters: feedforward filter and feedback filter.

3. GPS error de-correlation process using feedforward filter

INS is chosen to aid GPS to estimate and remove colour noises. We call this process as GPS error de-correlation process. INS measurements also contain noises. Feedback filter approach uses filter estimation to correct INS noises in real time. Feedforward filter approach leaves the INS noises without feedback correction. If the filter fails, INS is able to provide positioning independently in feedforward structure. In this section, we consider feedforward filter approach first. Feedback approach will be described in the next section.

3.1. Feedforward filter structure

Fig. 4 is a GPS/INS integration feedforward filter structure. INS indicated positioning information and GPS measurements are integrated into the system integration filter. The system filter estimates the velocity errors, position errors, shaping filter states, and outputs the position information. There is no feedback of position error estimation to sensors for real time correction.

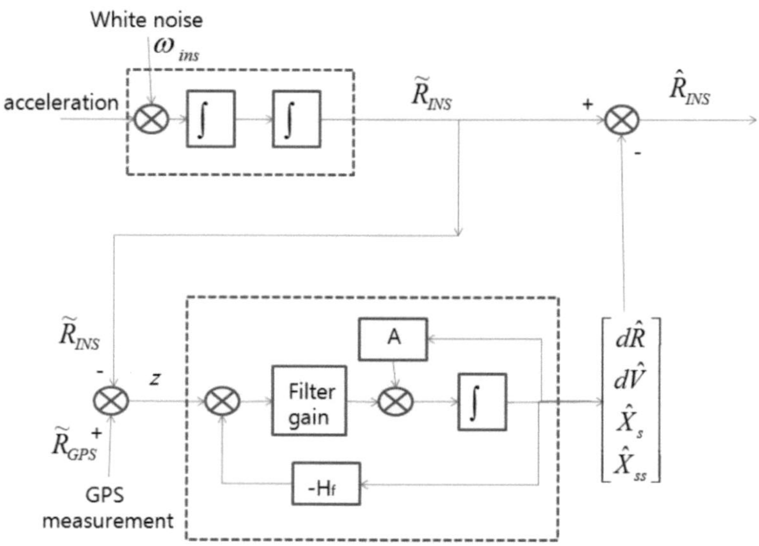

Figure 4. Feedforward filter structure

To focus on GPS error de-correlation process using this feedforward filter, we assume that the INS coordinate system is already transformed into the GPS coordinate system using gyros' information. The INS positioning reading is computed by integrating acceleration twice. The acceleration noise is assumed driven by white noise. Using the case of the second order shaping filter as example, Fig. 4 illustrates the filter inputs, outputs and state estimates.

The state of this feedforward filter is:

$$x(t) = [dR(t), dV(t), X_s(t), X_{ss}(t)]^T \tag{8}$$

The velocity error and position error in the filter state are defined as:

$$\begin{aligned} dR(t) &= \tilde{R}_{INS}(t) - R_{true}(t) \\ dV(t) &= \tilde{V}_{INS}(t) - V_{true}(t) \end{aligned} \tag{9}$$

where $R_{true}(t), V_{true}(t)$ are vehicle true position and velocity. $\tilde{R}_{INS}(t), \tilde{V}_{INS}(t)$ are INS computed position and velocity.

The filter observation $z(t)$ is the difference between the GPS output and INS output.

$$z(t) = \tilde{R}_{GPS}(t) - \tilde{R}_{INS}(t) \tag{10}$$

As described in the previous section, GPS outputs are modelled as the true position corrupted by colour noise *Xs* driven by white noise.

From Fig. 4, the following relationships can be derived:

$$\begin{aligned} d\dot{R}(t) &= dV(t) \\ d\dot{V}(t) &= \omega_{ins}(t) \end{aligned} \tag{11}$$

And

$$\begin{aligned} z(t) &= [R_{true}(t) + X_s(t) + \tau_{gps}(t)] - [R_{true}(t) + dR(t)] \\ &= -dR(t) + X_s(t) + \tau_{gps}(t) \end{aligned} \tag{12}$$

The de-correlation filter process model is therefore given by:

$$\dot{x}(t) = \begin{bmatrix} d\dot{R}(t) \\ d\dot{V}(t) \\ \dot{X}_s(t) \\ \dot{X}_{ss}(t) \end{bmatrix} = \begin{bmatrix} 0 & 1 & 0 & 0 \\ 0 & 0 & 0 & 0 \\ 0 & 0 & -2\alpha k & 1 \\ 0 & 0 & -\alpha^2 & 0 \end{bmatrix} \begin{bmatrix} dR(t) \\ dV(t) \\ X_s(t) \\ X_{ss}(t) \end{bmatrix} + \begin{bmatrix} 0 & 0 \\ 1 & 0 \\ 0 & \gamma \\ 0 & \gamma\beta \end{bmatrix} \begin{bmatrix} \omega_{ins}(t) \\ \omega_{sp}(t) \end{bmatrix} \tag{13}$$

The observation model of the filter is given by:

$$z(t) = H_f x(t) + \tau_{gps}(t) \tag{14}$$

where

$$H_f = \begin{bmatrix} -1 & 0 & 1 & 0 \end{bmatrix} \tag{15}$$

The covariance matrices of the process and observation noises are

$$\begin{aligned} Q &= E[w(t)w^T(t+\tau)] = \begin{bmatrix} q_{ins}^2 & 0 \\ 0 & 1 \end{bmatrix} \delta(\tau) \\ R &= E[\tau_{gps}(t)\tau_{gps}{}^T(t+\tau)] = \sigma_{GPS}^2 \delta(\tau) \end{aligned} \tag{16}$$

with the white noise variance q_{ins}^2 on the acceleration output and the variance σ_{GPS}^2 on the GPS position observation.

Equations (13) ~ (16) compose the feedforward de-correlation filter. In the next section, this feedforward filter will be analyzed in frequency domain and time domain respectively.

3.2. Analysis of de-correlation process using feedforward filter

The GPS/INS integration feedforward filter structure is described in the previous section. In this section, the positioning error de-correlation process is demonstrated in frequency domain and time domain. The de-correlation process is designed as the following procedure in Table 1.

Step	**Domain**	**Research Details**
Step 1	frequency domain	The filter transfer function from the observation $z(s)$ to the filter state estimates $x(s)$ is calculated.
Step 2	frequency domain	Tuning the filter noises levels, the gain of the filter transfer function is analysed using bode plots. INS noise levels are tuned during this step.
Step 3	time domain	The performance of the filter is analyzed at positioning system level. The shaping state time series and GPS error time series are compared. The system error level is examined.
Step 4	Iterative in both frequency and time domains	Continue tuning of the noise levels at Step 2 and analyzing time domain performance at Step 4. The de-correlation process is successful until the system error is tuned to the required level.

Table 1. Procedure of the positioning error de-correlation process in frequency domain and time domain

Using the feedforward filter structure in Fig. 4 and following the procedure in Table 1, the de-correlation process is demonstrated as below.

Step 1:

As illustrated in Fig. 4, in frequency domain, the transfer function from the observation z(s) to the filter state estimate is calculated by:

$$\frac{\hat{x}(s)}{z(s)} = (sI_{4x4} - A + KH_f)^{-1}K \tag{17}$$

where the filter-gain K is determined by the filter parameters and the noises of the filter process and observation.

and

$$A = \begin{bmatrix} 0 & 1 & 0 & 0 \\ 0 & 0 & 0 & 0 \\ 0 & 0 & -2\alpha k & 1 \\ 0 & 0 & -\alpha^2 & 0 \end{bmatrix} \tag{18}$$

Step 2:

The sensor noise levels are tuned in the filter. In the Case 1 for the filter in Fig. 4, a stand-alone GPS without differential correction is chosen. The variance of the position error is about $20 \times 20m^2$. The noise level of INS is tuned during this de-correlation analysis process before INS purchasing decision. For the example of Case 1, the range of accelerometer's noise variance is tuned from $1 \times 10^{-2}(m/s^2)^2$ to $1 \times 10^{-6}(m/s^2)^2$.

Fig. 5 is the bode plot by tuning an accelerometer with a variance of $1 \times 10^{-2}(m/s^2)^2$. The gain of the shaping state estimate $\hat{X}s$ is very small.

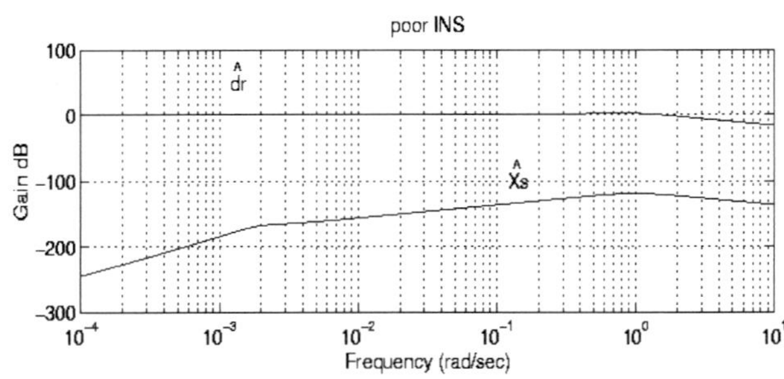

Figure 5. Bode plot of the feedforward filter using a low quality accelerometer

Step 3:

Switching to time domain, using the parameters in Step 2, the time series of GPS position estimates, error estimates, and shaping state estimates are illustrated in Fig. 6.

In the example of Case 1, the vehicle is in stationary state. The GPS measurement output consists of the entire colour noise. As in Fig. 6(a), the shaping state estimates do not follow the shape of the GPS measurements. Fig. 6(b) illustrates the system position estimates follow the INS position outputs. The position error estimates grow over time without bound in feedforward filter. De-correlation fails using this accelerometer quality level.

Step 4:

We iteratively tune the feedforward filter by adjusting sensor noise level using Step 2 and Step 3. Until in frequency domain, the gain of the shaping state in bode plot approaches 0dB, the de-correlation process is successful.

Fig. 7 is a successful de-correlation example of bode plot using an accelerometer with a variance of $1\times10^{-6}(m/s^2)^2$. The gain for the shaping state estimate $\hat{X}s$ within $10^{-4}\,rad/\sec$ to $10^{-2}\,rad/\sec$ is approaching 0dB. Examine the time domain performance for this case, Fig. 8 shows that the time series of the shaping state estimate $\hat{X}s$ follows the shape of the GPS output. The system position estimate errors are reduced to 4m.

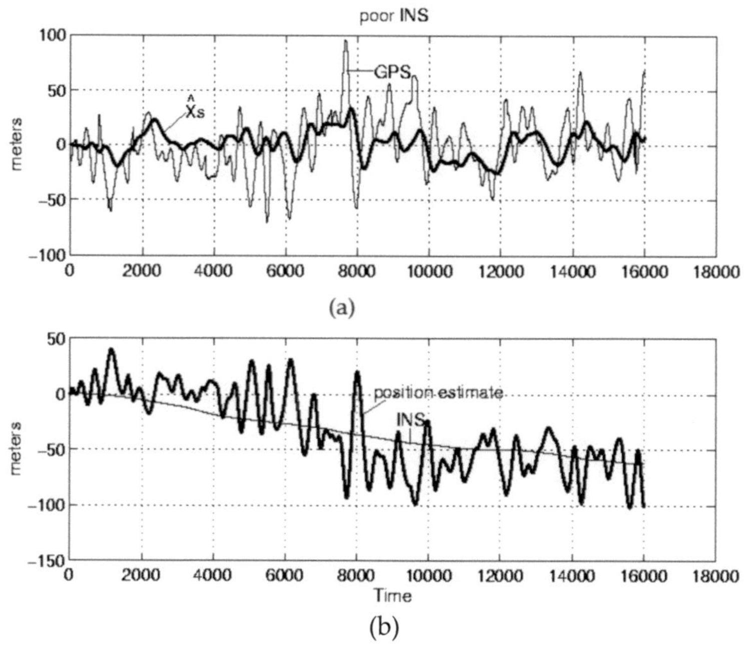

Figure 6. Time domain results using a low quality accelerometer (Unit: data iteration – meter) (a)Time domain performance of shaping filter estimate and GPS error measurement (b)Time domain performance of INS position estimate and system position estimate

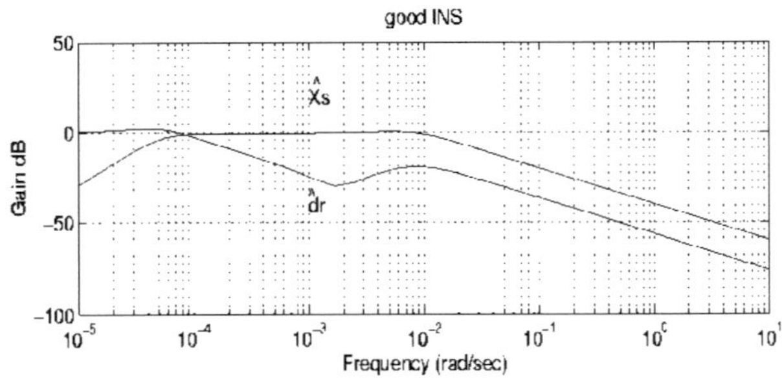

Figure 7. Bode plot of the feedforward filter using a high quality accelerometer

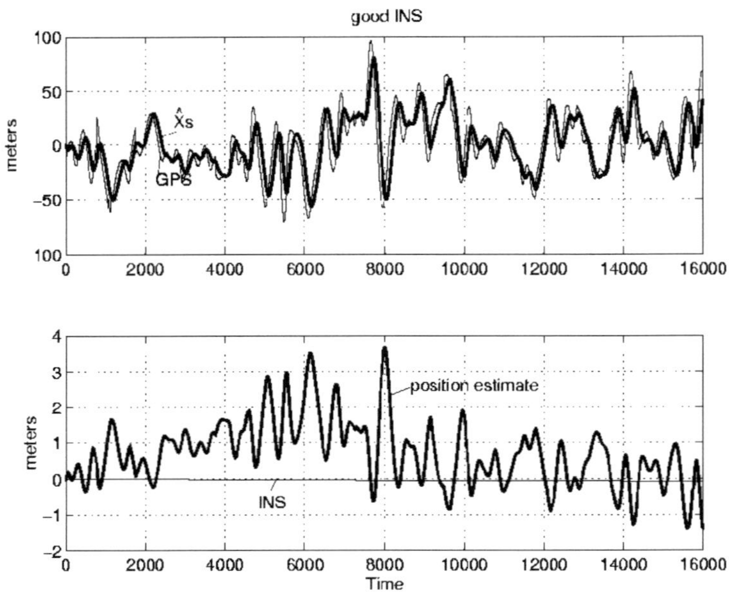

Figure 8. Time domain results using a high quality accelerometer (Unit: data iteration – meter)

During the filter tuning process using Step 2 to Step 4, the INS quality requirement for the successful de-correlation can be found from the bode plot of the filter transfer function. From the experiments we found that in order to use a feedforward filter to de-correlate GPS colour noises using an INS, the accuracy level of the INS should be above $3.16\times10^{-6}(m/s^2)^2$ on variance of the accelerometers.

The feedforward filter is able to output position estimates directly. The GPS and INS are integrated loosely. If the system integration filter fails, the INS is still able to provide positioning information without the filter. But there is no feedback to correct errors in INS in real time. The position outputs will grow over time due to the step by step integration of the inertial sensor's drift errors. Once a feedback scheme is introduced to the filter, the INS error could be reduced and the system position outputs could be bound. This feedback scheme is implemented in feedback filter.

4. GPS error de-correlation process using feedback filter

4.1. Feedback filter structure

A feedback scheme for inertial sensor correction is able to overcome the position error growth. Fig. 9 is a feedback filter structure. The GPS and INS are integrated tightly.

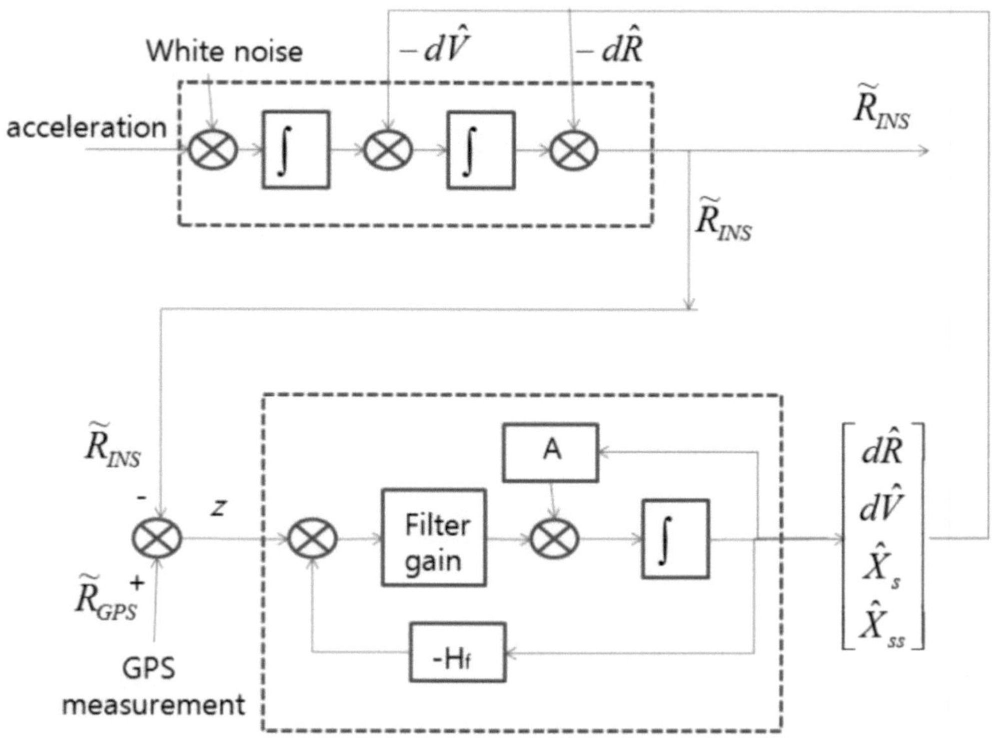

Figure 9. Feedback filter structure

In this feedback filter structure, the filter outputs the estimates of position error $d\hat{R}$, velocity error $d\hat{V}$, and shaping state estimation. Using the second order shaping filter as an example, the shaping states estimation include two states $\hat{X}s$ and $\hat{X}ss$. If using the first order shaping filter, the feedback filter will output one shaping state estimate $\hat{X}s$.

INS computes the velocity by integration acceleration. INS position is computed by integration the computed velocity. The filter then takes the input from the measurement difference between GPS and INS. The filter estimates the errors of velocity, position and GPS errors. The estimated velocity error and position error are feedback to INS computing process for correction as illustrated in Fig. 9. The filter states are still $[dR(t), dV(t), X_s(t), X_{ss}(t)]$ if using the second order shaping filter. The process model remains the same as the feedforward filter in Equation (13), and the observation model is the same as shown in Equations (14) and (15).

4.2. Feedback filter analysis in frequency domain and time domain

Following the same procedures in Table 1, the positioning system error de-correlation process using feedback filter is analyzed and implemented as follows.

In frequency domain, the filter transfer function from the observation to filter states is calculated. The filter is tuned using a range of accelerometers' error levels. Bode plots at each level are plotted. The gains of shaping states are analyzed. Fig. 10 shows the bode plot using a low quality accelerometer with a noise variance of $1\times10^{-2}(m/s^2)^2$.

Fig. 10 shows the gain of the shaping state estimate on the bode plot is very low. The ratio of the shaping estimate to the GPS error measurement is lower than 1. Using the parameters of this level to calculate the time series of shaping state, position estimates, error estimates, the time domain performance are illustrated in Fig. 11. Fig. 11(a) shows the shaping state estimates do not follow the shape of GPS errors. Fig. 11(b) shows the INS position estimate errors are bound over time due to the error feedback and INS correction. The system error de-correlation is unsuccessful using this quality level of accelerometer.

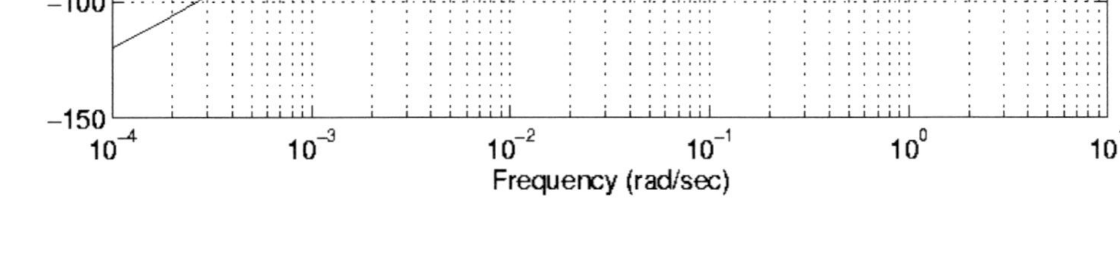

Figure 10. Bode plot of the feedback filter using a low quality accelerometer

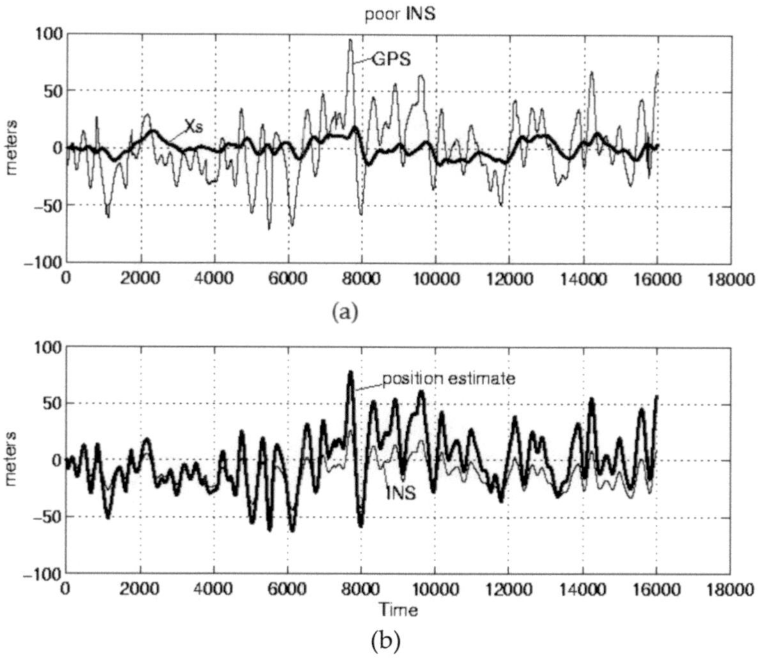

Figure 11. Feedback filter performance in time domain using a low quality accelerometer (Unit: data iteration -meter) (a)Time domain performance of shaping filter estimate and GPS error measurement (b)Time domain performance of INS position estimate and system position estimate

Continually tuning the feedback filter by trying different levels of accelerometer variance levels following Step 2 and Step 3 in Table 1, the best gain could be found in bode plot of the transfer function. Fig. 12 is a bode plot using an accelerometer variance of $1\times10^{-6}(m/s^2)^2$. The gain of the shaping state estimate to observation is approaching 0dB within $3\times10^{-4}rad/\sec$ to $10^{-2}rad/\sec$. As shown in Fig. 13 in time domain, the shaping state estimates track the GPS error shapes and match their magnitudes in time series.

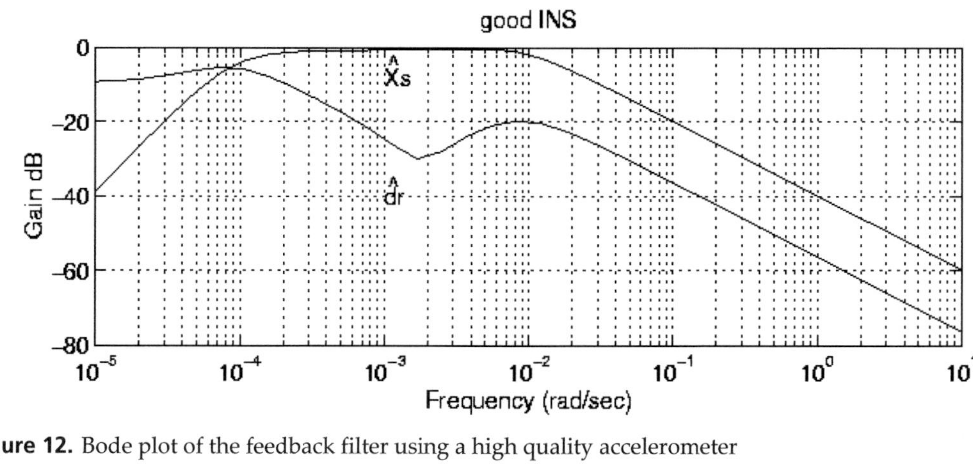

Figure 12. Bode plot of the feedback filter using a high quality accelerometer

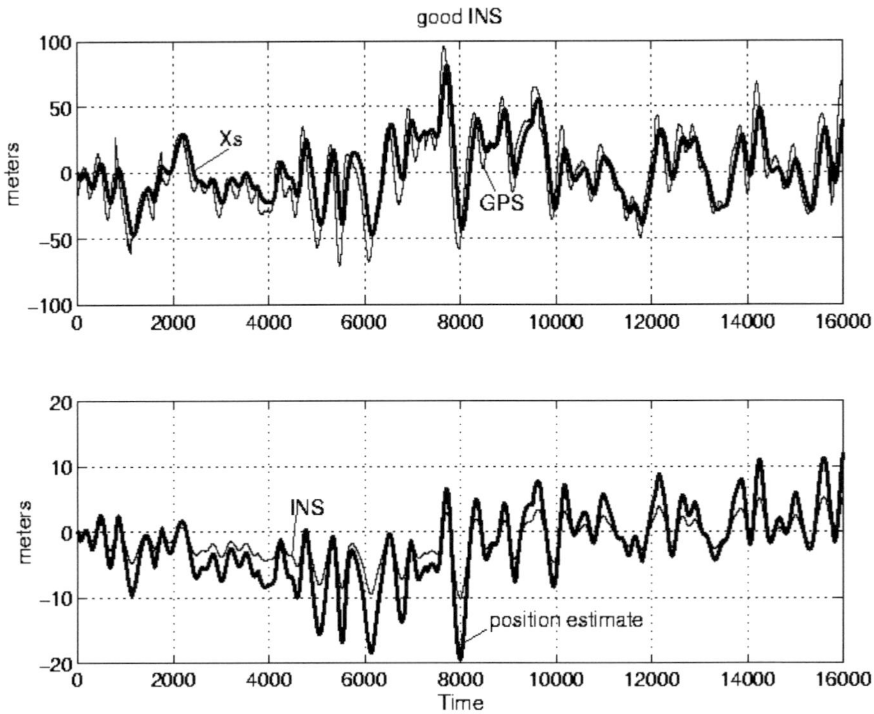

Figure 13. Feedback filter time domain performance using a high quality accelerometer (Unit: data iteration – meter)

In this feedback filter tuning process, we found that the accelerometer variance threshold for GPS error de-correlation using a stand-alone GPS with error variance of $20\times 20m^2$ is $3.16\times 10^{-6}(m/s^2)^2$.

Compare with the de-correlation threshold finding in the feedforward filter, the quality requirements based on accelerometer variance level are the same for both feedforward filter and feedback filter. The benefit of the feedback approach is that the position estimates are bound over time using the feedback filter structure.

5. Engineering process of the positioning error de-correlation approach

To effectively reuse the artifacts and minimize the designing efforts in both feedforward and feedback approaches, this GPS error de-correlation approach could be standardized using an activity-artifact flow to form a repeatable engineering process. Fig. 14 shows the activities and associated artifacts of this engineering process.

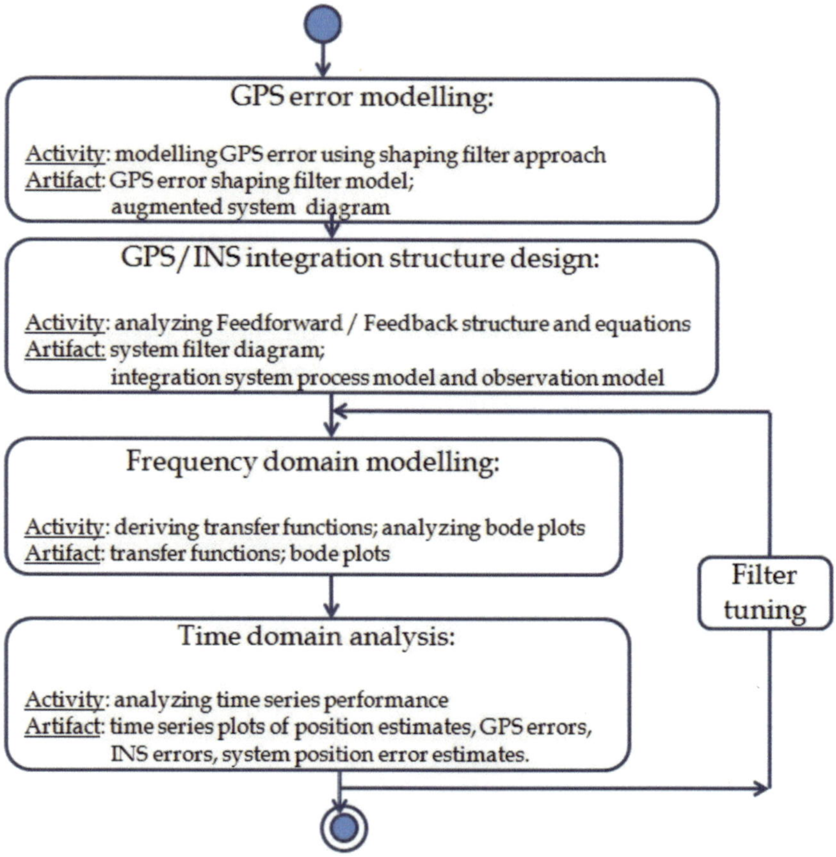

Figure 14. Activities and associated artifacts in the positioning error de-correlation process

As demonstrated in the previous sections of this chapter, this process includes the following major activities.

Activity 1: GPS error modelling using shaping filter approach. The associated artifacts of this activity include a GPS error shaping filter model and an augmented system diagram using this shaping filter.

Activity 2: GPS/INS integration design by analyzing the feebforward or feedback structures and their equations. The artifacts of this design activity are a system filter structure diagram, a set of filter equations of the process model and observation model.

Activity 3: Frequency domain modelling by deriving transfer functions and analyzing bode plots. The artifacts in frequency domain are transfer functions of the filter state estimates over filter measurement inputs, and bode plots using these transfer functions.

Activity 4: Time domain analysis by verifying the time series performance. The artifacts of this activity include a set of time series plots. These plots illustrate the GPS errors, INS errors, system errors, and system positioning estimates over time.

During this process, the artifacts in Activity 3 and Activity 4 will be iteratively analyzed using filter tuning with the trial of sensor noise levels. The system error de-correlation process stops once the threshold of the sensor noise level is found, and the final system accuracy level meets the system requirements.

6. Conclusions

This chapter presents a positioning system error de-correlation process approach. The aim of this approach is to solve the accuracy and availability issues of GPS based positioning system. In this chapter, this approach is standardized into an engineering process with a number of activities and associated artifacts. This approach starts from system component level by modelling GPS errors. GPS error is modelled using shaping filter approach. The artifact of GPS error modelling is an augmented system with the GPS error shaping filter. Once the component model is understood, the GPS/INS integration structures are analyzed at system high level. Feedforward filter and feedback filter structures are analyzed, designed and implemented for their benefits. Under each feedforward or feedback filter structure, the integrated system performance is analyzed both in frequency domain and time domain. The analysis activities are designed in an iterative filter tuning process. INS plays a role with its noise level in filter tuning process in both feedforward and feedback filter structures. Using the GPS/INS error de-correlation approach in this chapter, the experiments demonstrate that the final system accuracy level is improved.

This approach is designed for GPS/INS integration system. Further research could be extending this engineering process to other positioning sensor integration systems.

Author details

Xiaoying Kong
University of Technology, Sydney, Australia

Li Liu
The University of Sydney, Australia

Heung-Gyoon Ryu
Chungbuk National University, South Korea

7. References

Alban, S.; Akos, D.; Rock, S. & Gebre-Egziobher, D. (2003). Performance Analysis and Architectures for INS-Aided GPS tracking loops, Institute of Navigation - NTM, Anaheim, CA, 22-24 January, 2003, pp. 611-622

Committee on Oversight and Government Reform (2009). Hearing on GPS: can we avoid a gap in service? Last accessed: 13/4/2012. Available from: U.S. House of Representatives web site, http://oversight.house.gov/hearing/gps-can-we-avoid-a-gap-in-service/

Federal Aviation Administration (2012). WAAS Test Team Website, Last accessed: 13/4/2012. Available from: http://www.nstb.tc.faa.gov/

GAO (2009). Global positioning system, significant challenges in sustaining and upgrading widely used capabilities, United States Government Accountability Office report, GAO-09670T

Gaylor D.; Lightsey G. & Key K. (2005). Effects of multipath and signal blockage on GPS navigation in the vicinity of the International Space Station (ISS), Journal of The Institute of Navigation, Summer 2005, Vol. 52, No. 2, pp. 61-70

Geiger, W.; Bartholomeyczik, J.; Breng, U.; Gutmann, W.; Hafen, M.; Handrich, E.; Huber, M.; Jackle, A.; Kempfer, U.; Kopmann, H.; Kunz, J.; Leinfelder, P.; Ohmberger, R.; Probst, U.; Ruf, M.; Spahlinger, G.; Rasch, A.; Straub-Kalthoff, J.; Stroda, M.; Stumpf, K.; Weber, C.; Zimmermann, M.; Zimmermann, S. (2008). MEMS IMU for AHRS applications, 2008 IEEE/ION on Position, Location and Navigation Symposium, 5-8 May 2008; pp. 225 - 231; Monterey, CA

Groves, P. (2005). GPS signal-to-noise measurement in weak signal and high-interference environments, Journal of The Institute of Navigation, Summer 2005, Vol. 52, No. 2, pp. 83-94

Kong, X.; Nebot E. & Durrant-Whyte H. (1999). Development of a non-linear psi-angle model for large misalignment errors and its application in INS alignment and calibration, Proceedings of IEEE International Conference on Robotics and Automation, Detroit, MI, USA, May, 1999, pp. 1430-1435

Kong, X.; Liu, L. & Tran., T. (2010). Modeling satellite Positioning errors in different configurations from end user's viewpoint, Proceedings of the 16th Asia-Pacific Conference on Communications (APCC 2010), Oct. 31 -Nov. 3 2010, Auckland

Lashley, M.; Bevly, D. & Hung, J. (2010). Analysis of Deeply Integrated and Tightly Coupled Architectures, Proceedings of IEEE/ION PLANS 2010, pp.382-396, May 4-6, 2010

Maybeck, P. (1979). Stochastic models, estimation and control, New York, Academic Press, 1979

Nebot, E.; Durrant-Whyte, H. & Scheding S. (1996). Kalman filtering design techniques for aided GPS land navigation applications, Data Fusion Symposium, 1996. ADFS'96, First Australian.

Nebot E. & Durrant-Whyte H. (1999). Initial calibration and alignment of low cost inertial navigation units for land vehicle applications, Journal of Robotics Systems, Vol. 16, No. 2, Feb. 1999, pp. 81-92

Owen, J. & Wells, M. (2001). An advanced digital antenna control unit for GPS". IN: Look at the changing landscape of navigation technology; Proceedings of the Institute of Navigation 2001 National Technical Meeting, Long Beach, CA, Jan. 22-24, 2001, Alexandria, VA, Institute of Navigation, 2001, pp. 402-407

Rose, J.; Allain, D. & Mitchell, C. (2009). Reduction in the ionospheric error for a single-frequency GPS timing solution using tomography, Annals of Geophysics, Vol 52, No 5, 2009

Sahawneh, L. & Jarrah, M.A. (2008). Development and calibration of low cost MEMS IMU for UAV applications, 5th International Symposium on Mechatronics and Its Applications, 2008. ISMA 2008; 27-29 May, 2008

Scheding, S. (1997). PhD thesis, High Integrity Navigation, The University of Sydney, 1997

Scheding, S.; Nebot, E.; Stevens, M.; Durrant-Whyte, H.; Robots, J. & Corke, P. (1997). Experiments in autonomous underground guidance, IEEE International conference on Robotic and Automation, Albuquerque, NM, USA, Apr. 1997, pp. 1898-1903

Sukkarieh, S. (2000). Low Cost, High Integrity, Aided Inertial Navigation Systems for Autonomous Land Vehicles. PhD thesis, The University of Sydney, 2000

Wikipedia (2012). Differential GPS, Last accessed: 12/4/2012, Available from: http://en.wikipedia.org/wiki/Differential_GPS

Wikipedia (2012). Global Positioning System, Last accessed: 13/4/2012. Available from: http://en.wikipedia.org/wiki/Global_Positioning_System

Wu, J. & Hsieh, C. (2010). "Statistical modeling for the mitigation of GPS multipath delays from day-to-day range measurements", JOURNAL OF GEODESY Volume 84, Number 4, pp. 223-232, DOI: 10.1007/s00190-009-0358-6

Chapter 10

Inertial Navigation Systems and Its Practical Applications

Aleksander Nawrat, Karol Jędrasiak, Krzysztof Daniec and Roman Koteras

Additional information is available at the end of the chapter

http://dx.doi.org/10.5772/50578

1. Introduction

Nowadays one of the most common problems for science is the answer for the question how to improve our tools. Each year there are more attempts to solve the problem. Designed and created tools and prototypes' capabilities are increasing each year. In XXI century there is a trend to design tools that require as little human interaction as possible to fulfill their tasks.

Main intention of autonomous devices' designers is to develop tools used for implementation tasks, such as unknown territory exploration or performing tasks in strong radiation fields which are dangerous for human health and life. Unmanned flying objects used for military, mobile robots, space ships, exoskeletons or intelligent clothing monitoring body signals. These are only a few examples of useful devices that are being developed at the moment.

One of the main problems during autonomous mobile objects' development is the problem of precise navigation. In order to navigate the object it is required to know the exact position and orientation of the object in relation to the known environment. Creation of a sensor system capable of an environment perception as well as monitoring inner object parameters is an important problem in the unmanned mobile objects field. Control algorithms are designed on the top of available sensor data, therefore their flexibility and reliability are common requirements.

Recently, there is a lot of research being performed in the area of inertial measuring object orientation. Inertial measurement units (IMUs) are electronic devices used for detection of the current object orientation. Usually they measure changes in object's rotation and acceleration. As a measurement devices, they must fulfill a set of requirements e.g. smallest possible size and weight, configurable filtered output data. Sensor should be capable of working in the extreme temperature and pressure conditions. The main requirement is to

INTECH
open science | open minds

© 2012 The Author(s). Licensee InTech. This chapter is distributed under the terms of the Creative Commons Attribution License http://creativecommons.org/licenses/by/3.0), which permits unrestricted use, distribution, and reproduction in any medium, provided the original work is properly cited.

deliver high quality data with the highest speed possible. For mobile devices there is also an additional requirement of minimal electrical energy consumption.

In order to fulfill mentioned requirements, modern inertial measurement units are utilizing the newest technology achievements. Such sensors usually consist of at least two different types of subsensors. First type is an accelerometer measuring linear acceleration. Second is a gyroscope measuring angular acceleration value. Additionally, there are also different types of sensors mounted inside of IMU. One of the most popular additional sensors are reference magnetometers used for detection of north direction. Other types are temperature sensors used for temperature compensations algorithms and in case of UAV - altimeters.

Linear accelerometers measure object's linear acceleration and therefore detect direction of object's movement. In order to measure movement regardless of autonomous device's type there are usually three measurement axes X, Y and Z (fig. 1). For each axis there is a separate linear accelerometer. Modern sensors allow to measure data along all three axes by a single chip. Accelerometers are for instance used for collision detection, object's orientation measurement or as user interface. There are various methods of measuring linear acceleration. Due to this reason, there are sensors using capacity, piezoelectric, piezoresistant, magneto responsive, based on Hall effect and microelectromechanical systems (MEMS).

Gyroscopes are sensors which also measure acceleration. However it is not linear acceleration but angular. There are also three axes of measurement. They are traditionally called yaw, pitch and roll. Axes coordinate system are shown in fig.1. Gyroscopes are based on the law of conservation of momentum. In order to operate, they are required to preserve high rotation speed on the one side low and friction in bearings on the other side. There are three main types of gyroscopes: mechanical, optical and MEMS. Mechanical and optical sensors are much larger than MEMS, therefore the latter are usually used in practical solutions.

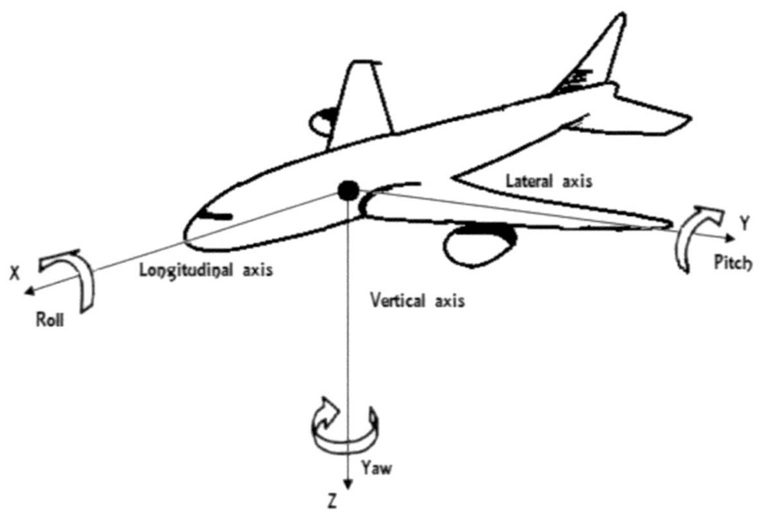

Figure 1. Three measurement axes for IMU.

Recently, most of the manufactured inertial measurement units are based on MEMS technology. Regardless of easy usability, the main advantage of such devices is considerably its small size. Electromicromechanical chips are made of silicon, glass and polymer materials. Masks for processors are made using microelectrical technologies as e.g. photolithography. MEMS sensors' circuits are divided into two parts. The first, mechanical part is responsible for measuring angular and linear velocity. The electronic part using analog-digital converter converts signals from the mechanical part. Because of relatively large surface to volume ratio there are dominating electrostatic phenomenon over the mass inertion.

Micromechanical systems are commonly used in navigation of mobile platforms and vehicles, for autopilots of aerial ships, boats or ground units. Another use is in aiding stabilization of opto-electrical gimbals. The IMUs are also used for seatbelts locking during detected crash moment, in PCs for HDD security during fall. It is also used in photo cameras for elimination of vibration effects. Modern toys for children also utilize inertial measurement units.

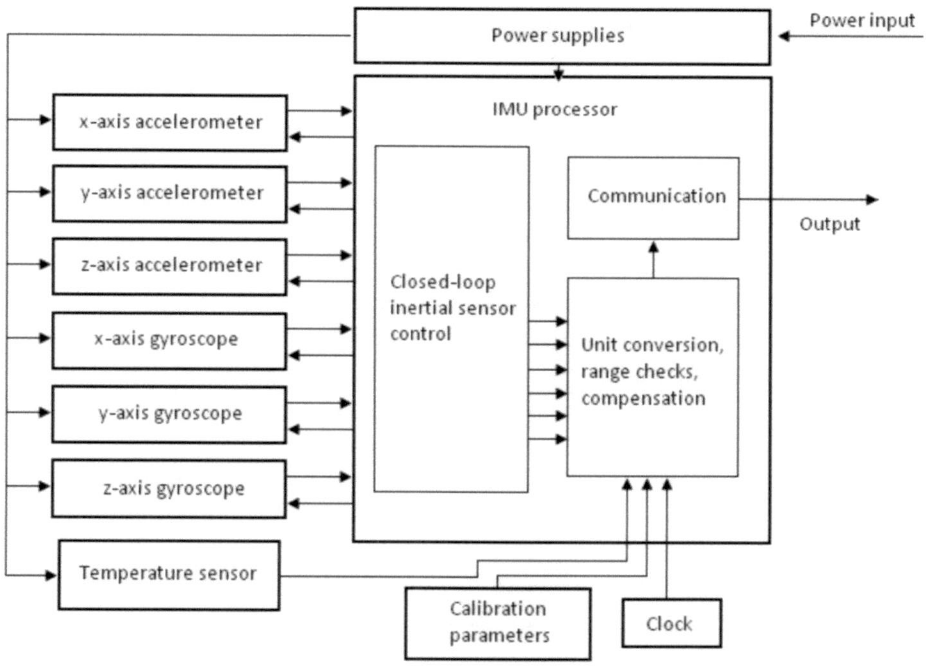

Figure 2. Inertial measurement unit operations schema.

An example of IMU working schema is presented in fig.2. Linear (x, y, z) and angular (fx, fy, fz) acceleration information is enough to estimate objects speed and further their position in space. Exemplary IMU is a 6-DOF type, which means it can measure signals in six dimensions. Raw data from sensors is converted to the known sensor's shell coordinate system. Measure values are converted to the respective physical units. Required parameters for transformation are computed during calibration phase of the sensor. It is useful to relate coordination systems to external, known reference coordinate system. Such reference coordinate system might be magnetic field of the Earth. Using magnetometers it is possible to measure the magnet response and after taking into account magnetic deviations depending on the current latitude and longitude, it is possible to estimate the north direction. Inertial measurement units also using magnetometers are called 9-DOF IMU.

IMUs internal processing frequency can be as high as 500 Hz. However output data rate is usually in range between 80 and 100 Hz.

Inertial measurement units are unfortunately subjects of a bias error. However, those errors can be measured in laboratories and computed fixes are stored in devices' flash memory. Sensor readings are also affected by the temperature of environment. Therefore each IMU is calibrated in climate chamber for full temperature operating range in order to establish compensation parameters. For temperature compensation installation of additional temperature sensor is required.

An affair of a great importance is devices' miniaturization. Nowadays, more and more research is being conducted all over the world in order to make current sensors smaller and lighter. Such miniaturized sensors could be used in micro military or medical robots. Micro-robots require smaller mechanical and propulsion parts. Size reduction is still a challenging task for scientists. It is essential for UAVs to reduce the total weight of mounted sensors. Possible applications of micro-sensors are tiny helicopters invisible for radars or capable of flying not only outdoors but also indoors. There is also a wide area for practical medical implementations. Recently, there is a lot of interest in the field of rehabilitation exoskeletons. Exoskeletons are electro-mechanical devices amplifying patients muscles' effective power. They are controlled by natural micro-electric signals in muscles measured by electromyography. For such devices it is important to deliver reliable data prom precisely chosen anthropomorphic points. Significantly reduced size of the sensor would allow to reduce the impact on the patient during acquisition. Therefore, measured movements would be closer to the natural patient's movement. Acceptable volume of such micro sensors should be around 2cm3 in order to not affect person's natural movements. Micro IMUs of a size not affecting movements could be used as a part of intelligent clothing. Knowledge about human's body parts' orientation would allow clothing to accommodate to person position. For instance, intelligent soldier uniform could enable masking mode every time the soldier starts crawling. Another implementation could monitor patients movements in order to detect e.g. Parkinson disease effects or measure pulse during everyday actions. It could be used as a tool for detection elderly people's potential threats in home environment.

2. Navigation aids

Navigation using only inertial measurement units is difficult or even impossible due to the accumulative character of possible errors. Typical errors of accelerometers and gyroscopes are biases, scale factors, triad non-orthogonalities and noise. Bias is defined as an independent and uncorrelated of the specific force and angular velocity experienced by the unit. The bias value affecting the unit consists usually of two or more sources of bias. The turn on bias commonly contains about 90% of the bias. Standard and temperature bias impact is estimated at about 10%. However the temperature bias in the MEMS IMU's can have similar magnitude as the turn on bias. A scale factor error is a ration of change in the output of the sensor with respect to true intended measurement. It can be estimated by the filter implemented in the inertial measurement unit. Another type of error is cross coupling errors which is the result from the non-orthogonality of the sensor triad. The error is estimated during the calibration and the required corrections are applied before using the sensor. Another type of error is existing within the gyroscopes which work due to use a spinning or vibratory mass. The source of the error is imbalance in the proof mass and the magnitude varies in the range from 1 to 100°/hr/g. It implies that units which are subjected to an acceleration of multiple [g] can be affected with an error of thousands of degrees per hour. The last type of error is a random noise error that comes from electrical and mechanical instabilities. It is known that during static operation the error follows Gaussian distribution however during unit movement the error distribution changes.

In order to aid the navigation the readings from the IMUs are compared against measurements from the independent sources. Various data fusion algorithms are used for combining the information from various external sources. One of the oldest of the still used navigation aids is using ground-based transmitting stations. Combination of bearings to at least two ground stations can be used to compute the exact position of the unit by a simple process of triangulation. In order to cover large areas it is desired to use low frequencies of electromagnetic waves. However the process of radio propagation is affected by the current condition of the atmosphere the readings may be unreliable. Hence, regardless the lower range higher frequencies are often used. Small corrections are always better than no corrections therefore a long-range navigation systems are also used. An example of such system is a Loran C system consisting of a master station and a chain of slave stations. The idea of operation of the system is measuring the time of arrival of the signal from the previous chain node. Using the time difference it computes the correction and broadcast it further to the next nodes and potential unit listeners.

It is well known that the stars may be used as a references for the purposes of celestial navigation. It is required to know the positions of at least two celestial objects in relation to the observer and the exact time of the observation. Knowledge of a position of a celestial object in relation to the observer allows to draw a circle centered on the point on the Earth directly below the observed object. There are pre-computed astronomical tables in which the center points of the circles can be found using the exact time of the measurement. It is assumed that an observer must be located at the circles intersection. Application of the following technique to the navigation using the INS requires development of an automatic measuring device. There are known such devices called star trackers however their accuracies are about three hundred meters on the surface of the Earth. Hence, such devices are rarely used in order to improve the quality of the INS.

Another type of a navigation aid is using a surface radar stations. It is possible to measure the line-of-sight from the ground station to the observed air ship. From the measurements the air ship vehicle's range (R), elevation, azimuth and bearing can be computed with respect to a local reference frame at the location of the radar ground station. The relative position of the object in the air and a known ground position of the station can be transmitted to the air object in order to compute the corrections and aid the INS.

$$R = \sqrt{x^2 + y^2 + z^2}$$
$$\psi = tan^{-1}(\frac{y}{x}) \tag{1}$$
$$\theta = tan^{-1}\{\frac{z}{\sqrt{x^2+y^2}}\}$$

Navigation aids can also be installed directly on the board of the unit e.g. air ship. One of such devices is a Doppler radar. It operates by transmitting a narrow beam of the microwave energy to the ground and measures the frequency shift that occurs in the reflected signal as a result of the relative motion between the aircraft and the ground. Let's assume that the air ship velocity is equal to V and the angle between the ground and the radar beam is θ, the frequency shift can be computed as:

$$\frac{2V}{\lambda}cos\theta, \tag{2}$$

where λ is the wavelength of the radar beam used for the transmission. Usually it is in the frequency range 13.24 – 13.4 GHz which stands for λ ≈ 0.22 dm. In order to estimate the velocity of an object capable of moving in three dimensions the number of radar beams used must be no smaller than three. Modern systems use a planar array of beams processed independently and computed by microprocessors.

Another navigation aid is based on measuring atmospheric pressure readings in order to estimate the height of the object. Altimeters due to their relatively high precision (typically less than 0.01 percent) are used to bound the growth of error in a vertical direction of the INS. The altimeter measurements can also be made by using sonic or ultra-sonic waves radar. Combination of height above ground and height above sea level values can be used in order to match a detailed map of a terrain. A sequence of such measurements accompanied by the orientation of the object from the INS can be fitted into the surface. However such application requires a map of terrain and that the air ship is moving above a terrain with a contour variation above 20%. It is difficult or impossible to use this technique over the sea. Recently laser rangefinders are becoming more and more popular in the field of measuring the distance however the technique is still limited to a short range over the ground.

Another technique possible to use as a navigation aid is a scene matching. A video camera mounted below the air ship delivers a sequence of images of the ground. If the scene visible in the ground can be recognized as a location with a known geographical coordinates it is possible to compute the corrections to the INS. The recognition is usually performed on the basis on two types of features. First are linear features detected in the image as edges and further extracted using techniques as Hough transform or active contours. Another type of features are SIFT like features consisting of a unique description of a point in the image that is scale, rotation, translation and illumination invariant. A combination of a terrain and a scene matching is map matching technique. First a geographical coordinates of a roads visible below are computed using scene recognition techniques. Next an air ship trajectory is fitted to the matching roads and finally any errors are corrected by the assumption that the object is moving only within the roads net. A variation of map matching technique is commonly used in car and pedestrian navigations.

One of the most popular navigation aids is using a magnetometers. The devices capable of measuring the magnetic field of the Earth. Regardless the type of the magnetometer it is measuring the magnetic field in a one, two or three dimensions. Two most commonly used as a navigation aid are 3D magnetometers. Almost any metal device has its own magnetic field that cannot be distinguished from the field of Earth, therefore a calibration of the magnetometers after mounting is required. After calibration it could be used as an 3D electronic compass with measurements expressed in a local object frame. The relationship between the magnetometer readings (M_x, M_y, M_z) and object's attitude can be expressed as follows:

$$\begin{bmatrix} M_x \\ M_y \\ M_z \end{bmatrix} = C_n^b \begin{bmatrix} Ecos\delta cos\gamma \\ Ecos\delta sin\gamma \\ Esin\delta \end{bmatrix}, \tag{3}$$

where C_n^b is a cosine matrix of the object with respect to the local geographic frame and E is the Earth's magnetic field. Magnetometers capability of measuring the magnetic field does not limit their usage only as an electronic compass. It is possible to measure local magnetic variation in the planet field. Such information could be stored in a map and used to navigate in a similar manner to terrain matching.

The most popular navigation aid is satellite navigation. Currently the most widely used system was names Global Positioning Systems Navigation Signal Timing And Ranging in short GPS. It consists of 31 satellites circling around the Earth. It is able to provide geographical location and time regardless the weather conditions nor the place on or near the Earth. In order to use the GPS information a GPS receiver is required. It listens for the information from the satellites. Each satellite transmits the time the message was transmitted and satellite position at the time of message transmission. Receiver using the acquired information computes by trilateration its distance to each satellite in the field of view. This distance is called pseudo-range. It is essential for very precise time measuring in the satellites because even small clock error is multiplied by the speed of light which results in a significant location miscalculations. A satellite's position and pseudo-range define a sphere, centered on the satellite with a radius equal to the pseudo-range. Therefore the position of the receiver is somewhere on the surface of the sphere. With at least four satellites the position of the GPS receiver should be at the intersection of the surfaces. However due to time measurement and atmospheric signal propagation errors the accuracy of the GPS is varies in the range of 0.5m to 100m depending of the receiver and the number of satellites visible. The computed coordinates are expressed using WGS 84 system.

3. Sensors miniaturization

The need for increasing miniaturization of electronic devices in order to use them in everyday tools, miniature robots and UAVs is foreseen to be stopped for a few years before jumping further from the micro to nano-sized devices. Therefore a currently available at the market inertial measurement units need a review. There are multiple commercial inertial measurement units available at the market. For instance 3DM-GX 1 from MicroStrain, MTi-G from Xsens Technologies, Crista IMU from Cloud Cap Technology, µNAV from Crossbow Technology, AHRS200AV2.5 from Rotomotion and ADIS 16400/405 from Analog Devices. In order to verify the capabilities of the mentioned solutions we have compared its characteristics with the current state of the art miniature MEMS IMU 5 developed by authors of the text. The comparison is visible in the tab. 1.

The current state of the art IMU 5 10-DOF allows for measuring angles in three dimensions, accelerations also in three dimensions, direction of the strongest magnet signal and the temperature of the surroundings. It utilizes MEMS technology in order to reduce both size and weight of the sensor. The sensor weighs only 1.13 [gram] without mounting and 3.13 [gram] with the standard mounting.

Measurement data from the IMU 5 sensor depending of the version can be sent through the USB or CAN bus or RS-232 which are typical industry standards of data transfer. The flexibility is required from modern sensors in order to integrate it with an existing systems. For the same reason the sensor has a flexible output data configuration. The output of the sensor can be acquired in three versions: as raw sensor data, as data after calibration and temperature compensation, or as the output of processing filter. Data from the sensors, may also be delivered in various forms to the user. The first form is the rotation matrix, which is generated based on data from the device. Data can also be supplied to the end user in the form of Euler angles and their values given in degrees or in radians. The third type of output data representation is by using quaternions.

Scheme of the operation of the micro measurement unit is presented in fig. 3 and the internal characteristics in tab. 2.

One of the fundamental inertial sensors is gyroscope measuring angular velocity Ω (in the schema GYRO(X), GYRO(Y), GYRO(Z)). Those sensors were oriented in such a way that their measurement axes create right handed Cartesian coordinate system. Analog MEMS type signals output is filtered by configurable low-pass filter (12.5, 25, 50, 110 Hz). After filtration the signal is converted into the digital form by analog-digital converter characterized by 16 bit resolution. The digital signal can be further filtered by configurable low and high-pass filters. The frequency of data from the gyroscopes can be set to 100, 200, 400 or 800 Hz. The resolution of the measurements can be configured with a modified precision in the range from 250 °/s to 2000 °/s.

Another important element in the schema is linear accelerations sensors block (in the schema ACC(X), ACC(Y), ACC(Z)). The sensors were oriented in such a way that the measurement axes also create right handed Cartesian coordinate system. Analog signals from the sensors MEMS output were redirected to the filter and further to the analog-digital converter. Maximum resolution of the measured accelerations can be configured to ±2/±4/±8 [g] ($g \approx 9.81 m/s2$). Accelerometers sensitivity was measured as 1 [mg].

Producer	MicroStrain	Xsens Technologies	Cloud Cap Technology

Model	3DM-GX1	MTi-G	Crista IMU
Internal sensors	accelerometers, gyroscopes, magnetometers, temperature sensor	accelerometers, magnetometers, gyroscopes, GPS	accelerometers, gyroscopes, temperature sensor
Gyroscopes range	± 300°/sec	± 300°/sec	± 300°/sec
Accelerometers range	± 5 g	± 5 g	± 10 g
Digital output	RS-232, RS-485	RS-232, USB	RS-232, CAN
Temperature range	-40°C to +70°C	-20°C to +60°C	-40°C to +70°C
Size with mounting [mm]	64x90x25	58x58x33	52.07x38.8x25.04
Weight [gram]	75	68	38.6
Producer	Crossbow Technology	Analog Devices	Rotomotion
			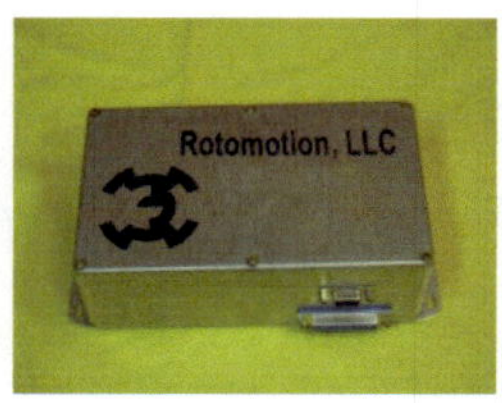
Model	μNAV	ADIS 16400/405	AHRS200AV2.5
Internal sensors	accelerometers, gyroscopes, magnetometers, temperature sensor, GPS	accelerometers, magnetometers, gyroscopes, temperature sensor	accelerometers, gyroscopes, magnetometers
Gyroscopes range	± 150°/sec	± 75-300°/sec	± 90°/sec
Accelerometers range	± 2 g	± 18 g	± 2 g

Producer	MicroStrain	Xsens Technologies	Cloud Cap Technology
Digital output	RS-232	SPI	RS-232, Ethernet
Temperature range	-5°C to +45°C	-40°C to +85°C	-5°C to +75°C
Size with mounting [mm]	57x45x11	31.9x23.5x22.9	>100x100x100
Weight [gram]	33	16	>100

Table 1. Comparison of currently available at the market miniature inertial measurement units.

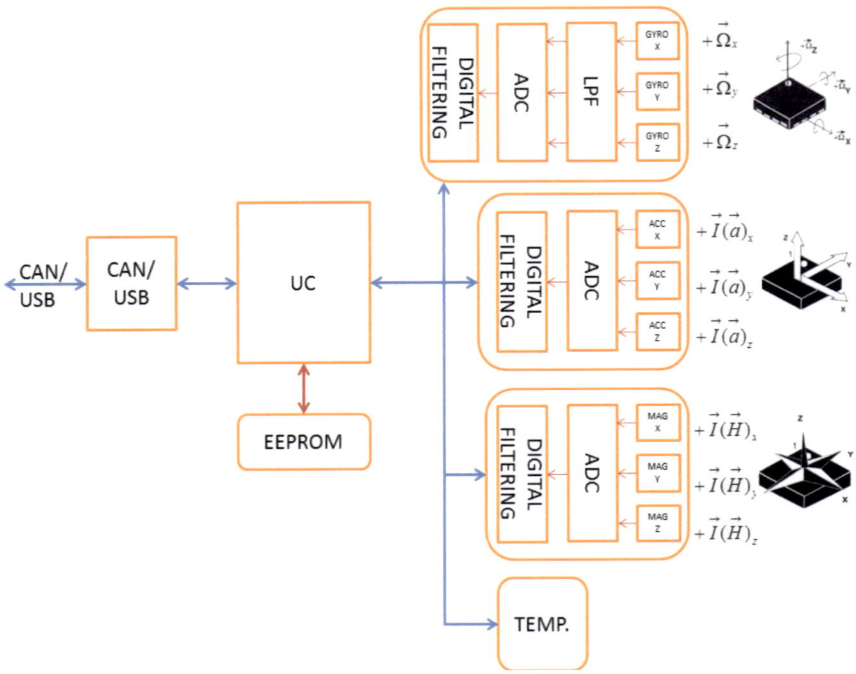

Figure 3. The IMU schema of operation.

Next functional block measures Earth magnetic field by using magnetometers (in the schema MAG(X), MAG(Y), MAG(Z)). Measurement axes of the sensors are oriented in right handed Cartesian coordinate system. The analog output from the sensors is connected with analog-digital converter and further with configurable filters block. Maximum values possible to measure are in the range from 1.3 to 81. [gauss] with resolution of 1/1055 [gauss].

The last functional block contains thermometer that measures temperature in the IMU's environment in order to allow for temperature compensation of the accelerometers, gyroscopes and magnetometers readings.

All of the functional blocks are connected with the central processing unit marked in the schema uProcessor. In order to store the required for algorithms parameters additional memory was added (marked in the schema as eeprom). The CPU performs the filtering before the data is redirected to the CAN or USB or RS-232 output. In order to establish the parameters of the filtering algorithm the calibration phase is required.

Internal sensors	accelerometers, gyroscopes, magnetometers, temperature sensor	Gyroscopes range	from ± 250 to 2000°/sec
Accelerometers range	from ± 2 to ± 8 g	Digital output	CAN, USB or RS-232

Table 2. Internal characteristics of the presented inertial measurement unit.

The presented IMU sensor volume is below 2cm3 with housing and weighs 3.13[gram]. Such size of the sensor can allow designers of robots to move from devices of considerable size to the designs of the micro scale. Maximum acceleration sensor is capable to withstand is 500 [g], while the input voltage is from 3.5 to 8V, and the current consumption is 35 mA. There are three different types of the housing that allows attachment of the sensor in various places. The smallest one Micro version (fig. 4) size with mounting is 18.6 x 14.7 x 7.3 [mm]. Version with the additional mounting holes Micro-Mounting version (fig. 5a) is 18.6 x 20.7 x 7.3 [mm]. The housings of Micro version and Micro-Mounting version output cable is appointed with the USB or RS-TTL connector depending of the version. There is also extended version with mounting holes and a LEMO plug embedded in the housing (fig. 5b). Its size is 32.0 x 18.0 x 16.5 [mm].

Physical measurements of all versions are presented in Tab.3.

Figure 4. The front and back side of the inertial measurement unit.

Calibrated axes orientations are engraved and colored on the top side as presented in Fig. 5. Size of the single mounting is 6.15 mm for Micro-Mounting Version, and 6.50 mm for Extended LEMO version.

Before the application of the IMU in the physical objects a comparison between the smallest available at the market sensors was made in order to verify the possibility of navigation of autonomous mobile vehicles. Comparative studies were designed to check the parameters set up the prototype, analyze the causes of measurement errors and to check how the sensors, which performed tests behave in extreme situations. For the tests two popular commercial miniature sensors were used: MTiG-28G from XSens Technologies and Crista IMU from Microstrain (fig. 6b).

IMU type	Dim. X	Dim. Y	Dim. Z	Units	Volume
Micro version	18.6	14.7	7.3	mm	1,996 cm^3
Micro-Mounting version	18.6	20.7	7.3	mm	2,811 cm^3
Extended version	32.0	16.5	18.0	mm	9,504 cm^3
Common parameters	Min.	Type	Max.	Units	
Weight (w/o housing)	-	1.13	-	gram	
Weight (w/ housing)	-	3.13	-	gram	
Operation temperature	-40	-	80	°C	
Storage temperature	-60	-	100	°C	
Maximal acceleration	-	500	-	g	
Input voltage	3.5	5.0	6.0	V	
Current	30	35	40	mA	

Table 3. Physical characteristic of the sensor.

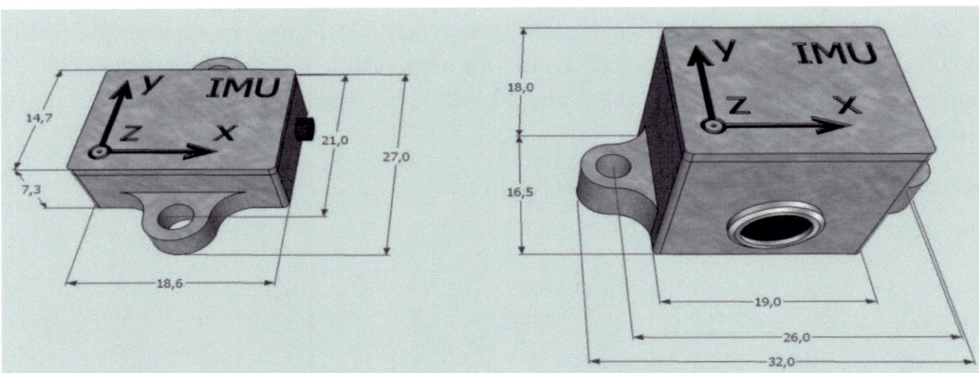

Figure 5. The micro IMU mounting schema. a) Micro-Mounting version, b) Extended LEMO version.

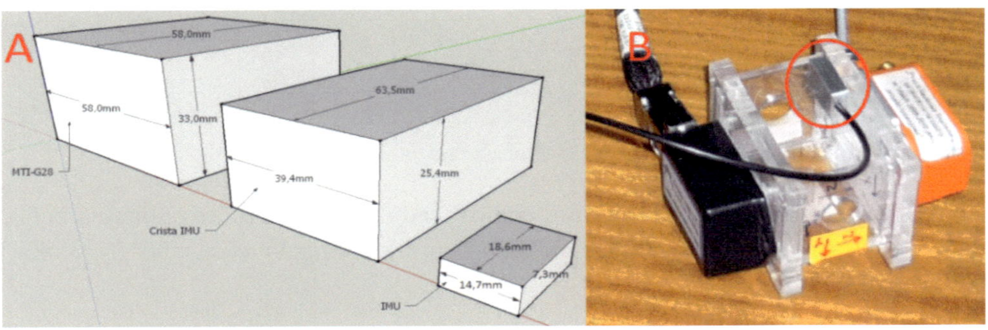

Figure 6. A) Comparison of the presented IMU size to the products available at the market, b) IMU comparison platform. The presented IMU is marked by the red ellipse. In the left side of the image Crista IMU can be seen. On the other side one can see MtiG Imu from XSense Technologies.

In order to perform comparative studies calibration platform was designed, produced and used (fig. 6b). All the tested IMU's were mounted to the platform and their raw data was calibrated to the common coordinate system for all the sensors using the presented calibration method. Several tests were performed: the raw data comparison, the filtered data comparison, time stability of the sensors and capability for temperature compensation. In the fig. 7a it is presented comparison of the readings from the accelerometers of the presented IMU sensor (CZK in the chart) and the MTiG sensor (MTI in the chart) and in the fig. 7b is presented the comparison of gyroscopes. Both comparisons were recorded during motion.

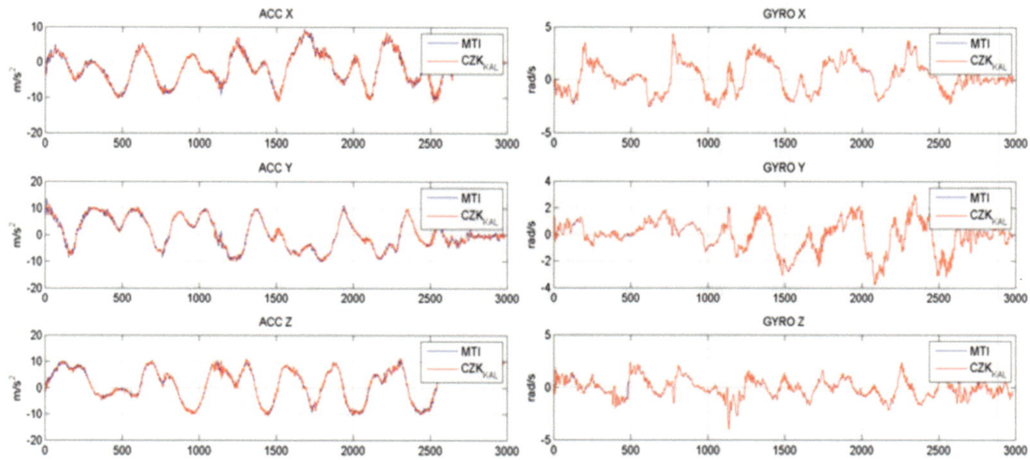

Figure 7. An example chart from comparative studies of the presented IMU sensor and the sensors available in the market. Presented IMU sensor is labeled as CZK and MTiG sensor from XSens Technologies is labeled MTI. Time units are [ms]. a) the readings from the accelerometers, b) the readings from the gyroscopes.

It can be noticed that the output data from the presented calibrated sensor and the reference sensor is almost the same. In order to measure the differences we have estimated the signal from raw data and separated it from the noise component. Only gyroscope Y measurement axe was chosen as an example for presentation. The chose was possible because achieved results are comparable regardless the axe and the sensor (magnetometers, accelerometers or gyroscopes). In the right part of the fig. 8 it is presented that standard deviation of the noise for MTi-G28 and our sensor is about 0.04 while the Crista IMU's result is about 0.12. The difference between standard deviation of MTi-G28 and our sensor is only 0.0035 which is usually undistinguishable by human eye.

Additional table collation of the signal to noise coefficient is presented in tab. 4.

An important problem for inertial measurement units is time stability of the output data. We have performed a series of stability tests with duration of 4h. The presented in the fig. 9 results were acquired after averaging from five samples of each IMU in the test. Two main observations can be made. First that MTi-G28 and our IMU results are comparable. The achieved average values difference is only 0.00027 and the difference between standard deviation values is 0.00226 which is even lower than during the test with movement. The second observation is that our sensor and MTi-G28 results are more than 20 times closer to the real 0 value.

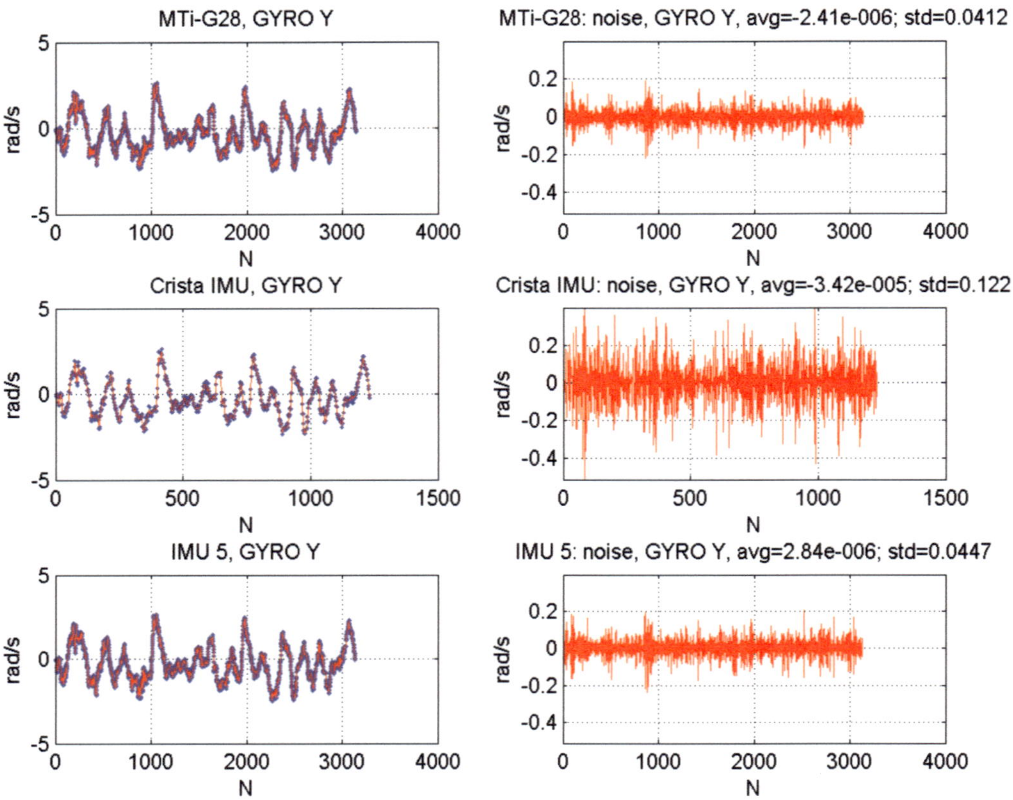

Figure 8. The comparison of the output data during movement from Y axe of gyroscopes for the tested sensors (MTiG-28, Crista IMU and our sensor). Time units are [ms].

	Crista IMU (S/NKAL)	MTiG (S/ NKAL)	IMU 5 (S/NKAL)
Accelerometers	20,3	38,1	43,2
Magnetometers	N/A	719,3	767,9
Gyroscopes	13,6	206,7	254,5

Table 4. The comparison of signal to noise coefficient for the tested IMUs. (The larger the value the better).

The global availability of the GPS, relatively high reliability of the readings and no need for additional infrastructure are the main reasons that most of the modern inertial navigation systems use GPS. The INS typically can be characterized by fast update rate and small but unbound error. GPS error is bounded however update time is slow and attitude estimation is not reliable. Because of the closed architecture of the most of GPS modules the most popular method type of fusion methods is called uncoupled or loosely coupled aiding. INS with GPS is used in vehicle safety systems for estimation of a vehicle sideslip. The standard usage involves vehicle guidance and navigation. GPS signal requires a clear visibility of the satellites which is often not possible in the canyons, especially in urban canyon environments. The GPS/INS systems are often utilized for navigation of a quadrocopter or missile guidance.

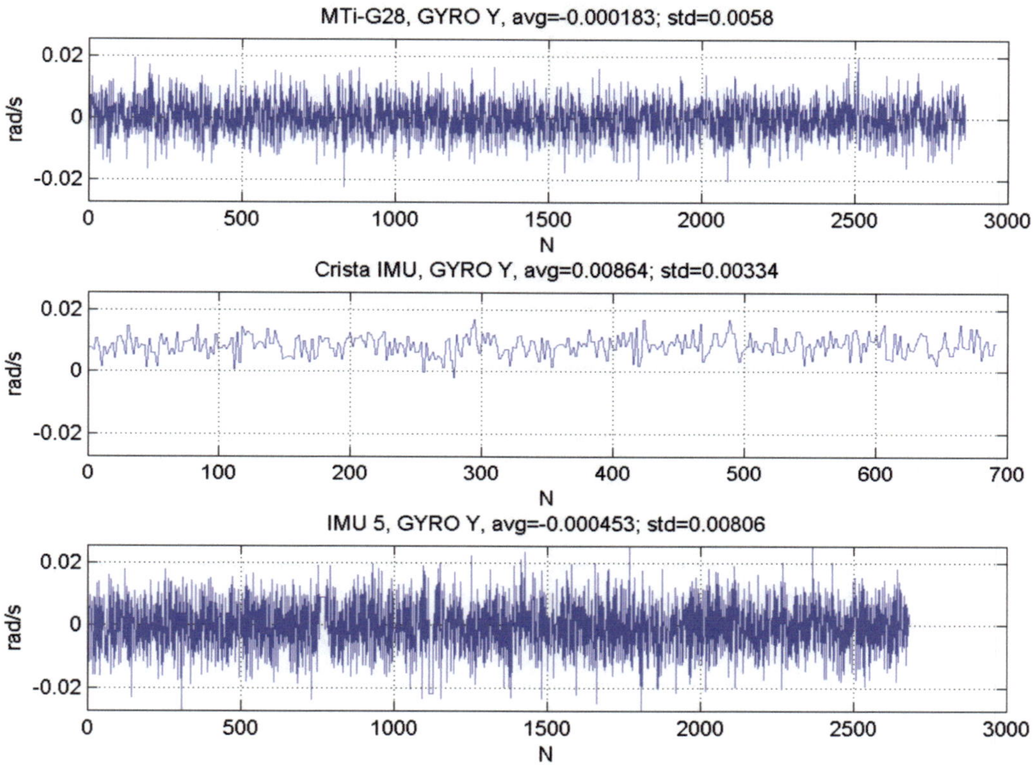

Figure 9. The comparison of the output data during stability test from Y axe of gyroscopes for the tested sensors (MTiG-28, Crista IMU and our sensor). Each value in the chart is an average value of 5 samples from the sensor.

We have developed an inertial navigation system as a complete navigation solution embedded in a single PCB. It contains 10-DOF inertial measurement unit presented above accompanied by Global Positioning System (GPS) module. In order to compensate for the low precision of the altitude estimation of the GPS additional barometer was embedded. The described INS allows for measuring angles in three dimensions, accelerations also in three dimensions, direction of the strongest magnet signal and the temperature of the surroundings. It utilizes MEMS technology in order to reduce both size and weight of the sensor. The INS printed circuit board with MCX connector weighs 3.67 gram without mounting and 8.86 gram with the standard mounting.

Measurement data from the sensor depending on the version can be sent through the USB or CAN bus which are typical industry standards of data transfer. The flexibility is required from modern sensors in order to integrate it with an existing systems. For the same reason the sensor has a flexible output data configuration. The output of the sensor can be acquired in three versions: as raw sensor data, as data after calibration and temperature compensation, or as the output of processing filter. Data from the sensors, may also be delivered in various forms to the user. The first form is the rotation matrix, which is generated based on data from the device. Data can also be supplied to the end user in the form of Euler angles and their values given in degrees or in radians. The third type of output data representation is by using quaternions. The output position coordinates from the GPS are sent in the Earth Centered Earth Fixed (ECEF) coordinate system. The unit of the estimated speed by the GPS is m/s. The used embedded atmospheric pressure altimeter is characterized by internal temperature compensation. It allows for measuring the atmospheric pressure in the range from 20 to 110 kPa with resolution of 1.5 Pa. It can be used to estimate the altitude of the sensor with a resolution equals to 30cm. Geographical localization of the sensor is computed using the GPS module. The GPS is connected to the microprocessor by serial communication bus. In order to improve the quality of the coordinates estimation it is possible to pass to the GPS differential corrections (DGPS). The output from the GPS is in Earth Centered Earth Fixed (ECEF) coordinate system and the frequency is 10 Hz.

The presented INS sensor volume is below 4cm3 with housing and weighs 8.86 [gram]. Such size of the sensor can allow designers of robots to move from devices of considerable size to the designs of the micro scale. Maximum acceleration sensor is capable to withstand is 500 [g], while the input voltage is from 3.5 to 8V, and the current consumption is 50 mA. There are two different types of the housing that allows attachment of the sensor in various places. The smallest one Micro version (fig. 10) size with mounting is 32.7 x 14.9 x 8.0 [mm]. Version with the MCX connector (fig. 3b) is 32.7 x 14.9 x 12.1 [mm]. The volume of the extended version is below 6cm3. The sensor can operate in a wide range of temperatures starting from -40°C to 80°C.

Physical measurements of all versions are presented in Tab.5.

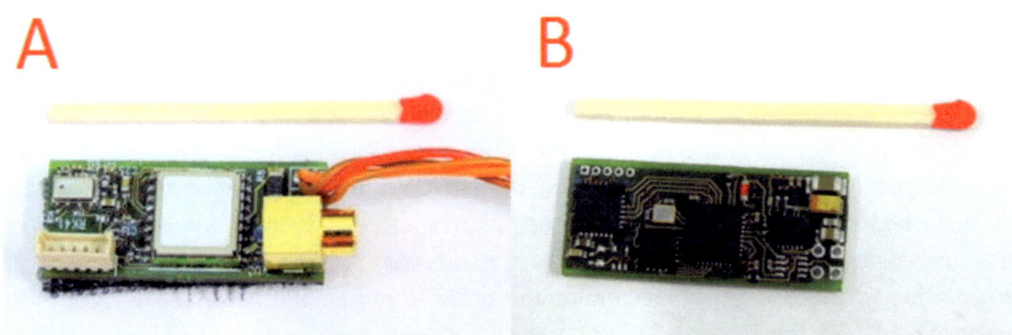

Figure 10. The front (a) and back (b) side of the inertial navigation system PCB with MCX connector.

INS type	Dim. X	Dim. Y	Dim. Z	Units	Volume
Micro version	32.7	14.9	8.0	mm	3.897 cm^3
Extended version	32.7	14.9	12.1	mm	5.895 cm^3
Common parameters	Min.	Type	Max.	Units	
Weight (w/o housing)	-	3.67	-	gram	
Weight (w/ housing)	-	8.86	-	gram	
Operation temperature	-40	-	80	°C	
Storage temperature	-60	-	100	°C	
Maximal acceleration	-	500	-	g	
Input voltage	3.5	5.0	6.0	V	
Current	40	50	60	mA	

Table 5. Physical characteristic of the sensor.

Calibrated axes orientations are engraved and colored on the top side as presented in Fig. 11.

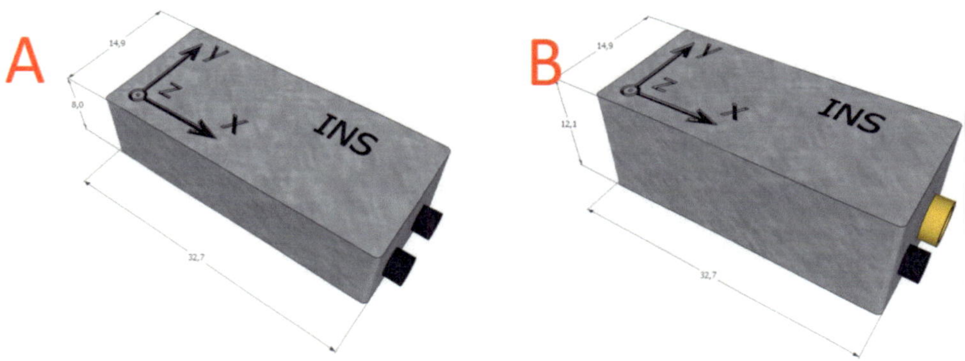

Figure 11. The INS mounting schema. a) Micro-Mounting version, b) Extended version.

The introduced INS was compared against the smallest IMU with a volume below 2 [cm3]. Both the measurement and size axes are marked in the fig 12ab. The X dimension of the micro IMU 5 is 18.6 [mm] which is about 57% of the INS size. The Y dimension is 14.7 [mm] which is about 99% of the INS size and the Z size of the IMU is more less equal to 91% of the Micro version of the INS and 60% of the Extended version. However the Extended LEMO version of the IMU is 18 [mm] which is 225% of the micro version of the INS. Comparison based on the volume only allows the statement that the micro version of the INS is less than two times larger (195% of the micro version of the IMU).

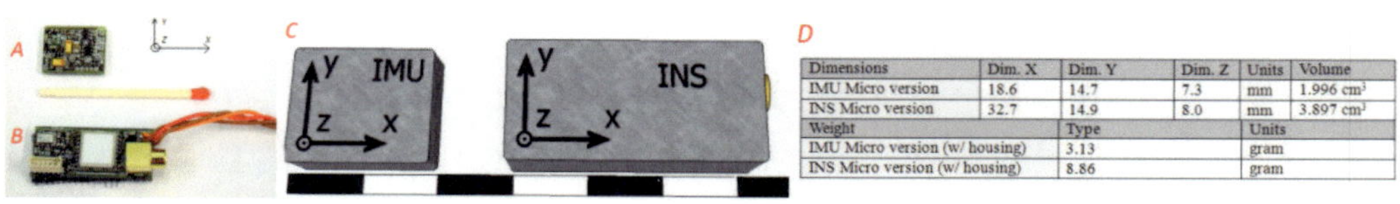

Figure 12. The comparison of the proposed INS (b) with the World's smallest IMU (a). The measurement axes are drawn in the image. c) The comparison of the mounting's size of the presented INS and IMU, d) The comparison of the external characteristics of the proposed INS and the IMU.

4. GPS and INS data fusion

Inertial measurement units are commonly used as a separate tools. However IMUs are also often used as a supplying unit for positioning systems e.g. Global Positioning System (GPS). GPS signal accuracy is not sufficient for object's precise navigation. Estimated position is biased with a large error especially on altitude signal's component. The error can be as high as hundred meters depending on the number of the satellites in the field of view of the GPS receiver. GPS is also not capable of working inside of buildings or in tunnels. There is no acceleration and orientation data. All the lacking information required for precise navigation can be obtained from IMUs. The integration of GPS and IMU into one system called inertial navigation system (INS) is usually implemented in one of two ways: loosely coupled and tightly coupled. Loosely coupled type consists of independent components with additional piece of software which allows sharing program resources. In such systems, application logic is implemented in loosely integrated group of communicating devices. Each device has its own local memory. The system is usually based on Kalman filter processing data from GPS and IMU altogether. Readings from GPS are treated as a reference fix information. As a counter measure for IMU error accumulation simplified schema is presented in fig. 13.

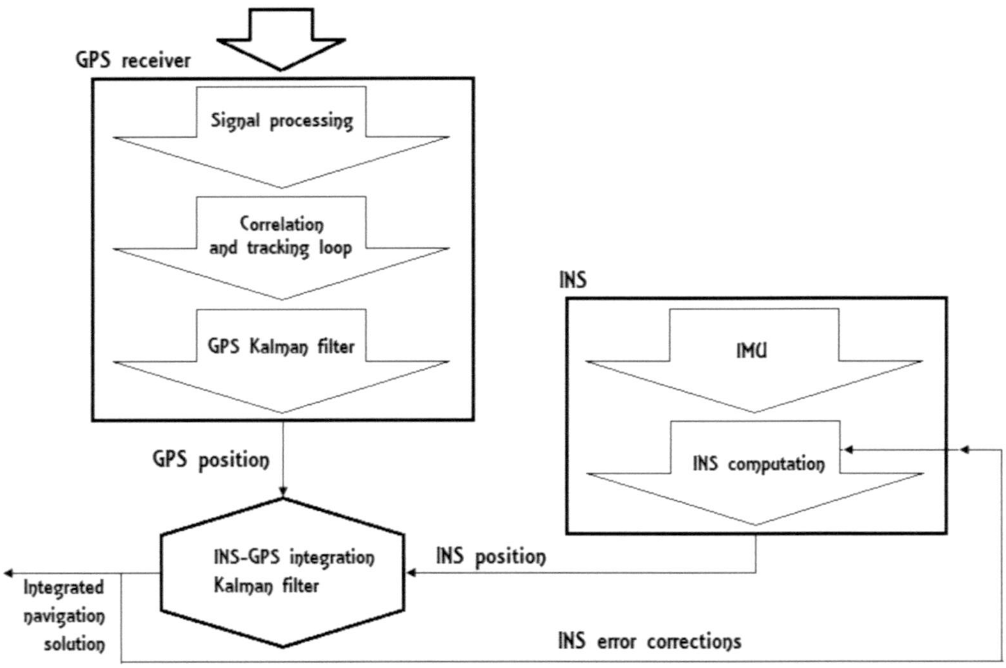

Figure 13. Loosely coupled integration type schema.

One of the most significant advantages of a loosely coupled system is a possibility to share resources among system nodes. Computation speed and reliability can be increased because of distributed processing. In case of error of one components the rest is still operating, therefore the overall reliability is increased. Communication network among the nodes allows to distribute information.

Tightly coupled systems are integrating GPS and IMU in one system with a central shared memory. Each of the devices can exist in common memory hierarchy. Inertial measurement units are used for measuring objects' orientation and acceleration and estimating velocity and position using navigation equations. Computed parameters aids readings, called pseudo-ranges, from the GPS receiver. After correlation, both signals are used by an integration Kalman filter. Simplified schema of the tightly coupled solution is presented in fig. 14. Due to INS corrections, GPS receiver in the system can work also when visible satellites are in range from one to four as good as it is working usually with five and more reference points visible. However accuracy of the systems increases with the number of visible satellites.

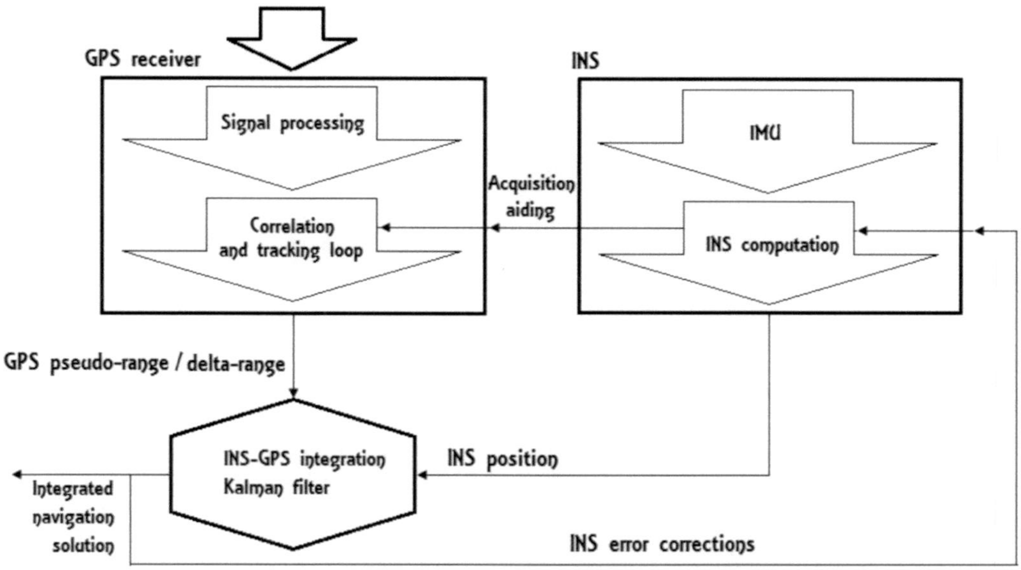

Figure 14. Tightly coupled integration type schema.

Many significant advantages of inertial measurement units can be found. The main advantages are high amount of possible applications, micro size and weight, high quality of data measurement. Due to those features IMUs are gaining more scientific attention each year and are implemented in various applications all over the world.

The dynamic model of the INS-GPS loosely coupled system can be formed from the differential equations of the navigation error. The time-varying stochastic system model is

$$\begin{gathered} \dot{x_{INS}}(t) = F_{INS}(t) + G_{LC}(t)b(t) + G_{LC}(t)w_{LC}(t), w_{LC} \sim N(0, Q_{LC}) \\ x_{INS} = [\delta L \quad \delta l \quad \delta h \quad \delta V_N \quad \delta V_E \quad \delta V_D \quad \theta_N \quad \theta_E \quad \theta_D]^T \quad , \\ w_{LC} = [w_{aX} \quad w_{aY} \quad w_{aZ} \quad w_{gX} \quad w_{gX} \quad w_{gX}]_T \end{gathered} \tag{4}$$

where the error state variable $x_{INS}(t)$ represents the position, velocity, and attitude errors of a vehicle, the bias term b(t) represents the inertial sensor bias errors, which consist of the accelerometer bias error and the gyro bias error, $w_{LC}$is the white noise of the inertial sensor with zero mean and covariance Q_{LC}. The system matrix FINS(t) and $G_{LC}(t)$ are

$$F_{INS} = \begin{bmatrix} F_{11} & F_{12} & F_{13} \\ F_{21} & F_{22} & F_{23} \\ F_{31} & F_{32} & F_{33} \end{bmatrix}, \quad G_{LC}(t) = \begin{bmatrix} 0_{3x3} & 0_{3x3} \\ C_b^n & 0_{3x3} \\ 0_{3x3} & -C_b^n \end{bmatrix} \tag{5}$$

where the information of F_{INS} (t) and G_{LC} (t) is referred in. The accelerometer error consists of white Gaussian noise and the accelerometer bias error, which is assumed as a random constant. The gyroscope error consists of white Gaussian noise and the gyroscope bias error, which is also assumed as a random constant. The measurement model of the INS-GPS loosely coupled system uses the position and velocity information of the INS and the GPS as

$$z(t) = H_{LC}x_{INS}(t) + v_{LC}(t) = \begin{bmatrix} P_{INS} \\ V_{INS} \end{bmatrix} - \begin{bmatrix} P_{GPS} \\ V_{GPS} \end{bmatrix}, \quad v_{LC} \sim N(0, R_{LC}), \tag{6}$$

where $H_{LC} = [I_{6x6} \quad 0_{6x3}]$ and v_{LC} is the white Gaussian noise with zero mean and covariance R_{LC}.

5. Experiments

In order to present the performance of the INS-GPS loosely coupled system, experiments were done on the stadium near the Silesian University of Technology. The differential GPS (DGPS) trajectory with 2 m accuracy was used as a reference trajectory for quantitative comparison. All measurements were provided by the previously introduced micro INS. In order to make the test comparable with existing solutions the output data frequency from the GPS was set to 1 Hz as most of commercial civilian GPS receivers instead of the possible 10 Hz frequency. The path the vehicle passed was recorded and visible in fig. 15f.

Most of practical applications in order to perform INS-GPS data fusion use the well-known Kalman filter. However the Kalman filtering technique requires complete specifications of both dynamical and statistical model parameters of the system. This assumption in most of practical cases is not fulfilled. The measured values often deviate from their nominate values or have an unknown time varying bias. Such conditions seriously affect the Kalman filter estimates. In order to reduce the impact of the deviations a two stage Kalman Filter (TKF) was suggested. In order to extend TKF to nonlinear systems a new two-stage extended Kalman filter was introduced TEKF. However TEKF was only an approximation of EKF. In order to solve this limitation, Caglayan proposed a general TEKF, which considers a general case in which the bias enters nonlinearly into the nonlinear system dynamics and measurements. Then, the bias is treated as not a random bias but a constant bias. An unknown random bias of a nonlinear system may cause a big problem in the TEKF because in a number of practical situations, the information of an unknown random bias is incomplete and the TEKF may diverge if the initial estimates are not sufficiently good. So the TEKF for a nonlinear system with a random bias has to assume that the dynamic equation and the noise covariance of an unknown random bias are known. To solve these problems, an adaptive filter can be used. The current state of the art algorithm is adaptive two-stage extended Kalman filter (ATEKF). The performance of the TEKF is the same as that of the EKF, as both filters are equivalent. However TEKF performance is significantly decreased when the bias is unknown or the sensor is temporary faulted which often happens in real life. Therefore we present the results acquired during the experiments when the bias was both known and unknown using both TEKF and ATEKF algorithms.

The navigation results using the ATEKF when there is no fault are presented in fig. 15.The partial figures show the position, velocity, attitude, an accelerometer bias, a gyroscopes bias, and a position error of the test vehicle used during the experiments. The position error is defined as the difference between a position of the DGPS and the computed value. The accelerometer bias and the gyroscope bias are both well estimated by the ATEKF. In order to perform the comparison of performance between the algorithms, the horizontal-axis average position error is defined traditionally as

$$e_T = \sqrt{\left(\frac{1}{T}\sum_{t_k=1}^{T}|e_N(t_k)|\right)^2 + \left(\frac{1}{T}\sum_{t_k=1}^{T}|e_E(t_k)|\right)^2}, \qquad (7)$$

where $e_N$and $e_E$mean the position errors of the north and east axis, respectively. T is the total navigation time of the vehicle. The TEKF shows the horizontal-axis average position error of 4.70 m and the ATEKF shows 4.90 m for the INS-GPS loosely coupled system without an unknown fault bias. It was confirmed that the TEKF and the ATEKF show almost an equivalent performance for the integrated navigation system when the bias was known.

In order to test the performance of the ATEKF for GPS-INS data fusion in a loosely coupled system when the bias is unknown two situations were taken into consideration: fault of accelerometer and fault of gyroscope. A fault biases (10, 30, 50, 100 [mg]) were inserted into measured data from the X-axis of the accelerometer and (50, 100, 200, 300 [deg/h]) gyroscope.

Fig. 16 shows the navigation results of the INS-GPS loosely coupled system using the ATEKF when the accelerometer sensor had a fault. These figures show the position error in the TEKF and the ATEKF for the integrated navigation system with an accelerometer fault bias. As expected, the position error of the ATEKF was smaller than that of the TEKF. Fig. 17 shows the bias estimation results of the TEKF and the ATEKF for several accelerometer fault biases.

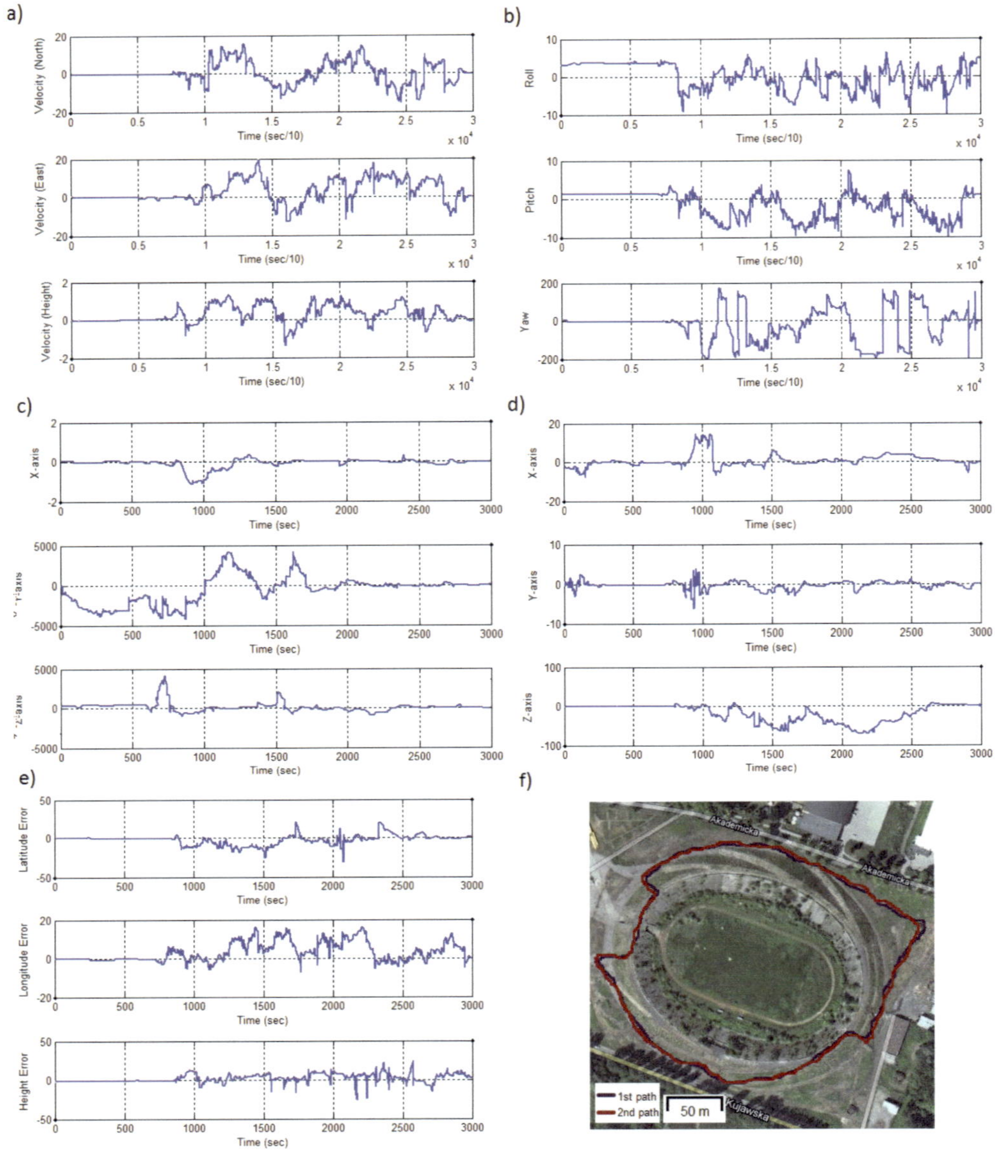

Figure 15. Results acquired using the ATEKF algorithm with known bias. a) Velocity, b) Attitude [deg], c) Accelerometer bias [μg], d) Gyroscope bias [deg/h], e) Position error [m], f) Path around the stadium.

It was evaluated that the tracking performance of the ATEKF was better than that of the TEKF. The ATEKF algorithm was capable of reliable tracking even with an unknown fault bias. The estimate of a fault bias was used for the compensation of the sensor fault, so the navigation error in the ATEKF was smaller than that of the TEKF. Fig. 18 shows the position error in the TEKF and the ATEKF for the integrated navigation system with a gyroscope fault bias. The total tendency was similar to the results in the accelerometer. Fig. 19 shows the bias estimation results of the TEKF and the ATEKF for several gyroscope fault biases. It can be observed in the figures 17 and 19 that the accelerometer bias converges faster than gyro biases.

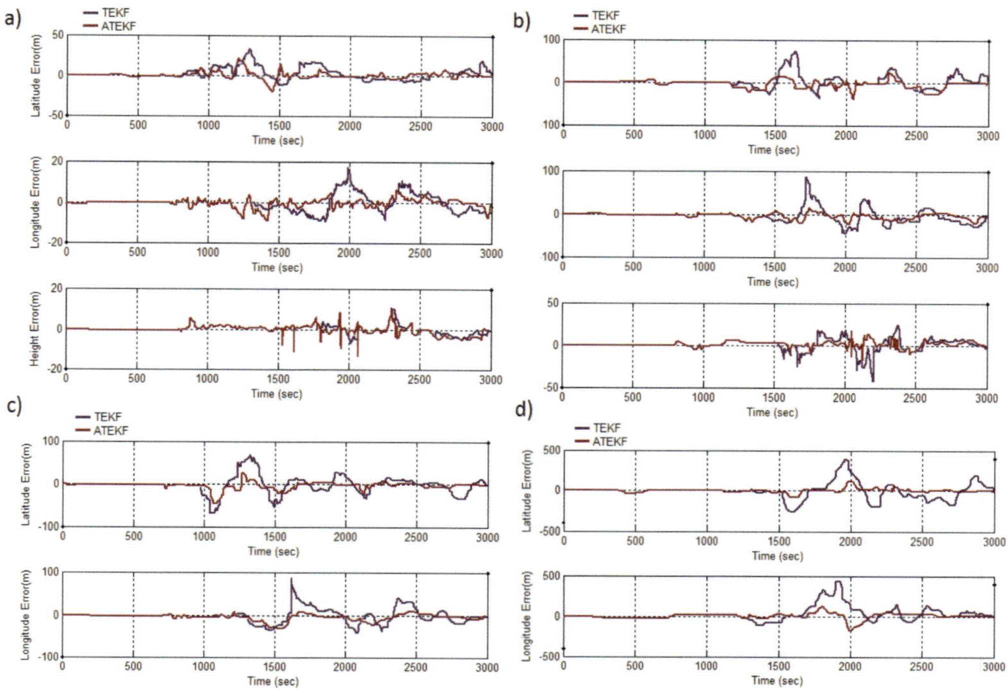

Figure 16. Comparison of the TEKF and ATEKF algorithms in case of accelerometer fault bias. a) X-axis bias 10mg, b) X-axis bias 30 mg, c) X-axis bias 50mg, d) X-axis bias 100mg.

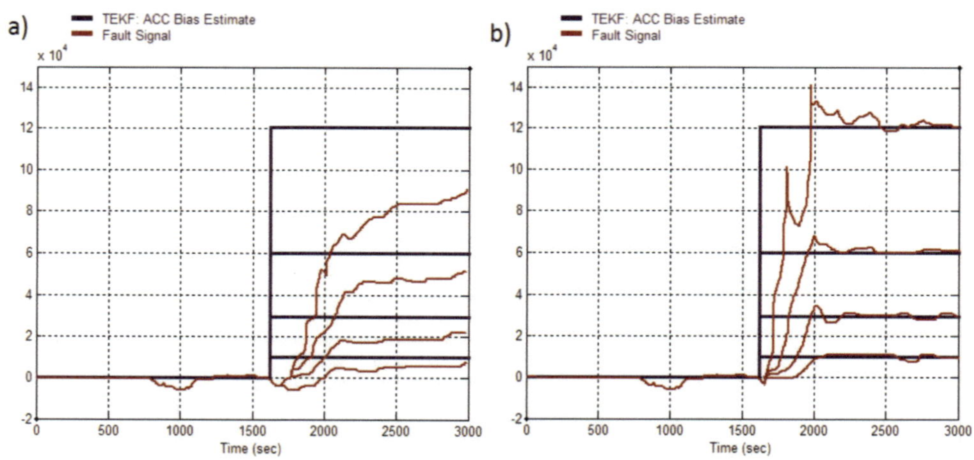

Figure 17. Comparison of the TEKF and ATEKF algorithms in case of several accelerometer fault biases. a) TEKF, b) ATEKF.

In the fig. 20 the response of the filter for a jumping fault bias are presented.

It was presented that the ATEKF filter successfully fuse together signals from accelerometers, gyroscopes and GPS in order to provide reliable and stable measurements required for many applications. Therefore it is justified to safely implement the algorithm in the sensors as it was done for the presented micro INS sensor.

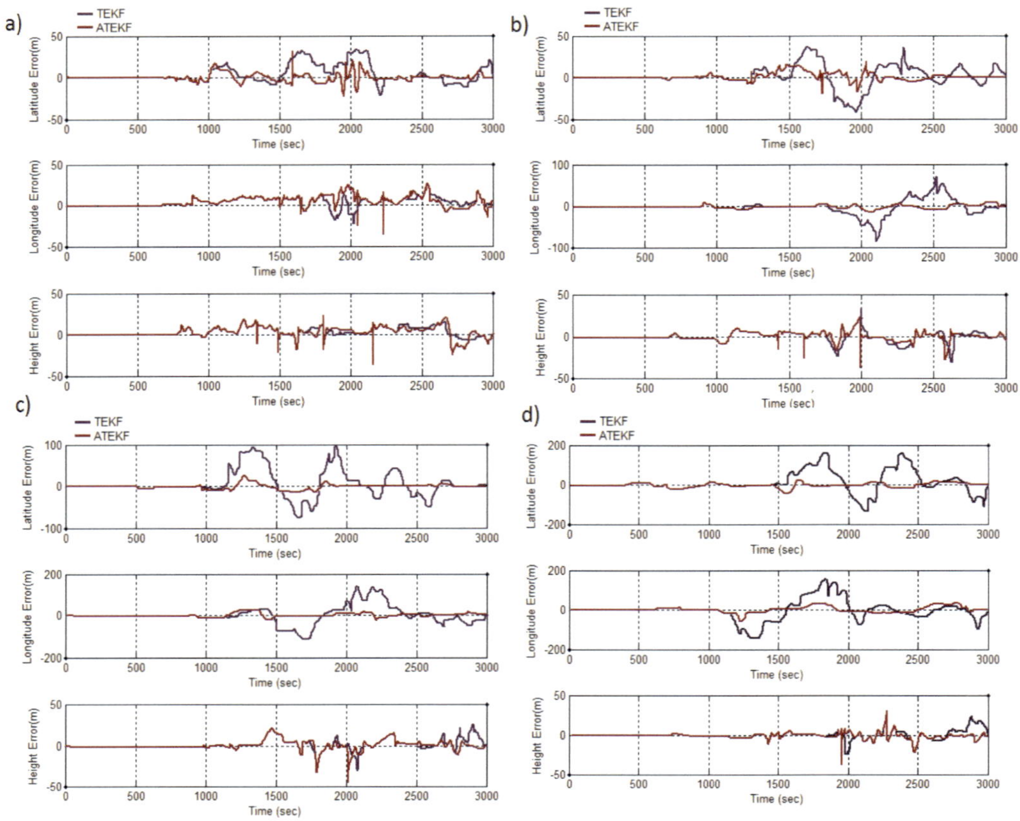

Figure 18. Comparison of the TEKF and ATEKF algorithms in case of gyroscope fault bias. a) X-axis bias 50deg/h, b) X-axis bias 100deg/h, c) X-axis bias 200deg/h, d) X-axis bias 300deg/h.

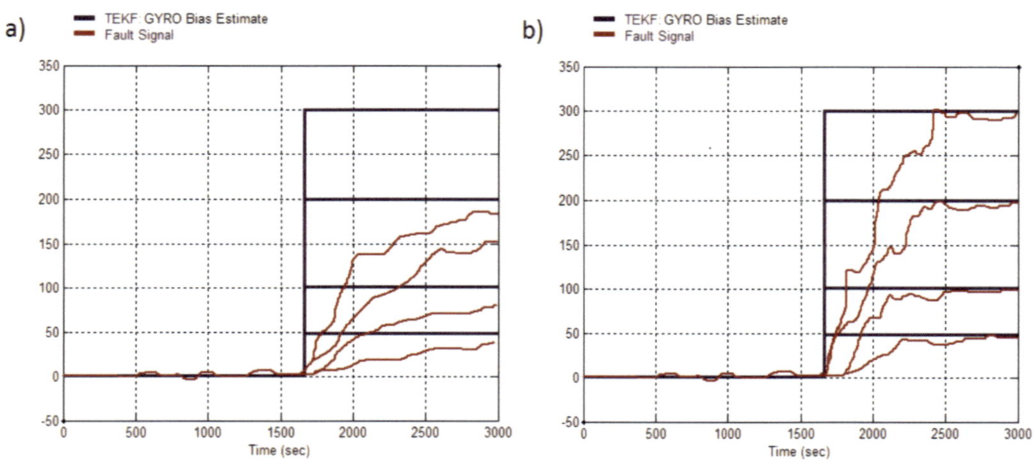

Figure 19. Comparison of the TEKF and ATEKF algorithms in case of several gyroscope fault biases. a) TEKF, b) ATEKF.

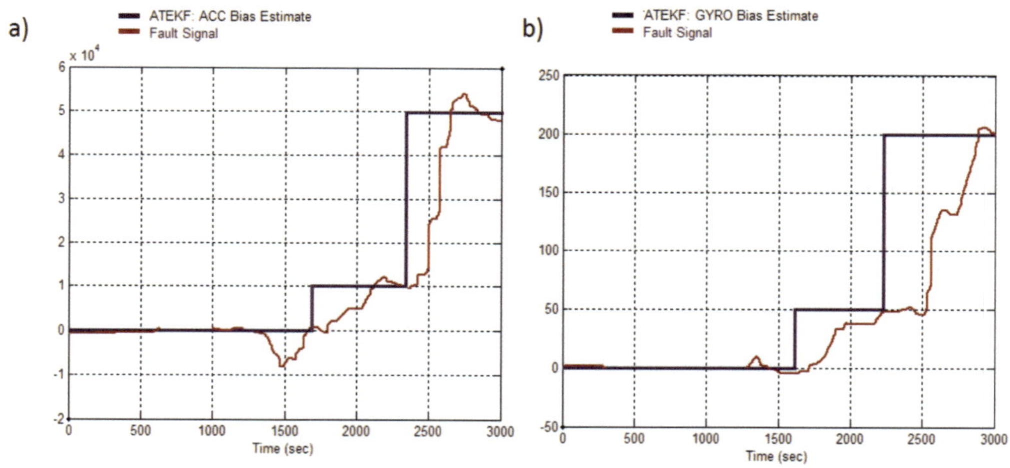

Figure 20. Results of the ATEKF algorithm acquired during fault bias jumping change. a) Accelerometer fault bias. b) Gyroscope fault bias.

6. Practical implementations

The most important in research is its practical application. After evaluation the quality of the inertial measurement units loosely coupled with the GPS receiver the next step was to implement it in real life application. IMUs are widely used in navigation various vehicles and objects. Application of the presented sensors for instance in autonomous navigation of unmanned flying or ground objects can be achieved by extending the approach presented in the experiments section. Main stream of current research in the field of inertial measurement units is to decrease the volume of the sensor in order to make it possible to use in different applications than navigation. Here we present few types of applications possible with the presented miniature sensors.

In order to verify the possibility of real-life application a challenging field was chosen. It was decided that the micro size of the IMU sensor will be used in sports. For instance in running. The sensor can be applied to each of the running shoes of the runner (fig. 21a-b). Alternatively the IMU can be placed on the wrist. It is worth mentioning that the runners were volunteers helping during the research. We have asked runners to test the application during the 4x100m relay race around the stadium. Each of four runners were equipped in the two IMUs. One on each of running shoes. The computed geographical trajectories are overlaid on the satellite view of the stadium. Using the output data from the system it was possible to record the exact path of each runner (fig. 21c). Therefore easily it was possible to calculate the time and velocity vector of each runner. The measured times are 11.89, 11.96, 12.10 and 12.01 and the average speed for all runners was 8,340 [m/s]. It can be seen that there is a movement of the runner while waiting for their turn (start without starting blocks) and after it. The start and stop of the recording was invoked remotely by radio communication and globally for all runners.

Encouraged by the success with the runner we have implemented a system for measuring statistics during football matches. It is important to be able to monitor the activity of each player during and after the match. The collected offline data can easily be used in order to profile the quality and activity of the players during the match and therefore polish the chosen tactic for the match. The statistics are also often used during players recruitment. However the most important in our opinion are two applications of the collected data. First analyzing the reasons behind the injuries during the match and second online monitoring of the match in order to polish the team tactic in the real time by the coaches. Thanks to Piast Gliwice sport club courtesy we were able to record a football match played by professional players. Two players were playing at central midfield, one player at external midfield and forward field. A similar approach to those previously tested with runners was used. Each recorded player was equipped in two micro INS mounted on the shoes and wireless communication board. The new part was a software which allows wireless online monitoring of the variables of each INS, therefore the current activity of selected players. The collected data was used in order to compute the total walking time, total running time, average time of non-activity, average, minimum and maximum speed of running. Regardless the statistics the whole motion path data was recorded for more detailed tactics offline processing. Current implementation was limited to four players however the system can be easily extended to all players. Fig. 21 presents acquired data during the match. An interesting extension of the presented system would be its integration with vision system in order to collect multimodal data.

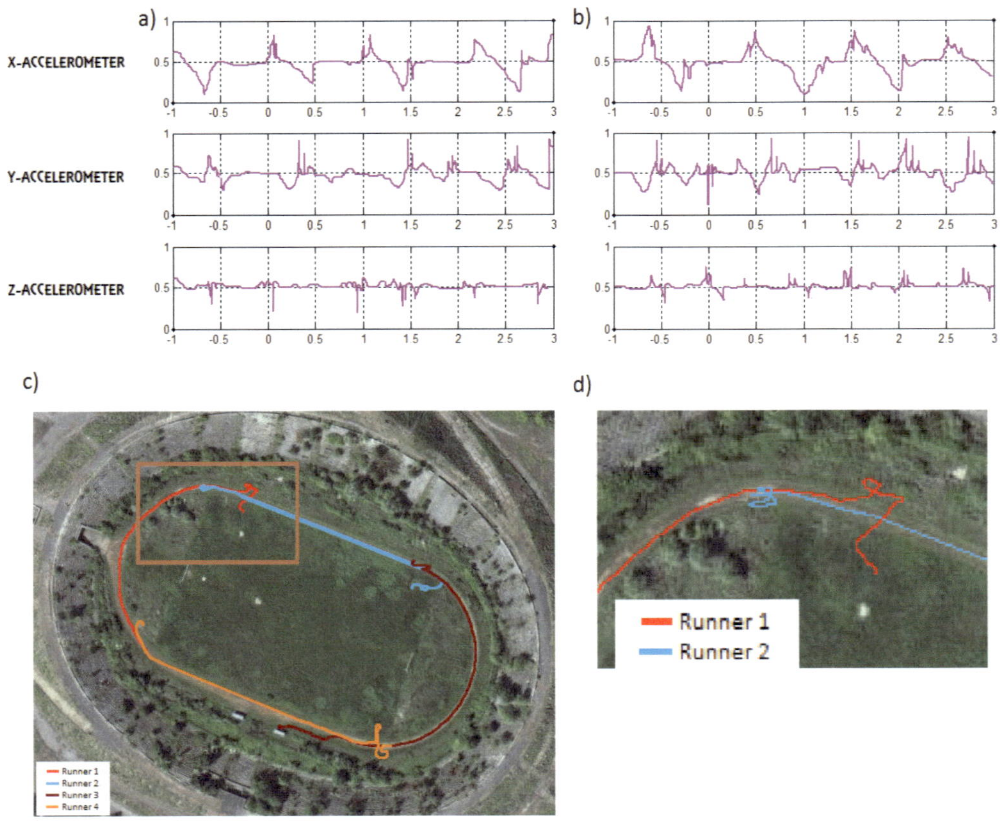

Figure 21. An example data collected during recording 4x100m relay race. a-b) accelerometer data during gait, c) the trajectory, d) zoom in the marked rectangle.

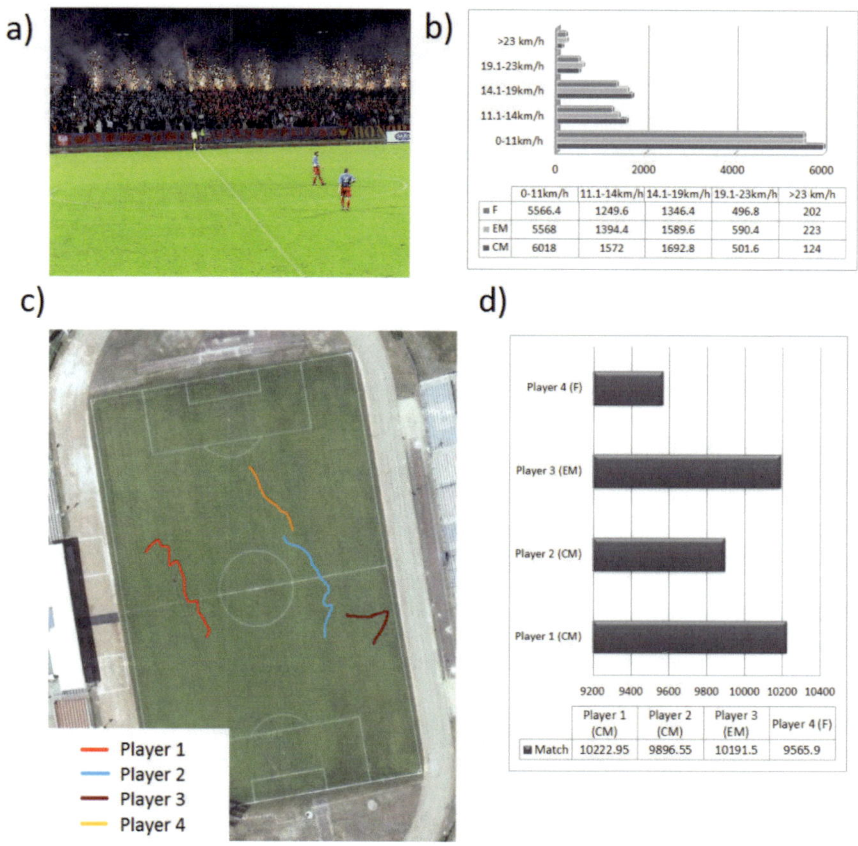

Figure 22. Motion paths acquired using the presented loosely coupled INS-GPS during the football match. a) Photograph from the match, b) distance covered by different types of players in different levels of activity, c) sample motion path overlaid on the satellite view of the stadium, d) total distance covered by different types of players.

7. Conclusions

The chapter presents the complete path of development of an example inertial measurement unit. The process of development includes: choosing algorithms of filtration and fusion, implementation and finally tests and practical applications. As a first step an analysis of existing solutions of algorithms for determining the orientation of the object in three dimensional space was performed. Acquired results allowed to select the technology capable of building a micro inertial measurement unit. Main criteria of selection was miniature size, therefore MEMS technology was chosen. Research team from Silesian University of Technology designed and produced two types of sensors. Micro inertial measurement unit IMU 5 for measuring object's orientation in the space and a second sensor INS extended with altimeter and the GPS receiver. Both presented devices can be distinguished by their many times smaller volume than the currently available at the market. However with smaller size comes no change in the quality of the solution. The next step of research was to design data fusion algorithm from data acquired from the inertial measurement unit and the GPS. The algorithm was implemented in hardware of the sensors and extensively tested in various conditions and applications. The considerations presented in the chapter allow to confirm the possibility of using miniature MEMS sensors in real applications without need to worry about loss of measurement's precision. It was proved that using the presented algorithm it is possible to acquire high quality output data signal. Such output data could be applied to the various real life applications, for instance sport.

Author details

Aleksander Nawrat, Karol Jędrasiak, Krzysztof Daniec and Roman Koteras
Silesian University of Technology, Department of Automatic Control and Robotics, Gliwice, Poland

Acknowledgement

This work has been supported by the National Centre of Research and Development funds in the years 2010 - 2012 as development project OR00 0132 12.

8. References

[1] M. Bao, Micro mechanical transducers: pressure sensors, accelerometers, and gyroscopes, Elsevier, 2000,

[2] W. Williamson, G. Glenn, V. Dang, J. Speyer, S. Stecko, J. Takacs, Sensor Fusion Applied to Autonomous Aerial Refueling, Journal of Guidance Control and Dynamics, Vol. 32, Issue 1, p. 262-275, 2009,

[3] I. Skog, P. Handel, Calibration of a Mems Inertial Measurement Unit, XVII Imeko World Congress, Metrology for a Sustainable Development, September 17-22, Rio De Janeiro, Brasil, 2006,

[4] K. H. Kim, J. G. Lee, C. G. Park, Adaptive Two-Stage Extended Kalman Filter for a Fault-Tolerant INS-GPS Loosely Coupled System, IEEE Transactions On Aerospace And Electronic Systems, Vol. 45, No. 1, January 2009,

[5] H. Junker, O. Amft, P. Lukowicz, G. Troster, Gesture spotting with body-worn inertial sensors to detect user activities, Pattern Recognition 41, 2010-2024, 2008,

[6] J. Moreno, E. Rocon De Lima, A. Ruiz, F. Brunetti, J. Pons, Design and implementation of an inertial measurement unit for control of artificial limbs: Application on leg orthoses, Elsevier, Sensors And Actuators B: Chemical, Vol. 118, Issues 1-2, p. 333-337, October 2006,

[7] D. H. Titterton, J. L. Weston, Strapdown Inertial Navigation Technology, London: Peregrinus, Ltd., 1997,

[8] O. Woodman, R. Harle, Pedestrian localization for indoor environments, Proceeding UbiComp '08, Proceedings of the 10th International Conference on Ubiquitous computing, New York, Usa, 2008,

[9] A. Nawrat, W. Ilewicz, Inertial Navigation using low cost IMU – calculations of UAV orientation in 3-D space, Technological Advancement in Defence and Security of a State, p. 45-61, ISBN: 978-83-62652-06-8, 2010,

[10] A. Nawrat, Modelling and Control of Unmanned Aerial Vehicles, ISBN: 978-83-7335-580-4, 2009.

Chapter 11

The Design and Realization of a Portable Positioning System

Xin Xu, Kai Zhang, Haidong Fu, Shunxin Li, Yimin Qiu and Xiaofeng Wang

Additional information is available at the end of the chapter

http://dx.doi.org/10.5772/50483

1. Introduction

With the rapid development of positioning techniques, the design and realization of a positioning system is inclined to introduce user-friendly platform which can provide easy way to assist end users to locate themselves by utilizing either sensor-based or satellite-based measurements [4].

In the early 1960s, the Inertial Navigation System (INS) was developed by Litton Industries. INS is a sensor-based, self-contained, dead-reckoning positioning system which uses a computer, motion sensors and rotation sensors to perform continuous positioning. The computation of position, velocity and attitude information of moving objects in INS is different from GPS-based methods. This information is calculated by an inertial measurement unit (IMU), where the difference relative to a known starting position, velocity and attitude can be obtained [20]. However, INS suffers from integration drift and leads to unbounded accumulation of errors when calculating the time varied information, due to the needed integrations of small errors in the measurement of acceleration and angular velocity [28], [18]. Therefore, standalone INS-based positioning is unsuitable for accurate positioning over an extended period of time.

Then, in the early 1970s, the US Department of Defense (DoD) created the first Global Positioning System (GPS). GPS is a typical example of space-based satellite navigation system that provides sufficiently accurate information including time, position and velocity [11]. However, GPS cannot work under specific conditions where there lacks an unobstructed line of sight to four or more GPS satellites [3], [29]. In the GPS-denied environments such as urban canyons and tunnels, it is difficult for GPS-based measurements to perform continuous and reliable positioning [15]. Due to the lack of line-of-sight between the receiver's antenna and the satellites in these circumstances, blockage, interference, jamming and multipath effects may easily influence the GPS to perform its function of locating the position of receiver [7].

As mentioned above, either GPS-based or sensor-based measurement cannot provide a continuous and reliable solution. An alternative solution can be obtained by integrating

INTECH
open science | open minds

© 2012 The Author(s). Licensee InTech. This chapter is distributed under the terms of the Creative Commons Attribution License http://creativecommons.org/licenses/by/3.0), which permits unrestricted use, distribution, and reproduction in any medium, provided the original work is properly cited.

measurements from sensor-based systems with GPS-based measurements. The integration is reasonable and can improve the performance by mitigating each others' problems. As the characteristics of sensor-based measurements are complementary with those of GPS-based positioning systems.

In the past, several previous efforts have been made to realize the integration of GPS-based and sensor-based positioning systems. In [12], the GPS was integrated with a full IMU, where three accelerometers and three gyroscopes were contained. In [23], a reduced inertial sensor system (RISS) was proposed, which integrated the GPS with only one gyroscope and an odometer or wheel encoders, to provide positioning information. In [24], Abdelfatah realized this RISS and GPS integrated positioning system. The developed system can compute the data-fused positioning on a field programmable gate arrays (FPGA).

Most of the recent positioning systems can provide effective and efficient solution. However, these products may face many difficulties like long development cycle, high development cost, and short life cycle. In addition, the size and weight of most positioning systems are not easy for end user to use and to carry. The design and realization of portable positioning systems (PPS) is important, which can provide easy way for consumers to locate themselves anywhere. Fortunately, the System On Programmable Chip (SOPC) technology can integrate various modules in a single FPGA, including CPU, memory storage, I/O interface, and etc. And this will provide a great flexibility in hardware configuration. Besides the requirement of PPS hardware, the requirement in PPS software is also need to be considered. The micromation and portability requirement in the design of PPS hardware will inevitably result in more limitation in the storage capacity and processing speed of PPS software. As a result, few storage space and rapid transmission of cartographical is needed in the design of PPS software to ensure the end user experience and runtime efficiency.

The goal of our work described in this chapter is to develop and implement a SOPC based PPS which is capable of handling the data-fused positioning and can provide user a feasible way to obtain their position anywhere.

2. Framework of PPS

The PPS generally includes two main components: receiving terminal and monitoring center. As illustrated in figure 1, the receiving terminal can receive the GPS signal from satellite, which is then fused with the signal from built-in digital compass to perform a combined positioning. Then, the resulting information is transferred to the monitoring center through wireless network and several base stations. In monitoring center, the received positioning information is analyzed and matched with high precise digital map to display the location of the receiving terminal in real time. Finally, the resulting location can be sent back with the compressed digital map to display the location of receiving terminal on its own screen.

In this chapter, we will introduce the design and realization of the PPS, in which three core contents are included.

1. The design and realization of receiving terminal (PPS hardware). A solution integrating measurements from GPS and digital compass is proposed to overcome the problems that arise from using GPS positioning systems in standalone mode. This method can provide

continuous and reliable positioning, even under the circumstances of urban canyons, tunnels, and other GPS-denied environments.

2. The design and realization of monitoring center (PPS software). The processing speed is important for the long-term stable operation of PPS. In order to improve runtime efficiency of the PPS, we compare the performance of native table file and memory table in MapX by using some frequently used operation. By calculating each response time, a suitable storage type can be found out.

3. Digital map compression for PPS. A new digital map compression algorithm is presented to provide low cost storage and rapid data transmission in PPS. This algorithm overcomes the problems that arise from the three traditional compressing algorithms, and can provide reliable and efficient performance.

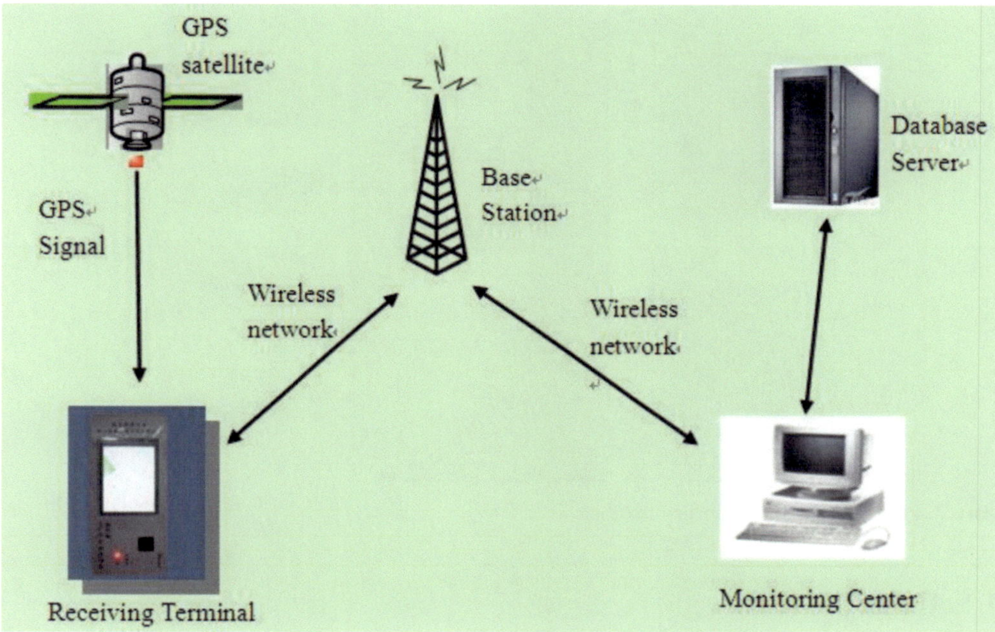

Figure 1. Framework of PPS.

3. Design and realization of receiving terminal

Several hardware platforms can be used for the development of receiving terminal, such as microcontrollers (μC), digital signal processors (DSP), field programmable gate arrays (FPGA) and application specific integrated circuits (ASIC) [16], [9], [22]. The choosing of one hardware platform over others should depend not only on the requirements of PPS such as the performance, power consumption, cost per chip; but also on the easy to use of the tools for producing PPS within the constrained system cost and project time.

The design and realization of receiving terminal in PPS uses FPGA as the development platform. The single FPGA can be integrated, using SOPC technology, with various modules including CPU, memory storage, I/O interface, and etc. And this will provide great flexibility in hardware configuration. The specific hardware description of receiving terminal is illustrated in figure 2.

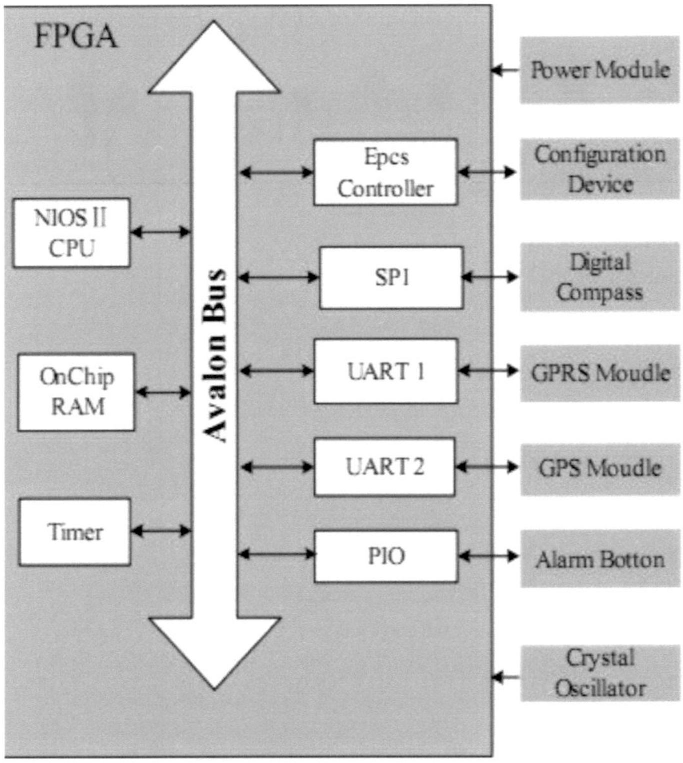

Figure 2. Framework of Receiving Terminal.

3.1. Soft core processor

The receiving terminal is implemented using soft-core processor. A soft-core processor is a microprocessor fully described usually in a Hardware Description Language (HDL), which can be synthesized in programmable hardware. Soft-core processors implemented in FPGAs can be customized easily according to the needs of a specific target application.

Nios II, developed by the Altera Corporation, is a 32-bit general-purpose Reduced Instruction Set Computing (RISC) processor-core [1]. Compared with traditional portable positioning system, Nios II embedded receiving terminal has many advantages:

1. High performance and low price. Nios II soft-core processor has full 32-bit instruction set, data path, and address space. It can perform beyond 150 DMIPS, but only costs 35 cents to implement in a low-cost Altera FPGA device [2]. Besides, the integration implementing of processors, peripherals, memory, and I/O interfaces on a single FPGA can also contribute to the reducing in cost.

2. High flexibility in design. SOPC development tools enable developers to customize the exact peripherals, memory storage, and interface component freely. This is somewhat similar to the component software development, developers can download and configure the needed components in a single FPGA according to different applications.

3. Easy-to-update. By implementing a soft-core processor in a FPGA, in-field hardware component can be updated conveniently. As a result, new features and latest standards can be incorporated easily.

3.2. FPGA

As illustrated in figure 2, the GPS and digital compass modules were both connected with I/O interfaces in FPGA. The signal from satellite can be received by the GPS module, and then fused with the signal from digital compass to perform integrated positioning. In the receiving terminal, GPS module took use of GSU-40 developed by Nihon Kohden Corporation. The size of GSU-40 is 26*26*97 mm3, which meets the portability requirement of the PPS; besides, the positioning data is updated every second to ensure accurate positioning over an extended period of time. The data format in GSU-40 complies with the NMEA-0183 standard.

NIOS II Soft-core CPU is connected with other IP cores via Avalon on-chip bus, which defined time sequence of port and communication between master units and slave units. The hardware design can be accomplished conveniently by using the SOPC Builder system development tools. According to function requirement in this system and high flexibility of NIOS II Soft-core processor, IP cores are designed in FPGA.

Besides, other components were also integrated in this ALTERA CY1C12Q240C8 based on SOPC technology, including NIOS II Soft-core processor, On-Chip RAM, Timer, digital interface and etc. NIOS II/e "economy" soft-core processor: Three kinds of optional processor modes are provided by NIOS II, in this system, NIOS II/e "economy" core is employed which is designed to achieve the simplest possible core at the smallest possible size; Serial Peripheral Interface (SPI): A communication interface used to obtain the positioning data from digital compass; Universal Asynchronous Receiver Transmitter (UART): The interfaces for GPRS module and GPS module to communicate with CPU; On Chip RAM: CY1C12Q240C8 FPGA provides as many RAM memories as 239,616 bits, which are enough for the requirement of PPS hardware system; Parallel Input/Output (PIO) module is used as alarming module, which enable the connection with CPU via PIO to send alarming message to the monitoring center; Epcs Controller module is used to control hardware configuration files and download programs from serial configuration device to FPGA when the system is power up; Timer module is used to provide system time interruption; Power module uses Lithium battery and provide power to PPS hardware system; Crystal Oscillator module provides system clock to PPS hardware system; and An EPCS4 chip is used to store the configuration and software data.

3.3. Digital compass

The signal of digital compass is not from the traditional magnetic needle compass or the gyrocompass, but from an electronic solution (3-axis magnetic sensor and tilt sensor) which adopts the principle of earth magnetic field. The 3-axis magnetic sensor collects the 3-axis earth magnetic field signal; while the tilt sensor is used for compensating "roll" and "pitch". This electronic solution can provide the advantages of a solid-state component without moving the parts and the convenience to interconnect with other electronic systems [6]. The structure of digital compass is shown in figure 3.

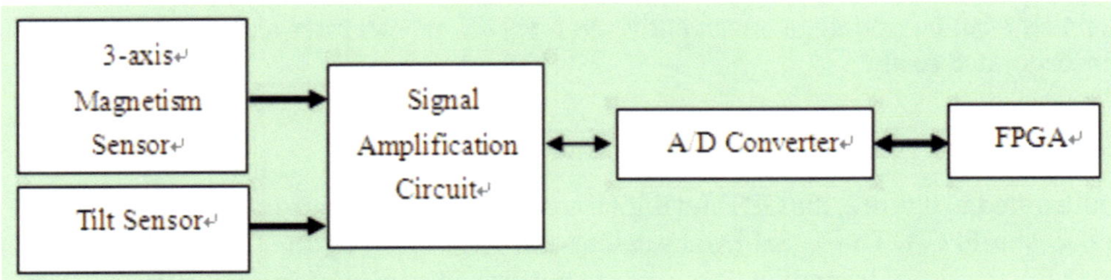

Figure 3. The structure of digital compass.

3-axis magnetic sensor collects the 3-axis earth magnetic field signal and the tilt sensor is used for compensating "roll" and "pitch", and then all of them are amplified and convened to digital signals through A/D converter, at last transmitted to NIOS II by PIO interface.

3.4. Integrating positioning algorithm

GPS positioning signal may be blocked in underground park, tunnel or other GPS-denied places. To solve the problem, this system adopts digital compass to realize combined positioning. An adaptive dead reckoning algorithm is used for fusing the two positioning information.

Dead reckoning is a typical positioning technology in the two-dimensional plane and widely used in vehicle positioning. Once the start location (x_0, y_0) and azimuth (θ_0) angle are given, the current location could be calculated from the prior traveled distance (s_i) and shift angle $(\Delta\theta_i)$. Equation (1)-(3) gives the formula for dead reckoning:

$$Lat_{compass} = x_k = x_0 + \sum_{i=0}^{k-1} (s_i * sin\theta_i) \tag{1}$$

$$Lon_{compass} = y_0 + \sum_{i=0}^{k-1} (s_i * cos\theta_i) \tag{2}$$

$$\theta_i = \theta_{i-1} + \Delta\theta_{i-1} \tag{3}$$

Dead reckoning position error is accumulating with the procedure continues because the current calculated position during each sampling period depends on the previous calculation cycle. So in PPS, we adopts the adaptive dead reckoning algorithm proposed in [21] to improve the precise of this portable position system. The azimuth information is collected by digital compass. When the PPS located not in GPS-denied environment, the two position devices are both used to calculate the location, not only the DR system status are amended constantly, but also exact start location and azimuth information can be provided by the combined positioning output.

The characteristics of INS-based measurements are complementary with those of GPS-based positioning systems. The positioning information from GPS can be used as the initial value and the error correction of digital compass; on the other hand, the signal from digital compass can be used to compensate the random error in GPS to provide accurate positioning, and also can smooth the positioning trajectory. Thus, the resulting positioning information can be given by the following formula:

$$Latitude = (1 - \beta) * Lat_{compass} + \beta * Lat_{GPS} \tag{4}$$

$$Longitude = (1 - \beta) * Lon_{compass} + \beta * Lon_{GPS} \tag{5}$$

Latitude in (4) and Longitude in (5) are the results of positioning coordinates. $Lat_{compass}$ and $Lon_{compass}$ are the coordinates of digital compass; while Lat_{GPS} and Lon_{GPS} are the coordinates of GPS. β is the weight value and varies according to different positioning environments, which can be modeled by fitting the empirical data.

3.5. Experimental results

Figure 4 and figure 5 are the tracking results of the two positioning modes in the monitoring center. As illustrated in figure 4, the standalone GPS based method fails to perform its function in locating its position when it is in the GPS-denied environment (urban tunnel). However, this problem was successfully tackled using the integrated GPS/INS based positioning. The continuous positioning trajectory can be found in figure 5.

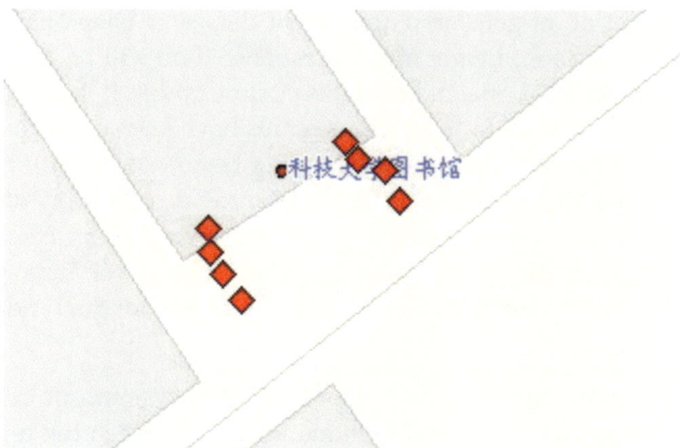

Figure 4. Result of standalone position solution.

By comparing these two tracks of the same object moving on the same route, we can find that standalone GPS based positioning fail to work in the GPS-denied environment regularly, where the line-of-sight between the receiver's antenna and the satellites is lacked. On the contrary, the proposed integrating positioning solution which adopts the positioning information both from GPS and digital compass works well, which can provide continuous and reliable positioning solution under the circumstances of urban tunnels.

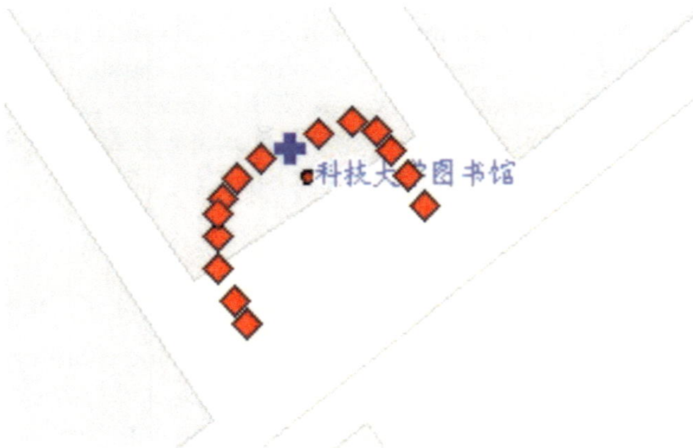

Figure 5. Result of integrated position solution.

Besides, we also conducted a number of testing in different GPS-denied environment. All of them indicate that a good positioning effect can be gained by adopting this integrating positioning technology, which provides a feasible way for the development of the PPS.

4. Design and realization of monitoring center

Real-time monitoring of multiple targets is crucial in PPS. However, due to the frequent status change of each monitoring target, large amount of data was generated during the whole runtime of system. As a result, longer system response time will be taken to process these data and display them on the digital map. In this circumstances, it may incur a problem that follow-up data may have to wait until previous points have been processed. This will worse the runtime efficiency of system. In our monitoring center, MapX component is used for the processing and displaying of digital map. MapX provides a variety of object, properties and methods, among which the Animation Layer can be used as an efficient mean for real time monitoring. However, the main objective of Animation Layer aims to avoid the vision fluttering bring by the fast refresh rate, which is different from our goal and cannot effectively solve this problem.

Conventional approach for reducing the response time of system can be divided into two categories. The first method is to extend the time interval between the receiving of locating information, and realize the real-time positioning through the use of some technology, such as dead reckoning etc, which can assist for positioning. However, it is not a reliable and continuous positioning solution. Another method for real-time positioning is to optimize the runtime efficiency of system by improving the speed of information processing. Specifically, we use memory table instead of native table file to process the locating information, which can avoid frequent I/O process, thereby improve efficiency of the system. In order to illustrate the effectiveness of this approach, we respectively use native table file and memory table in MapX to add and modify the map objects. Then a comparison is made to find the difference in efficiency of the system by calculating their response time.

4.1. Testing methods selection

The MapX Object Model is hierarchical tree structure. Every object, properties and methods in MapX are derived from the Map Object, which located in the top-level of the tree. Maps are the basic building blocks for MapInfo MapX. Each map is defined by three objects: Layer, Dataset and Annotations objects and collections [19].

The first important object is Layer Object and Layers Collection.

Each Map has a collection of layers. The Layers collection is made up of Layer objects. The Layers collection has methods and properties used to add and remove Layer objects from the collection. Computer maps are organized into layers. Think of the layers as transparencies that are stacked on top of one another. Each layer contains different aspects of the entire map. Each layer contains different map objects, such as regions, points, lines and text. For example, one layer may contain state boundaries, a second layer may have symbols that represent capitals, a third layer might consist of text labels. By stacking these layers one on top of the other, you begin to build a complete map. You can display multiple layers at a time. Map layers form the building blocks of maps in MapX. Once you have created your map of layers, you can customize the layers in a variety of ways, add and delete layers, or reorder them.

Map layers display in a particular order. It is important to order your layers correctly. For example, you have a layer of customer points and a layer of census tracts. If the layers are incorrectly ordered in the map window, MapX might draw the customer points first and then display the census tract layer second. Your points would be obscured beneath the census tract layer.

Another object is Dataset object and Datasets collection.

Each Map has a collection of Datasets. The Datasets collection has methods and properties used to add and remove Dataset objects from the collection. Datasets enable you to bind user data to your maps. For example, if you have a Microsoft access database of sales by county and you had a Lotus Notes database of the location of your sales force, you could bind that data to a map and spot trends or notice correlations between the two sets of data. There are many different types of databases in businesses today; therefore, MapX lets you bind to several different types of Data Sources. In MapX, the data is represented as a Dataset object.

The last one is Annotation Object and Annotations Collection.

Each Map has a collection of Annotations (Map.Annotations property). Annotations are either symbol or text objects, and are drawn on top of the map. Annotations are typically used to add messages (text) to a map or to put symbols on a map. These annotations will scale with the map as you zoom in and out. They are not necessarily associated with a particular layer of the map and are always on top.

The greatest impact on the runtime efficiency of system lies in the operation which processes and displays large amount of data to the digital map. According to the preceding discussion on the MapX Object Model, we can confirm that this operation mainly involved in two objects: Layer and Annotations objects and collections. The Dataset object and Datasets collection does not seriously affect this operation. Therefore, we decide to use AddFeature, AddAnnotation and Update these three methods for testing.

4.2. Experimental results

The efficiency of the system can be measured by their response time at runtime. The less response time system needed the higher efficiency system worked in. On the other hand, the more response time system needed, that means longer system response time needed to be taken to process the locating information, and therefore took a toll on efficiency of system. In order to evaluate the efficiency of system, we respectively used native table file and memory table in MapX to add and modify the map objects. Then their response time was calculated. Firstly, we created a temporary layer for testing:

layerinfo.Type = miLayerInfoTypeTemp

Then the storage type, TableStorageType in MapX, of this layer was set up to native table file and memory table using the AddParameter method. For example, we used the following command to set the storage type to native table file:

layerinfo.AddParameter "TableStorageType", "Native"

After the AddFeature, AddAnnotation and Update methods were used into this layer, their respective response time was calculated.

Specifically, we at first calculated the response time needed by AddFeature method to add M*N nodes. And the Timer component was used in telling time. Calculating the response time of AddAnnotation method is similar to AddFeature.

Then the response time can be calculated by using the Update method to operate the M*N nodes. Likewise, the Timer component was used in telling time. And after these operations, their respective response time to memory table and native table file was calculated.

The experiment result was shown in table 1. AddFeature, AddAnnotation and Update in the first line of table corresponded to the operation of adding nodes, adding annotations and updating nodes status. The mem and file in the second line of table corresponded to memory table and native table file. The Nodes in row 1 stood for the number of nodes. The data in table 1 was the result of response time for corresponding operation on the nodes. Let us took the first data in table 1 for example. The number 0.1725 stood for the response time when adding 900 nodes into the memory table.

The experimental results indicate that the use of memory table could avoid frequent I/O process, thereby improved the efficiency of system. Due to large amount of data brought by the frequent status change of each monitoring target, overlong response time would result in the deteriorating of system runtime efficiency. Accordingly, it is suitable to use memory table as storage type in personal portable positioning system. This result would be helpful for other research in which the status of monitoring target needed to change frequently, such as industrial valve monitoring, telecommunications terminal monitoring etc.

5. Digital map compression for PPS

Low cost storage and rapid cartographic transmission is important for PPS. And whether or not the digital map can be entitled low cost storage and rapid transmission is subjected to the efficient compression of the digital map [31].

	AddFeature	AddAnnotation	Update
30*30 mem	0.1725	0.09375	0.28125
30*30 file	7.687625	0.09375	9.313
40*40 mem	0.344375	0.203125	0.593875
40*40 file	14.98094	0.203125	17.43406
50*50 mem	0.6095	0.40625	1.062875
50*50 file	24.65269	0.421875	27.94931
60*60 mem	1.031875	0.75	1.8445
60*60 file	36.87169	0.765625	40.98056
70*70 mem	1.844375	1.40625	3.266
70*70 file	51.12194	1.46875	55.93431

Table 1. Comparison of AddFeature, AddAnnotation, and Update methods on memory table (mem) and native table (file)

Digital map can be divided into two categories: vector structures and grid structure [26]. And they have their own advantages and disadvantages in expressing different geographical phenomenon. The compressing of grid map is not involved in this paper, for it is similar to image compression. On the other hand, vector data compression can be simply defined as the process of eliminating redundancy. As stated by ZHENG [30], vector data compression was extracting a subset from data sets with possibly little data, and at the same time reflecting the original appearance of it as far as possible. Depending on the nature of the application, two classes of compression methods exist: lossless methods and lossy methods. In the lossless methods, data is compressed in such a way that upon reconstruction (decompression), the original data is exactly restored. In the lossy methods, an error measure between the original and the reconstructed data is used to allow a tolerable data distortion. Most vector map compression techniques, including the algorithm introduced in this paper, are of this last category, and eventually a combination of the two categories. Structural tables, namely geometry data, were usually used to store the graphical and topological information of vector map in current applications of GIS, such as MapInfo data. Other data, attribute data for example, was saved in other files. Moreover, attribute data could be saved in files or database, while geometry data was saved in files in most of the time. Still, it organized the files through different layer according to their geographical information. Generally, geometry data, compared to attribute data, took the most of the vector map's storage. For instance, MapInfo vector map's layer, which was saved in file format, generally had two textual file, namely *.MAP for geometry data's storage and *.TAB for attribute data's storage. And this paper will focus on the compression of the *.MAP file.

Specifically, vector map compression is based on the principle of selecting a reduced set of dominant points (DPs) on the curve from the map according to an adequate strategy. Currently, there are many traditional algorithms for compressing vector data. However, they have their own problems when applying the compression to vector map. As a result, an enhanced vector data compressing algorithm was put forward in this work, whereafter, the reliability of this algorithm was tested through experiment. The result of the experiment shows that the enhanced vector data compressing algorithm introduced in this paper could efficiently tackle the problems caused by the classical compressing algorithms, especially effective when aiming at the vector map. Besides, it will hopefully serve as useful feedback information for related works.

5.1. Traditional compressing algorithms for digital map

Three types of geographic information have been developed to represent the vector map over the past years, namely point element, curve element and surface element [13]. Point element, similar to point in geometric feature, stands for location of the target in ground. And two-dimensional coordinates was used to express its spatial information. For instance, we can use point element to express hospital, station, pier, shop and so on. Curve element, similar to curve in geometric feature, can stand for road, river, railway and etc. Surface element, similar to polygon or closed region in geometric feature, usually has the character of area and perimeter. For example, we can use surface element to express farmland, grassland, Public Park, Public Square and so on. Among them, compression towards point element will not have significant effect in saving storage space. Surface element compression can be seen as the compression towards a family of curves [27]. Therefore, compressing towards vector map can be treated as the compressing towards curve element, namely curve simplification.

5.2. Definitions of curve simplification

Curve simplification [5]: Given a polygonal curve $C_{A,B} = (P_1 = A, P_2, ..., P_N = B) = P$, where A and B are the curve endpoints and N is the number of samples, the curve simplification of $C_{A,B}$ consists in computing another polygonal curve $C'_{A,B} = P'(A = Q_1, ..., Q_M = B) \in C_{A,B}$, with $M < N$ and $[P, P'] < \epsilon$, with $\epsilon > 0$, a preset tolerable error of simplification.

5.3. Limitation of traditional curve simplification algorithms

Traditional compressing algorithms can be successfully applied to vector data. However, they had their own problems when applying the compression to vector map.

In James Algorithm [32], due to its' lack of consideration of the curves' global features, the overall direction of the curve may be distorted when there was a continuous small angle change. As shown in figure 6, we calculated the angle change α_2 between line $L_{1,2}$ and line $L_{2,3}$ through the point 2. Then the angles change α_2 was less than the tolerance band α_0. As a result, the point 2 should not be preserved. And similar statements could be made about the points 3, 4, 5 and 6. At last the curve $C_{1,2,3,4,5,6,7}$ could be simplified to line $L_{1,7}$ with a distortion of the overall direction.

Douglas - Peucker Algorithm [8], which is the most popular method to reduce the number of vertices in a digital curve, can keep the overall direction well. The algorithm iteratively selects new points for inclusion in the thinned output curve based on their deviation from a baseline connecting two neighboring points already chosen for inclusion. A review of relevant GIS literature, in journals and online, indicates that implementations often incorporate an error in the method used to calculate the distance between baselines and intermediate data points.

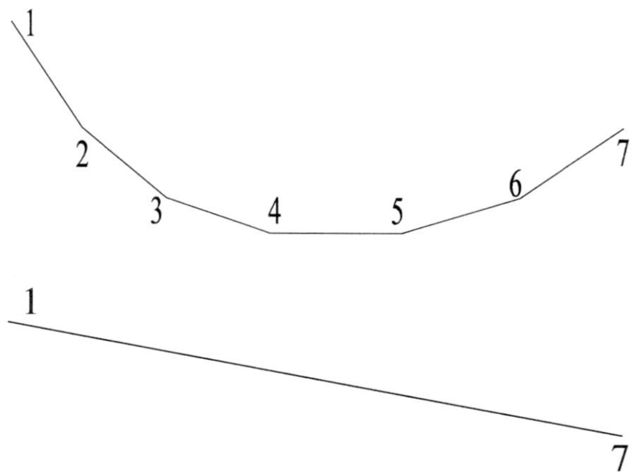

Figure 6. Distortion of the overall direction.

This problem exists because the original Douglas-Peucker paper is somewhat ambiguous in its definition of the distance criterion for selection of a point [8]. On pages 116 and 117 of their paper we see: (1) The perpendicular distance of C from the segment A-B, (2) distance from the straight line segment, (3) the furthest point from the straight segment, (4) maximum perpendicular distance from the segment, (5) the greatest perpendicular distance between it and the straight line defined by the anchor and the floater. Of these definitions, (2) and (3) are correct, but (5) seems to be the most widely used by programmers [10]. We need the distance from the segment, not the distance from the line or the perpendicular distance from the segment. To see the effect of this mistake, consider the situation shown in figure 7. When the curve witnessed a significant angle change, some characteristic bending points (point P in figure 7) might be lost after compression and resulted in terrain distortion.

Light Column Algorithm calculated through ever-dwindling caliber to detect the point outside the fan-shaped region [14]. Owing to the increasingly stringent conditions of detection, the compressing was not that efficient. Still, similar to Douglas - Peucker Algorithm, when the curve saw a significant angle change, some bending feature points might be lost and resulted in terrain distortion. A simpler example, which could be used to test whether an implementation of the Light Column Algorithm suffers from this mistake, may be seen in figure 8. The modified line segment will miss the removed point during the line $L_{B,P}$ and $L_{P,A}$.

In spite of the reduction in tolerance band could solve the above problems; however there, at the same time, will inevitably be a significant reduction in the compressing ratio. Accordingly, some improvement on compressing algorithm for vector map was investigated in this paper.

5.4. Proposed enhanced vector data compressing algorithm

In order to solve the limitation of classical algorithms and extract as less data as possible without reduction in accuracy, it was considerable to integrate the merits of the above classical algorithms.

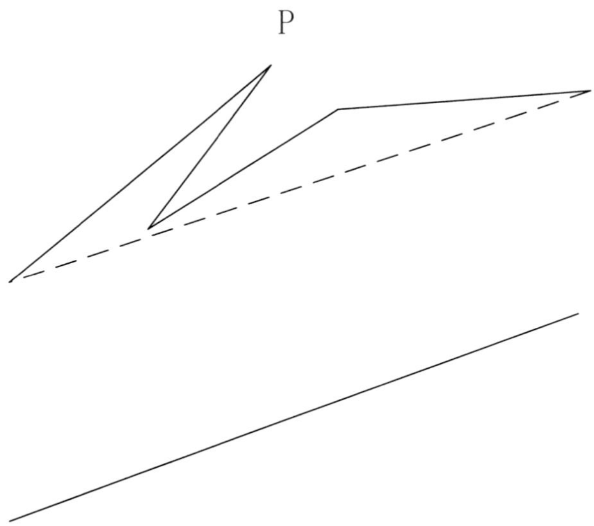

Figure 7. Terrain distortion.

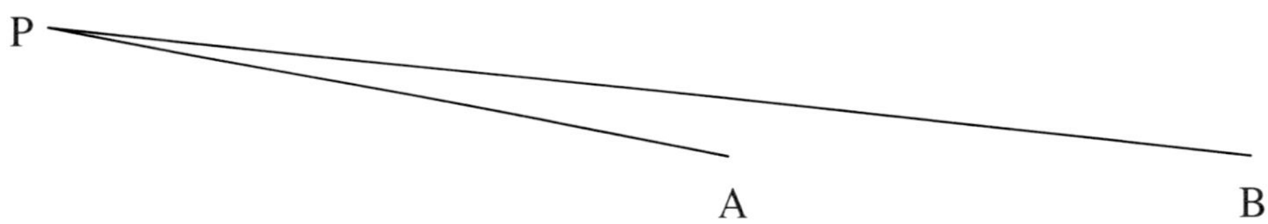

Figure 8. Polyline with three vertices.

Procedure of this algorithm:

(1) Using James Algorithm to extract the characteristic point with a significant angle change, and compressing the curves by Light Column Algorithm at the same time.

For all points in each curve, take P_i for example, the angle change α_i between line L_{P_{i-1},P_i} and line $L_{P_i,P_{i+1}}$ was calculated. Afterward, whether or not the point P_i witnessed a significant angle change and should be preserved was subjected to whether or not the angle change α_i was more than the tolerance band α_0.

On the other hand, we drew a straight line perpendicular to the line L_{P_{i-1},P_i}, which intersect the edge of the fan-shaped region defined by Light Column Algorithm at b_1 and b_2. Then, a_1 and a_2, which was located in this straight line and d/2 apart from the point P_i, was connected separately to P_{i-1} for defining a new sector. Besides, if the variable flag was equal to the true value, a_1 or a_2 should be replaced by b_1 or b_2 when they were outside the sector. As a result,

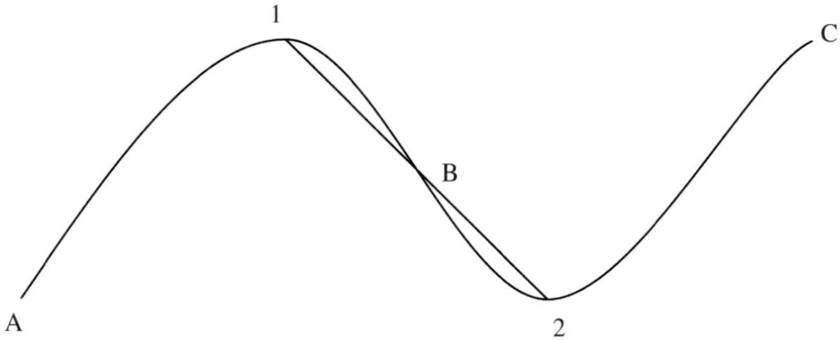

Figure 9. Endpoint redundancy.

the sector was updated. Afterward, if P_{i+1} was inside this sector, P_i should be replaced by P_{i-1}, and set flag to true. Else flag should be set to false.

Then the first step of compression was completed, and the point with a significant angle change was marked at the same time.

(2) Cutting the compression curve through the point with a significant angle change, and then compressing them separately by Douglas - Peucker Algorithm.

There were some problems, which had been discussed in reference [27], should be pay attention to in using Douglas - Peucker Algorithm to compress the vector map.

Closed curve should be divided into two curve end to end, and then compressing them separately by Douglas - Peucker Algorithm. The principle for selecting the split points lay in three criterions.

a The split point should be one of the DPs.

b The other split point should be the point with the greatest distance between it and the split point.

c The split curve should be the longest one of DPs' connection.

Redundant data may follow the compression to a family of curves linked end to end, for the initial point and the endpoint were extracted as the characteristic point for each curve. As shown in figure 9, point 1 and 2 respectively stand for the nearest vertices to endpoint B, therefore the curve $C_{1,B}$ and $C_{B,2}$ can respectively be replaced by subtense $L_{1,B}$ and $L_{B,2}$. However, the entire curve $C_{1,B,2}$ can be replaced by subtense $L_{1,2}$ with reduction of point B.

Accordingly, additional procedure is needed: detecting whether or not the endpoint (point B in figure 9) can be deleted.

a If the two succession curves were the public edge of a polygon, then the public point should be preserved.

b Else if the curve between the two points which were extracted from the two succession curves satisfied the criterion of Douglas - Peucker's tolerance band, then the joint point can be deleted.

c Else the point should be preserved.

In this program, depth-searching algorithm was adopted to find the public edge of the polygon [25].

After these two steps of compression above, the curve was compressed. Also, the procedure associated to this strategy was given in Algorithm 1, where $C_{A,B}$ is the curve needed to be simplified and J, G and D were the tolerance band of James, Light Column and Douglas-Peucker Algorithm. Besides, α_k stood for the angle change between line $L_{k-1,k}$ and line $L_{k,k+1}$ through the point P_k; LCA(P_i, P_k, G) stood for a sector defined by points P_i, P_k and tolerance band G; DP($C_{A,B}$, D) stood for procedure of the Douglas-Peucker Algorithm.

Algorithm 1: CurveSimp($C_{A,B}, J, G, D$)

Input: J, G and D were the tolerance band of James, Light Column and Douglas - Peucker Algorithm, $C_{A,B}$ is the curve needed to be simplified.

Output: P' is the set of dominant points after compression, M is the number of the points.

```
P' ← ∅ ∪ P_1, i ← 1, k ← 2, flag ← false;
while P_k ∈ C_{A,B} do
    if α_k ⩾ J then
        P' ← P' ∪ P_k;
    else
        LCA(P_i, P_k, G);
        if flag ≠ false then
            a_1 ← b_1, a_2 ← b_2;
        end
        if P_{k+1} ∈ LCA(P_i, P_k, G) then
            P_k ← P_i, i ← i + 1, flag ← true;
        else
            P' ← P' ∪ P_k, flag ← false;
        end
    end
    k ← k + 1;
end
if P_1 == P_N then
    CurveSimp(C_{A,C}, J, G, D); CurveSimp(C_{C,B}, J, G, D);
end
M ←| P' |;
DP(C_{A,B}, D);
return (P', M);
```

5.5. Performance evaluation

The reliability of algorithm can be tested through the comparison of the data, including the total length and average coordinate of the curve, with corresponding data before compression. The less difference in corresponding data, the higher fidelity was.

Name	Length (km)	Difference
Original curve	9.023876126834	0
James Algorithm	8.951782492573	7.98921E-3
Douglas-Peucker Algorithm	9.013681028737	1.12979E-3
Light Column Algorithm	9.015937028463	8.7979E-4
This Algorithm	9.016173945937	8.5353E-4

Table 2. Length Comparison

Name	Coordinates (km)	Difference
Original curve	114.385484755	0
	30.6358562641	0
James Algorithm	114.385866870	3.3406E-6
	30.6349333413	3.0126E-5
Douglas-Peucker Algorithm	114.385768397	2.4797E-6
	30.6353483255	1.6580E-5
Light Column Algorithm	114.385682864	1.7319E-6
	30.6354687928	1.2648E-5
This Algorithm	114.385515924	2.7249E-7
	30.6356739586	5.9507E-6

Table 3. Average Coordinates Comparison

In order to evaluate the reliability, the implementation of the algorithm was put forward in Windows XP platform, using the Visual C++ 6.0 as the Integrated Developing Environment. Besides, MapInfo data was adopted to show the vector map with the precision of 1:10000, from which the 3rd trunk road of Qingshan District in Wuhan was selected for experiment. Based on the requirement of the application, threshold value was set to 2 meters for corresponding tolerance band. An improvement on the result shown below could be made by based on the data provided in Table 2 and Table 3.

As shown in Table 2, the enhanced vector data compressing algorithm introduced in this paper had the least difference (8.5353E-4 in Table 2) in the comparison of the length value after the compression.

Similar statements could be made about Table 3, where the corresponding data of average coordinates had the least difference (2.7249E-7 difference in longitude and 5.9507E-6 difference in latitude in Table 3) after the compression by the enhanced vector data compressing algorithm.

Furthermore, the enhanced vector data compressing algorithm could process during the digitalization of the map with linear temporal complexity and linear spatial complexity in most of the time [17]. Thus, it had a fast compression speed.

6. Conclusion

This work addresses the design and realization of PPS. In PPS, a integrating positioning solution is proposed to overcome the problems both in standalone modes and in traditional integrating measurements. This method, which integrates the data both from GPS and digital compass, can provide continuous and reliable positioning, even under the circumstances of urban canyons, tunnels, and other GPS-denied environments. In order to improve runtime efficiency of PPS, we conduct a testing at the monitoring centor where the performance of native table file and memory table in MapX are compared using some commonly used operations. In addition, a new digital map compression algorithm is presented to provide low cost storage and rapid data transmission in PPS. This algorithm overcomes the problems that arise from the three traditional compressing algorithms, and can provide reliable and efficient solution.

There are a few improvements can be made in future research. First, despite the use of a combination of switching power supplies and linear power can improve the efficiency of power usage, it is still necessary to further improve the power component design. Second, the proposed PPS can only be positioned in the horizontal plane. It could be further investigated with tilt compensation based 3D positioning scheme.

Acknowledgments

The authors are thankful to Professor Kangling Fang from the College of Information Science and Engineering at Wuhan University of Science and Technology for her support and useful discussion. This work was supported in part by the Young Scientists Foundation of Wuhan University of Science and Technology (2012xz013), the Project (2008TD04) from Science Foundation of Wuhan University of Science and Technology, the Program of Wuhan Subject Chief Scientist (201150530152), the Educational Commission of Hubei Province (Q20101101, Q20101110), the projects (2009CDA034, 2009CDA136, 2008CDB345) from the Natural Science Foundation of Hubei Provincial, the project from Hubei Provincial Natural Science Funds for Distinguished Young Scholar of China (No. 2010CDA090), the project from Wuhan Chen Guang Project (No. 201150431095), and the Natural Science Foundation of China (60803160, 60975031, 61100055).

Author details

Xin Xu, Kai Zhang, Haidong Fu, Shunxin Li, Yimin Qiu and Xiaofeng Wang
School of Computer Science and Technology, Wuhan University of Science and Technology, China

7. References

[1] Altera, C. [2005a]. *Embedded Processor Solutions Overview*, http://www.altera.com.cn/technology/embedded/emb-index.html.

[2] Altera, C. [2005b]. *NIOS II Processor Reference handbook*, http://www.altera.com.cn/literature/litnios2.jsp.

[3] Bahr, A. [2009]. *Cooperative localization for autonomous underwater vehicles*, PhD thesis, Massachusetts Institute of Technology, Cambridge, MA, USA.

[4] Bekir, E. [2007]. Introduction to modern navigation systems, *Technical report*, World Scientific Publishing Company.

[5] Boucheham, B. & Ferdi, Y. [2006]. Recursive Versus Sequential Multiple Error Measures Reduction: A Curve Simplification Approach to ECG Data Compression, *Computer Methods and Programs in Biomedicine* 81: 162–173.

[6] Caruso, M. [1997]. Applications of magnetoresistive sensors in navigation systems, *Technical report*, SAE Technical Paper 970602.

[7] D. Gaylor, G. L. & Key, K. [2005]. Effects of multipath and signal blockage on gps navigation in the vicinity of the international space station (iss), *Journal of The Institute of Navigation* 52: 61–70.

[8] Douglas, D. H. & Peucker, T. K. [1973]. Algorithm for the Reduction of the Number of Points Required to Represent a Digitized Line or Its Caricature, *The Canadian Cartographer* 10(2): 116–117.

[9] Dubey, R. [2008]. *Introduction to Embedded System Design Using Field Programmable Gate Arrays*, Springer, Berlin, Germany.

[10] Ebisch, K. [2002]. A Correction to the Douglas - Peucker Line Generalization Algorithm, *Computers and Geosciences* 28: 995–997.

[11] El-Rabbany, A. [2002]. *Introduction to GPS: The Global Positioning System*, Artech House Publishers.

[12] Farrell, J. [2008]. *Aided Navigation: GPS with High Rate Sensors*, McGraw-Hill Professional.

[13] Fu, W. [1997]. Expression of Knowledge and Organization of Knowledge base in Geographical Expert System, *Journal of Applied Science* 15(4): 482–489.

[14] Hu, P., Huang, X. & Hua, Y. [2002]. *Tutorial of Geographical Information System*, Wuhan University Press, Wuhan.

[15] Kehrer, D. & Bachu, D. [2011]. New generation high linearity navigation front-end devices covering gps and glonass, *Microwave Journal* 54: 86–96.

[16] Lapsley, P. [1997]. *DSP Processor Fundamentals: Architectures and Features*, IEEE Press, New York, USA.

[17] Liu, X. & Li, S. [2005]. Study on Subsection Douglas Algorithm with the Goniometry in Generalization, *Journal of Surveying and Mapping* 28(2): 51–52.

[18] M. Valtonen, L. Kaila, J. & Vanhala, J. [2011]. Unobtrusive human height and posture recognition with a capacitive sensor, *Journal of Ambient Intelligence and Smart Environments* 3: 305–332.

[19] MapInfo, C. [2004]. MapX online help.

[20] Nassar, S. [2003]. *Improving the inertial navigation system (INS) error model for INS and INS/DGPS applications*, PhD thesis, The University of Calgary.

[21] Q. Chang, D. Y. [2005]. *Vehicles Navigation Positioning Methods and Applications*, China Machine Press.

[22] Stringham, G. [2009]. *Hardware/Firmware Interface Design: Best Practices for Improving Embedded Systems Development*, Newnes: Waltham, MA, USA.

[23] U. Iqbal, A.F. Okou, A. N. [2008]. An integrated reduced inertial sensor system-riss/gps for land vehicle, *In Proceedings of the Position, Location and Navigation Symposium*, pp. 1014–1021.

[24] W. F. Abdelfatah, J. Georgy, I. U. N. A. [2012]. FPGA-based real-time embedded system for RISS/GPS integrated navigation, *Sensors* 12: 115–147.

[25] Wang, J. & Jiang, G. [2003]. Researching and Realization of the Quick Compression Method aimed at the Non-Topology Vector Data, *Journal of Surveying and Mapping* 32(2): 173–177.

[26] Wu, L. [2002]. *Geographical Information System: Principles and Applications*, Science Press, Beijing.

[27] Wu, L. & Shi, W. [2003]. *Theory and Algorithm of Geographical Information System*, Science Press, Beijing.

[28] Wu, Q. [1998]. Survey of gps/ins integrated navigation systems, *Navigation* 8: 1–10.

[29] Z. Lichuan, L. Mingyong, X. D. Y. W. [2009]. Cooperative localization for underwater vehicles, *4th IEEE Conference on Industrial Electronics and Applications*, pp. 2524–2527.

[30] Zheng, H. [1997]. *Theory of Mapping and Cartography Based Computer*, Institute of Surveying and Mapping, Zhengzhou.

[31] Zhong, S. & Gao, Q. [2004]. An Efficient Lossless Compression Algorithm for a kind of Two-Dimension Vector Maps, *Journal of System Simulation* 16(10): 2189–2194.

[32] Zhu, Z. [2006]. Geographical Information System Technology.
URL: *http://www.geog.ntu.edu.tw/course/gistech/*